MAPPING CHINA AND MANAGING THE WORLD

From the founding of the Qin dynasty in 221 BCE to the present, the Chinese have been preoccupied with the concept of order (*zhi*). This cultural preoccupation has found expression not only in China's highly refined bureaucratic institutions and methods of social and economic organization but also in Chinese philosophy, religious and secular ritual, and a number of comprehensive systems for classifying every form of human achievement, as well as all natural and supernatural phenomena. Richard J. Smith's *Mapping China and Managing the World* focuses on several crucial devices employed by the Chinese for understanding and ordering their vast and variegated world, which they saw as encompassing "all under Heaven."

The book begins with discussions of how the ancient work known as the *Yijing* (Classic of Changes) and maps of "the world" became two prominent means by which the Chinese in imperial times (221 BCE to 1912) managed space and time. Smith goes on to show how ritual (*li*) served as a powerful tool for overcoming disorder, structuring Chinese society, and maintaining dynastic legitimacy. He then develops the idea that just as the Chinese classics and histories ordered the past, and ritual ordered the present, so divination ordered the future. The book concludes by emphasizing the enduring relevance of the *Yijing* in Chinese intellectual and cultural life as well as its place in the history of Sino-foreign interactions.

This selection of essays by one of the foremost scholars of Chinese intellectual and cultural history will be welcomed by Chinese and East Asian historians, as well as those interested more broadly in the cultures of, and interactions between, China and East Asia.

Richard J. Smith is George and Nancy Rupp Professor of Humanities and Professor of History at Rice University, Houston, Texas.

Asia's Transformations

Edited by Mark Selden, *Cornell University, USA*

The books in this series explore the political, social, economic and cultural consequences of Asia's transformations in the twentieth and twenty-first centuries. The series emphasizes the tumultuous interplay of local, national, regional and global forces as Asia bids to become the hub of the world economy. While focusing on the contemporary, it also looks back to analyse the antecedents of Asia's contested rise.

This series comprises several strands:

Asia's Transformations

Titles include:

1. **Debating Human Rights***
   Critical essays from the United States and Asia
   Edited by Peter Van Ness

2. **Hong Kong's History***
   State and society under colonial rule
   Edited by Tak-Wing Ngo

3. **Japan's Comfort Women***
   Sexual slavery and prostitution during World War II and the US occupation
   Yuki Tanaka

4. **Opium, Empire and the Global Political Economy***
   Carl A. Trocki

5. **Chinese Society***
   Change, conflict and resistance
   Edited by Elizabeth J. Perry and Mark Selden

6. **Mao's Children in the New China***
   Voices from the Red Guard generation
   Yarong Jiang and David Ashley

7. **Remaking the Chinese State***
   Strategies, society and security
   Edited by Chien-min Chao and Bruce J. Dickson

8. **Korean Society***
   Civil society, democracy and the state
   Edited by Charles K. Armstrong

9. **The Making of Modern Korea***
   Adrian Buzo

10. **The Resurgence of East Asia***
    500, 150 and 50 year perspectives
    Edited by Giovanni Arrighi, Takeshi Hamashita and Mark Selden

11. **Chinese Society, second edition***
    Change, conflict and resistance
    Edited by Elizabeth J. Perry and Mark Selden

12. **Ethnicity in Asia***
    Edited by Colin Mackerras

13. **The Battle for Asia***
From decolonization to globalization
Mark T. Berger

14. **State and Society in 21st Century China***
Edited by Peter Hays Gries and Stanley Rosen

15. **Japan's Quiet Transformation***
Social change and civil society in the 21st century
Jeff Kingston

16. **Confronting the Bush Doctrine***
Critical views from the Asia-Pacific
Edited by Mel Gurtov and Peter Van Ness

17. **China in War and Revolution, 1895–1949***
Peter Zarrow

18. **The Future of US–Korean Relations***
The imbalance of power
Edited by John Feffer

19. **Working in China***
Ethnographies of labor and workplace transformations
Edited by Ching Kwan Lee

20. **Korean Society, second edition***
Civil society, democracy and the state
Edited by Charles K. Armstrong

21. **Singapore***
The state and the culture of excess
Souchou Yao

22. **Pan-Asianism in Modern Japanese History***
Colonialism, regionalism and borders
Edited by Sven Saaler and J. Victor Koschmann

23. **The Making of Modern Korea, second edition***
Adrian Buzo

24. **Re-writing Culture in Taiwan***
Edited by Fang-long Shih, Stuart Thompson, and Paul-François Tremlett

25. **Reclaiming Chinese Society***
The new social activism
Edited by You-tien Hsing and Ching Kwan Lee

26. **Girl Reading Girl in Japan***
Edited by Tomoko Aoyama and Barbara Hartley

27. **Chinese Politics***
State, society and the market
Edited by Peter Hays Gries and Stanley Rosen

28. **Chinese Society, third edition***
Change, conflict and resistance
Edited by Elizabeth J. Perry and Mark Selden

29. **Mapping Modernity in Shanghai**
Space, gender, and visual culture in the Sojourners' City, 1853–98
Samuel Y. Liang

30. **Minorities and Multiculturalism in Japanese Education**
An interactive perspective
Edited by Ryoko Tsuneyoshi, Kaori H. Okano and Sarane Boocock

31. **Japan's Wartime Medical Atrocities**
Comparative inquiries in science, history, and ethics
Edited by Jing-Bao Nie, Nanyan Guo, Mark Selden and Arthur Kleinman

32. **State and Society in Modern Rangoon**
Donald M. Seekins

33. **Learning Chinese, Turning Chinese***
Becoming sinophone in a globalised world
Edward McDonald

34. **Aesthetic Constructions of Korean Nationalism**
Spectacle, politics and history
Hong Kal

35. **Popular Culture and the State in East and Southeast Asia**
Edited by Nissim Otmazgin and Eyal Ben Ari

36. **Japan's Outcaste Abolition**
The struggle for national inclusion and the making of the modern state
Noah Y. McCormack

37. **The Market and Temple Fairs of Rural China**
Red fire
Gene Cooper

38. **The Role of American NGOs in China's Modernization**
Invited influence
Norton Wheeler

39. **State, Society and the Market in Contemporary Vietnam**
Property, power and values
Hue-Tam Ho Tai and Mark Sidel

Asia's Great Cities
Each volume aims to capture the heartbeat of the contemporary city from multiple perspectives emblematic of the authors' own deep familiarity with the distinctive faces of the city, its history, society, culture, politics and economics, and its evolving position in national, regional and global frameworks. While most volumes emphasize urban developments since the Second World War, some pay close attention to the legacy of the *longue durée* in shaping the contemporary. Thematic and comparative volumes address such themes as urbanization, economic and financial linkages, architecture and space, wealth and power, gendered relationships, planning and anarchy, and ethnographies in national and regional perspective. Titles include:

1. **Bangkok***
Place, practice and representation
Marc Askew

2. **Representing Calcutta***
Modernity, nationalism and the colonial uncanny
Swati Chattopadhyay

3. **Singapore***
   Wealth, power and the culture of control
   Carl A. Trocki

4. **The City in South Asia**
   James Heitzman

5. **Global Shanghai, 1850–2010***
   A history in fragments
   Jeffrey N. Wasserstrom

6. **Hong Kong***
   Becoming a global city
   Stephen Chiu and Tai-Lok Lui

Asia.com is a series which focuses on the ways in which new information and communication technologies are influencing politics, society and culture in Asia. Titles include:

1. **Japanese Cybercultures***
   Edited by Mark McLelland and Nanette Gottlieb

2. **Asia.com***
   Asia encounters the Internet
   Edited by K. C. Ho, Randolph Kluver and Kenneth C. C. Yang

3. **The Internet in Indonesia's New Democracy***
   David T. Hill and Krishna Sen

4. **Chinese Cyberspaces***
   Technological changes and political effects
   Edited by Jens Damm and Simona Thomas

5. **Mobile Media in the Asia-Pacific**
   Gender and the art of being mobile
   Larissa Hjorth

Literature and Society

Literature and Society is a series that seeks to demonstrate the ways in which Asian Literature is influenced by the politics, society and culture in which it is produced. Titles include:

1. **The Body in Postwar Japanese Fiction**
   Douglas N. Slaymaker

2. **Chinese Women Writers and the Feminist Imagination, 1905–1948***
   Haiping Yan

Routledge Studies in Asia's Transformations

Routledge Studies in Asia's Transformations is a forum for innovative new research intended for a high-level specialist readership. Titles include:

1. **The American Occupation of Japan and Okinawa***
Literature and memory
Michael Molasky

2. **Koreans in Japan***
Critical voices from the margin
Edited by Sonia Ryang

3. **Internationalizing the Pacific**
The United States, Japan and the Institute of Pacific Relations in War and Peace, 1919–1945
Tomoko Akami

4. **Imperialism in South East Asia***
'A fleeting, passing phase'
Nicholas Tarling

5. **Chinese Media, Global Contexts***
Edited by Chin-Chuan Lee

6. **Remaking Citizenship in Hong Kong***
Community, nation and the global city
Edited by Agnes S. Ku and Ngai Pun

7. **Japanese Industrial Governance**
Protectionism and the licensing state
Yul Sohn

8. **Developmental Dilemmas***
Land reform and institutional change in China
Edited by Peter Ho

9. **Genders, Transgenders and Sexualities in Japan***
Edited by Mark McLelland and Romit Dasgupta

10. **Fertility, Family Planning and Population Policy in China***
Edited by Dudley L. Poston, Che-Fu Lee, Chiung-Fang Chang, Sherry L. McKibben and Carol S. Walther

11. **Japanese Diasporas***
Unsung pasts, conflicting presents and uncertain futures
Edited by Nobuko Adachi

12. **How China Works***
Perspectives on the twentieth-century industrial workplace
Edited by Jacob Eyferth

13. **Remolding and Resistance among Writers of the Chinese Prison Camp**
Disciplined and published
Edited by Philip F. Williams and Yenna Wu

14. **Popular Culture, Globalization and Japan***
Edited by Matthew Allen and Rumi Sakamoto

15. **medi@sia***
Global media/tion in and out of context
Edited by Todd Joseph, Miles Holden and Timothy J. Scrase

16. **Vientiane***
Transformations of a Lao landscape
Marc Askew, William S. Logan and Colin Long

17. **State Formation and Radical Democracy in India**
Manali Desai

18. **Democracy in Occupied Japan***
The U.S. occupation and Japanese politics and society
Edited by Mark E. Caprio and Yoneyuki Sugita

19. **Globalization, Culture and Society in Laos***
Boike Rehbein

20. **Transcultural Japan***
At the borderlands of race, gender, and identity
Edited by David Blake Willis and Stephen Murphy-Shigematsu

21. **Post-Conflict Heritage, Post-Colonial Tourism**
Culture, politics and development at Angkor
Tim Winter

22. **Education and Reform in China***
Emily Hannum and Albert Park

23. **Writing Okinawa: Narrative Acts of Identity and Resistance**
Davinder L. Bhowmik

24. **Maid in China***
Media, mobility, and a new semiotic of power
Wanning Sun

25. **Northern Territories, Asia-Pacific Regional Conflicts and the Åland Experience**
Untying the Kurillian knot
Edited by Kimie Hara and Geoffrey Jukes

26. **Reconciling Indonesia**
Grassroots agency for peace
Birgit Bräuchler

27. **Singapore in the Malay World***
Building and breaching regional bridges
Lily Zubaidah Rahim

28. **Pirate Modernity***
Delhi's media urbanism
Ravi Sundaram

29. **The World Bank and the post-Washington Consensus in Vietnam and Indonesia**
Inheritance of loss
Susan Engel

30. **China on Video**
Smaller screen realities
Paola Voci

31. **Overseas Chinese, Ethnic Minorities and Nationalism**
De-centering China
Elena Barabantseva

Critical Asian Scholarship

Critical Asian Scholarship is a series intended to showcase the most important individual contributions to scholarship in Asian Studies. Each of the volumes presents a leading Asian scholar addressing themes that are central to his or her most significant and lasting contribution to Asian studies. The series is committed to the rich variety of research and writing on Asia, and is not restricted to any particular discipline, theoretical approach or geographical expertise.

1. **Southeast Asia***
   A testament
   George McT. Kahin

2. **Women and the Family in Chinese History***
   Patricia Buckley Ebrey

3. **China Unbound***
   Evolving perspectives on the Chinese past
   Paul A. Cohen

4. **China's Past, China's Future***
   Energy, food, environment
   Vaclav Smil

5. **The Chinese State in Ming Society***
   Timothy Brook

6. **China, East Asia and the Global Economy***
   Regional and historical perspectives
   Takeshi Hamashita
   Edited by Mark Selden and Linda Grove

7. **The Global and Regional in China's Nation-Formation***
   Prasenjit Duara

8. **Decoding Subaltern Politics***
   Ideology, disguise, and resistance in agrarian politics
   James C. Scott

9. **Mapping China and Managing the World***
   Culture, cartography and cosmology in late imperial times
   Richard J. Smith

* Available in paperback

MAPPING CHINA AND MANAGING THE WORLD

Culture, cartography and cosmology in late imperial times

Richard J. Smith

Routledge
Taylor & Francis Group
LONDON AND NEW YORK

First published 2013
by Routledge
4 Park Square, Milton Park, Abingdon, Oxon OX14 4RN
605 Third Avenue, New York, NY 10017

Routledge is an imprint of the Taylor & Francis Group, an informa business

British Library Cataloguing in Publication Data
A catalogue record for this book is available from the British Library.

Library of Congress Cataloging in Publication Data
Smith, Richard J. (Richard Joseph), 1944–
Mapping China and managing the world: culture, cartography and cosmology in late imperial times / Richard J. Smith.
p. cm.
Includes bibliographical references and index.
1. Cartography—China—History. 2. Human geography—China.
3. Political geography—China. 4. Cosmology, Chinese.
5. China—Social life and customs. I. Title.
GA1121.S65 2012
526.0951—dc23
2012012457

ISBN: 978-0-415-68509-2 (hbk)
ISBN: 978-0-415-68510-8 (pbk)
ISBN: 978-0-203-09475-4 (ebk)

Typeset in Bembo
by Book Now Ltd, London

CONTENTS

FIGURES

PREFACE AND ACKNOWLEDGEMENTS

In the essays that follow I have converted all Wade-Giles transliterations to Pinyin, updated each of the chapters in various ways (primarily by addressing new scholarship and adding new bibliographical citations), and indicated avenues of further research. Chinese characters for most of the names, book titles and technical expressions mentioned in these chapters can be found in the orginal publications and/or the various glossaries at http://chaocenter.rice.edu/Content.aspx?id=601.

The first essay of the collection, "The languages of the *Yijing* and the representation of reality," is based loosely on R. Smith 1998d, amplified with some entirely new material, as well as excerpts and paraphrases from several of my other writings on the *Changes* (esp. R. Smith 1991, 2003, 2005, 2009–10, 2011 and 2012). Chapter 2, "Mapping China's world: cultural cartography in late imperial China" updates R. Smith 1998a, with additional material drawn from R. Smith 2001b and 2006a. Chapter 3, "Ritual in Qing culture" updates R. Smith 1990a, and Chapter 4, "The teachings of ritual and the rectification of customs: echos of tradition in the political culture of modern China" updates R. Smith 1998b. Chapter 5, "Divination in the Qing" combines material from R. Smith 1993a, 2008b and 2011, and Chapter 6, "Jesuit interpretations of the *Yijing* in global perspective" is an updated version of R. Smith 2001a, supplemented with material from R. Smith 2002, 2003, 2006b, 2008a, 2011 and 2012. I have, however, cut a good deal of information from the last essay that I have addressed in the first one.

I would like to thank the following publishers for permission to use their printed materials, and to allow me to modify them in the ways indicated above: Global Scholarly Publications in New York (R. Smith 2008b), the Institute of East Asian Studies in Berkeley (R. Smith 1998a), The Institute of Modern History at the Academia Sinica (Zhongyang yanjiu yuan) in Nankang (Nangang), Taiwan (R. Smith 1998b), the *Journal of Chinese Philosophy* in Honolulu (1998c and 2006b), the Oxford University Press in New York (R. Smith 1996a), the Princeton University Press

(R. Smith 2012), the University of California Press in Berkeley (R. Smith 1990a), the University of Hawaii Press in Honolulu (R. Smith 1993a), and the Westview Press in Boulder, Colorado (R. Smith 1991). I am also grateful to Watkins Publishing in London for permission to reproduce several excellent illustrations from Aylward 2007 (Figures 1.1–1.7).

I would like to thank the following individuals for their contributions to this volume. First, the three people whom I consider to be my mentors in the China field: Liu Kwang-Ching, John Fairbank, and Joseph Needham. I owe each of them, particularly K.C., more than I could ever express. Next, sincere thanks to my many friends and colleagues at Rice, most particularly Allen Matusow and Nanxiu Qian, who have offered support and invaluable assistance over a long period of time. And, of course, my family, as always. I would also like to thank a number of other individuals whose scholarship has been particularly challenging and/or inspiring to me in the process of writing or revising the articles for this book: Catherine Bell, Cao Wanru, Pingyi Chu, Paul Cohen, Cordell Yee, Pamela Crossley, Nicola Di Cosmo, Patricia Ebrey, Mark Elliott, Benjamin Elman, Beverly Foulks, John Henderson, James Hevia, Ho Peng-Yoke, Tze-ki Hon, Laura Hostetler, Huang Yinong, Marc Kalinowski, Michael Lackner, Mark Lewis, Li Ling, Liu Dajun, Benjamin Wai-ming Ng, Peter Perdue, Evelyn Rawski, Edward Shaughnessy, Nathan Sivin, Donald Sutton, Claudia von Collani, John Wills Jr., David Woodward, Cordell Yee, Zhang Qicheng, Zhang Longxi and Angela Zito. I would also like to thank Susan Dunsmore for her heroic editorial work on this diverse collection of essays.

Finally, I am especially grateful to Mark Selden for encouraging me to put these essays together for the Routledge series, and for providing a truly excellent example of stimulating and challenging scholarship and editorship.

NOTE ON THE COVER

The cover image is a detail from the upper middle section of a mid-nineteenth century map of Hainan Island. It shows two members of the Li minority offering local tributary products to a Qing dynasty official. The individuals in black and red hats are the Chinese official's bodyguards.

Source: Library of Congress, Geography and Map Division.

To my uncle, Robert D. Livingston, a fellow traveler on the scholarly path.

INTRODUCTION

For my entire academic life I have been interested in constructions of "Chineseness" and "Otherness"—perhaps because I am an "Other" who studies "China." What I have learned over time, something that, in retrospect, should have come as no surprise, is that these categories are remarkably fluid. But in my years as a graduate student in the late 1960s and early 1970s they remained largely unproblematical. So were a host of other broad categories, such as "culture," "tradition" and "modernization," even though, unbeknownst to me at the time, they were highly contested in most academic circles. I do not blame my teachers in graduate school for any failure to encourage reflection about these terms; I came late to the China field (my ambition was to be a professional baseball player, which might have come to pass if my senior year had been a good one at the plate and if graduate school had not become an attractive alternative to military service in Vietnam), and so my teachers spent much of their time tutoring me in Classical Chinese, teaching me the fundamentals of historical research, and showing me how to appear intelligent and well-prepared in the gladiatorial events they called seminars.

My dissertation topic, delivered to me in the fashion of an imperial edict, eventually became my first book, *Mercenaries and Mandarins: The Ever-Victorious Army in Nineteenth Century China* (1978). It focused on the policies of the Qing dynasty (1644–1912) in managing foreign employees in the Chinese service during the Taiping Rebellion and other anti-dynastic uprisings in the period from 1850 to 1874. Research on the historical and cultural dimensions of these policies led me to write a small book on Qing dynasty culture in 1978, and another more substantial one in 1983, which I subsequently revised in 1994 as *China's Cultural Heritage: The Qing Dynasty, 1644–1912.*

One of my motives in writing the 1994 edition of this book was to respond to the move toward "disaggregation" in Chinese studies—advocated, for instance, by Paul Cohen in his masterful critique of American scholarship on China. A key feature of what Cohen calls a "China-centered" approach to Chinese history is that "it seeks to cope with the immense variety and complexity of the Chinese world by breaking it down into smaller, more manageable spatial units, thereby facilitating close scrutiny of the whole range of local variation (including religious, ethnic, and social difference)."[1] Although there is obviously great value in this approach,[2] as well as in studies of what Cohen calls the "lower" levels of society, I believe that the dialectic between holistic and particularistic studies must be sustained. How, after all, can we discuss dissidence, diversity, and change in China without an understanding of consensus, conformity, and continuity? What, in the end, makes terms such as "China" and "Chinese" meaningful? I shall return to this vexing question.

During the 1980s and early 1990s my research took two rather different and yet related forms. One was a continuation of my interest in Qing policies toward foreign employees. Professor John K. Fairbank, the mentor of my mentor, shared this interest, and so he and I collaborated on three books: *Entering China's Service: The Journals of Robert Hart, 1854–1863* (1986), *Robert Hart and China's Early Modernization: His Journals, 1863–1866* (1991), and *H. B. Morse: Customs Commissioner and Historian of China* (1995).[3] One of the most fascinating features of this work, which involved substantial archival research on my part in the massive collection of Chinese and Western documents contained in the Robert Hart Papers at Queen's University, Belfast, was the light these documents shed on Hart's personal life (he had a Chinese mistress who bore him three children) and on his interactions with high-level Qing bureaucrats in Beijing. John had a particular interest in Hart's personal affairs, while I was more interested in the many dimensions of his "official" life, but in both cases we learned a great deal about Sino-foreign political, social and cultural interactions from Hart's papers and his private journals.

In part, because of my interest in these interactions, and also because of my earlier research on the Ever-Victorious Army and Qing dynasty culture, I began to explore in a systematic way certain phenomena that seemed particularly relevant to an understanding of Chinese political, social and intellectual life in the nineteenth century. Ritual, both civil and military, loomed large at every turn, as did divination—whether undertaken by official diviners in the Imperial Bureau of Astronomy (*Qintian jian*), private scholars, military personnel, or professional fortune-tellers. I also came to appreciate the extraordinarily important role of official calendars (*shixian shu*) and popular almanacs (*huangli*, *tongshu*, etc.) in the regulation of Chinese society—not least by their identification of specific activities regarded as auspicious or inauspicious for every day of the year.[4]

In the course of my research at the British Library, the Library of Congress, the National Library of China, and the Vatican Archives—all of which happen to have extraordinary map collections—I acquired a deep personal interest in Chinese cartography, ranging from the large-scale representations of "China," the world and the cosmos contained in almanacs and encyclopedias, to representations of provinces and counties (*xian*) in gazetteers, not to mention strategic and tactical maps used in campaigns against anti-dynastic rebels. I also discovered in these collections a wealth of individual maps from the Ming and Qing periods, both printed and in manuscript form, many of them stunningly beautiful and all of them useful as indices of the way Chinese scholars envisioned and represented space in late imperial times.

This research yielded four books: a full-length study of divination titled *Fortune-tellers and Philosophers: Divination in Traditional Chinese Society* (1991; revised 1993), a volume that I co-edited with D.Y.Y. Kwok, titled *Cosmology, Ontology and Human Efficacy: Essays in Chinese Thought* (1993), and a couple of short monographs for the Oxford "Images of Asia" series, *Chinese Almanacs* (1992) and *Chinese Maps: Images of "All under Heaven"* (1996). As these projects came to fruition, I became ever more interested in the political, social and cosmological significance of ritual in Chinese cultural life. And the more I looked into the cosmological dimensions of ritual, divination, almanacs and even maps, the more I began to question the basic conclusion of John Henderson's pioneering book, *The Development and Decline of Chinese Cosmology* (1984): that vigorous critiques by late Ming and early Qing scholars associated with the "school of evidential learning" (*kaozheng xue*) had thoroughly undermined the traditional Chinese view of the cosmos. It just didn't seem to be the case, and it still doesn't, for reasons that will become clear below.

My growing interest in the many dimensions and ramifications of Chinese cosmology led to a fateful (yes, fateful) decision to study the *Yijing* or *Classic of Changes*. It was, to say the least, a serious and long-term commitment. As I put the matter in my preface to *Fathoming the Cosmos and Ordering the World: The Yijing (I-Ching or Classic of Changes) and Its Evolution in China* (2008), "The study of the *Changes*, or *Yixue*, is a black hole within the China field, a dense and immense space that allows no possibility of escape for anyone drawn by its powerful pull." For more than a decade I devoted myself to an exploration of the evolution and eventual "globalization" of the *Changes*. My most recent monograph, titled *The I Ching: A Biography* (2012), for Princeton's series on "The Lives of Great Religious Books," examines not only the origins and development of the *Changes* in China but also its travels to other parts of Asia (Japan, Korea, Tibet and Vietnam) and eventually to the West.

What, then, connects these seemingly disparate research projects? The main unifying theme is that of order (*zhi*). For some three thousand years, Chinese thinkers have been preoccupied with this notion—a preoccupation expressed throughout the imperial era (221 BCE–1912 CE) not only in China's highly refined bureaucratic institutions and methods of social and economic organization but also in Chinese philosophy, religious and secular ritual, standards of literary and artistic achievement, and in various comprehensive systems of classifying all natural and supernatural phenomena.[5] This preoccupation has also been evident in what some have called the "cultural psychology" of the Chinese—notably in their attitudes toward social conformity, consensus, collective responsibility, and their almost pathological fear of disorder (*luan*). The philosopher Xunzi (*c.* 312–230 BCE) tells us that in ancient times the Chinese sage kings "hated disorder," and therefore "they established ritual principles in order to curb it, to train men's desires and to provide for their satisfaction." The famous nineteenth-century statesman, Zeng Guofan, put the matter succinctly: "The mind of Heaven hates disorder."[6]

But the effort to impose order on all aspects of the known world was at once China's greatest cultural strength and its most critical weakness, for the very factors that contributed to an unparalleled record of cultural cohesiveness and continuity also posed serious administrative problems for the Chinese state—especially in contending with the unprecedented challenges of the nineteenth and twentieth centuries.

The essays in hand focus on four critical means by which the Chinese in imperial times sought to order their vast and variegated world, which literally encompassed "all under Heaven." All deal in one way or another with what may be called applied cosmology.

One of the most important and enduring ordering devices in China was the *Yijing*, an ancient work widely if not universally believed to contain the symbolic resources for understanding all of human experience, both physical and metaphysical.[7] Originating as a divination manual some three thousand years ago, the *Changes* acquired a set of increasingly sophisticated commentaries over time and became a classic in 136 BCE. From that point onward it occupied a central place in virtually all realms of Chinese discourse, from politics, philosophy, religion, art and literature to law, medicine, mathematics and science. Into the twentieth century the *Yijing* remained a work of enormous and unchallenged scriptural authority; everyone in Chinese society esteemed it and employed it in some way, from emperors and officials to artisans and peasants. From the Han period onward it came to be regarded as the "first of the [Confucian] classics."

Intellectuals saw the *Changes* as the key to unlocking the mysteries of the universe. The great Song dynasty scholar Cheng Yi (1033–1107), for example, believed that the work "conjoins everything, from the darkness and brightness of Heaven and Earth to the minuteness of insects, grasses, and trees," and Wang Fuzhi (1619–1692), another of China's greatest intellectuals, described the classic in the following terms:

> [The *Yijing*] is the manifestation of the Heavenly Way, the unexpressed form of nature, and the showcase for sagely achievement. *Yin* and *yang*, movement and stillness, darkness and lightness, contraction and expansion—all are inherent in it. Spirit operates within it; the refined subtlety of ritual and music is stored in it; the great utility of humaneness and right behavior issues forth from it; and the calculation of order and disorder, good and bad fortune, life and death is in accordance with it.[8]

According to the editors of the massive eighteenth-century compilation known as the *Siku quanshu* (Complete collection of the Four Treasuries), the most important literary compilation in Chinese history: "The way of the *Changes* … [encompasses everything, including] astronomy, geography, music, military methods, phonetics, numerical calculations, and even alchemy."[9]

The opening chapter of this volume, "The languages of the *Yijing* and the representation of reality," seeks to show how the *Changes* came to be viewed in these ways. An analysis of its language and symbolic logic will help us to understand and appreciate other devices by which the Chinese sought to order their existence, for example, through the manipulation of space.

Mark Lewis, in *The Construction of Space in Early China* (2006), discusses this process as it unfolded from Neolithic times to the Han dynasty (206 BCE–220 CE), providing a systematic analysis of various Chinese conceptions of space—in ascending order from homes and lineages to cities, regions, states, the world and the cosmos—all of which were the products of "conscious and continuous human endeavor."[10] For well over two thousand years the elite cultural myth in China was one of unity and harmony within the realm of "all under Heaven." But the reality, as Lewis's book makes abundantly clear, was that the Chinese empire, like all other forms of "ordered human space," was often divided against itself. Each spatial entity, from the family to the state, had its own organizing principles and priorities, and although the goal of Chinese statecraft was always to bring all of these forms into some sort of harmonious order, the task was extremely difficult. How, for example, could the "private" (*si*) concerns of the household be completely reconciled with the "public" (*gong*) interests of the state?

As Lewis points out in Chapters 4 and 5 of his book, one of the important divisions that existed in the world of ancient China was between the "Chinese people" and "others." In these two chapters, Lewis describes the process by which "Chinese writers" in the Warring States period (*c*. 475–221 BCE) "increasingly established what was distinctive about themselves by describing alien peoples," and he also seeks to show how the Qin and Han dynasties sought to unify areas that were marked by significant differences in language, dress, religious traditions, and other "local" customs.[11] This process continued throughout the imperial era, as we shall see.

Of particular interest are Lewis's discussions of early models of "the world" as represented in the nine-part grid known as the "Tribute of Yu" (*Yugong*), the places and people described in the *Shanhai jing* (Classic of mountains and seas), and the geometrical "Hall of Light" (*Mingtang*), which served as the ritual embodiment of Chinese conceptions linking space and time. Archaeological excavations in China over the past few decades have revealed efforts on the part of several Han rulers to build such a hall, and they have also unearthed a great many ritual and divinatory devices designed to reflect the structure of the universe, and thus to provide guidance for the harmonization of heavenly patterns and earthly affairs.[12]

By late imperial times, Chinese techniques for representing large-scale space had gone well beyond earlier geometric constructions. From at least the Song dynasty (960–1279 CE) onward, Chinese cartographers began producing extraordinarily accurate maps that depicted the contours and major geographic features of China—territory that would be instantly recognizable by any well-informed viewer today. At the same time, however, they created other types of maps that emphasized significantly different features and gave far less attention to mathematical accuracy. A number of these more "cultural" cartographic productions continued to refer to earlier conceptual models such as the *Yu gong* and to reflect in their cartouches long-standing Chinese assumptions about the "other" from late Zhou sources such as the *Shanhai jing*. How do we account for the breadth and tenacity of these ideas?

Chapter 2, "Mapping China's world: cultural cartography in late imperial China," focuses on two fundamental questions: How did changing conceptions of "the world" shape the contours of Chinese cartography; and how did changing (as well as enduring) cartographic practices affect Chinese conceptions of the world? It also pays attention to other forms of representation, including depictions and descriptions of foreigners in Ming and Qing times. These depictions and descriptions reveal themes of both continuity and change. Among the specific questions I address in this chapter are: How were foreign peoples and their living spaces conceived and represented by cartographers and ethnographers in late imperial times? What motivated these conceptions and representations? And to what extent were "non-Chinese" distinguished from one another across time and space? In short, what were the concrete images that the Chinese of late imperial times had in mind when they thought about "foreign" people and "foreign" places and what were the implications of these perceptions?

The third major theme of this book is ritual (*li*).[13] As Mark Lewis has noted in *The Construction of Space in Early China*, a nearly obsessive fear of chaos (*luan*) "haunted the Chinese imagination for millennia," leading to a perpetual search for ways of establishing and maintaining order. One of these was naturally ritual. In Lewis's words:

> Early discussions of ritual repeatedly asserted that it maintained order through imposing divisions. It was through separating men from women, senior from junior, ruler from subject, or civilized from barbarian that ritual constituted social roles and groups. Without ritual's constant guidance to create and maintain these divisions, society would collapse back into undifferentiated chaos or animal savagery.[14]

No major aspect of life in imperial China was devoid of ritual significance, and ritual specialists were ubiquitous at all levels of society, especially in the Ming and Qing periods. Everyone from emperor to peasant recognized the importance of ritual in preserving status distinctions, promoting social cohesion, sanctifying ethical norms, and transmitting tradition. Ritual was also closely connected with the conduct of China's foreign relations, including the so-called tributary system (see below). Closely linked to both cosmology and law, ritual in China quite literally performed the function Clifford Geertz assigns to "sacred symbols" in synthesizing moral values, aesthetics, and worldview.[15]

Angela Zito's book *Of Body and Brush: Grand Sacrifice as Text/Performance in Eighteenth Century China* (1997) provides an excellent illustration of this point. Zito's emphasis is on the complex ways in which the ruler, the imperium, and the cosmos itself were symbolically constructed in eighteenth-century China. In her view, the emperor, as the "filial" Son of Heaven, ruled by means of both ritual texts (the brush) and ritual performance (the body). The two together formed a complementary set. In the ritualist metaphysics of eighteenth-century China, "subjectivity and language were organized to operate within the world rather than upon it," and human consciousness "constantly performed itself into being through actions of significance, the set of practices called *wen* [translated as 'text-patterns;' 'signs of cosmic order accessible to those who can discern them'].”[16] Zito thus defines *li* broadly as “ways of being human that are considered necessary to the workings of the cosmos as well as its embedded social order, including everything from how to dress to how to venerate ancestors.”[17]

In many ways, ritual played a role in pre-modern Chinese social and bureaucratic life analogous to the role played by the *Yijing* in Chinese intellectual life. Both were “all-encompassing” ordering devices, which hinged on “centering” (and the need to determine, and then to take advantage of, the proper time and place for human action). They provided a means by which what was hidden could be made manifest, what seemed separate could be united, and what was unbalanced could be rectified and re-centered. Relying on images, numbers, and “spirituality,” as well as *yin/yang* logic, “five agents” symbolism, and other notions of cosmological correspondence, both the *Changes* and Chinese ritual practices extended the Dao visibly into the world, thus contributing to cosmic harmony. Like the lines, trigrams, and hexagrams of the *Yijing*, ritual texts and practices imitated the patterns of the cosmos. Chapter 3, “Ritual in Qing culture,” discusses this in detail.

Chapter 4, “The teachings of ritual and the rectification of customs: echoes of tradition in the political culture of modern China” extends the discussion of *li* chronologically, offering a comparative analysis of the way that Chinese leaders attempted to use ritual as a means of achieving the common goal of moral transformation and social harmonization. The case studies in this chapter come from three state-sponsored “morality movements” that occurred at different times in Chinese history—one focusing on the reading of the “Sacred Edict” in the Qing dynasty (1644–1912), one focusing on the “New Life Movement” in the Republican era (1912–49), and one focusing on the “Five Emphases and Four Beautifications” campaign in the People's Republic of China (1949–present). In each era, the Chinese state considered self-cultivation to be the first step in the moral transformation of society, and in each case it relied upon literate elites to spread its official orthodoxy and combat heterodoxy. But in all three instances the attempt to “rectify customs” met severe and ultimately insurmountable challenges. What went wrong and why?

The fourth theme of this book is the way that the Chinese at all levels of society sought to manage the future. Just as history ordered the past and ritual ordered the present, divination ordered the future. From earliest times into the twenty-first century, divination has been a prominent way of knowing in China, with far-reaching cultural implications. Fortune-telling has existed in all societies, past and present, but its different manifestations and social purposes naturally reflect distinctive cultural characteristics and preoccupations. Thus, the more pervasive the phenomenon of divination, the more revealing it will be as a window to culture. And in China it was widespread indeed. Like drama and ritual, with which it shows remarkable affinities, divination can often be “read” as a social performance, as well as evaluated on the basis of its own written and visual materials—whether in the form of esoteric manuals or simply-worded oracles. Like the practice of medicine, it is both diagnostic and prescriptive, and it invites a wide range of reactions, from wholehearted acceptance to skepticism and outright denial. It is for some a game, and for others, an important, perhaps even vital, means of gaining psychological insights. Among other things, divination is a device for constructing and articulating social meanings.[18]

Most of the material in Chapter 5, "Divination in the Qing" grew out of my work on *Fortune-tellers and Philosophers*. In both the book and the original essay I tried to answer a series of questions designed to address issues of unity and diversity as well as continuity and change in pre-modern times.[19] Among these questions were: What forms of divination prevailed in late imperial China, and what sorts of cosmologies and other conceptual systems did they reflect? How does the vocabulary of divination shed light on the way the Chinese organized their experience and interpreted "reality?" How was fortunetelling viewed by the various sectors of traditional Chinese society in the Qing period? What social and cultural factors affected the popularity of divination? What special knowledge did diviners possess, how did they acquire it, and how did they use it? What social or political power did they enjoy? Who were the clients of diviners? What in particular did they want to know? How were the messages of divination conveyed? What rituals, symbols, and modes of discourse were employed, what specific values were expressed, and how did various fortune-telling techniques relate to one another? What did people do with the knowledge they acquired through divination? How, in other words, did they try to deal with the future, and how did the imperial Chinese state react to their activities?

Although its particular manifestations and social significance varied from time to time, place to place, and group to group, divination touched every sector of Chinese society, from emperor to peasant. Nearly everyone in China believed that certain cosmological factors affected human destiny, and all used a similar symbolic vocabulary to express these cosmic variables. Although the notion of "fate" may have been differently conceived, and despite the fact that certain cosmological symbols could be interpreted in different ways, there nonetheless existed a shared "grammar" in the discourse of divination, a common ground of cultural understanding. In other words, fortunetelling, like religion and culture itself, was capable of generating an infinite number of statements, but it was nonetheless constrained by its own internal "logic." Furthermore, like magic, divination had the potential of empowering everyone in Chinese society—a notion particularly appealing to the traditionally disadvantaged. Knowledge of the future could be in some respects more efficacious than conventional knowledge, at least if one could do something about it.

The final chapter in this volume returns to the *Classic of Changes* in an effort to locate it within the larger historical framework of Sino-foreign interactions and the process of globalization. Chapter 6, "Jesuit interpretations of the *Yijing* in global perspective" seeks to show how the *Changes* traveled across time and space and what happened to it in the course of its travels.

The *Yijing*'s great prestige and multifaceted cultural role in China naturally commended it to several civilizations on the Chinese periphery—notably Korea, Japan, and Vietnam—each of which had long been influenced significantly by Chinese philosophy, religion, art, literature, statecraft, law, and social customs. In all these environments, the *Changes* enjoyed an exalted reputation, and in each it was employed in a variety of cultural realms, as it had been in China. The process of transmission in East

Asia was relatively uncomplicated, in part because the classical Chinese language in which the *Yijing* was written served as the literary lingua franca of virtually all educated Korean, Japanese, and Vietnamese elites until the late nineteenth century. Despite this powerful cultural common denominator, however, over time the *Changes* came to be used and understood in ways that reflected the particular needs and interests of the host environment, and in the process the document became domesticated.[20]

Similar processes of appropriation and adaptation took place much later in the West, but for somewhat different reasons and with sometimes radically different results. In the first place, the *Yijing* had to be translated into various Western languages by scholars who had different levels of language ability and different political, religious, or personal agendas. Often these translators turned out to be fierce competitors. And whereas in East Asia the *Changes* remained a part of the dominant culture into the twentieth century, in Europe and the Americas it served completely different functions. One was as a "bridge" between Chinese and Western culture and the other was as a countercultural document, designed to challenge the establishment rather than to support it.[21]

Let me turn now to some of the specific controversies that the chapters in this volume address. One of the most important and enduring issues is how to talk about "culture," "Chinese" culture in particular. The introduction to Paul Cohen's *China Unbound: Evolving Perspectives on the Chinese Past* (2003) raises this issue at several points—usually when he is addressing the inadequacy of old paradigms of Western scholarship, such as the "impact-response" model or the binary notion of "tradition-modernity." He writes:

> When I initially advanced the notion of a China-centered approach, I observed that one of the approach's more important concomitants was a gradual shift away from *culture* and toward *history* as the dominant mode of structuring problems of the recent Chinese past (by which I meant chiefly the nineteenth and twentieth centuries). During the 1950s and 1960s, when the impact-response and tradition-modernity paradigms held sway in American scholarship, enormous explanatory power was invested in the *nature* of China's "traditional" society or culture—and, of course, in the ways in which this society-culture differed from that of the West (or Japan).[22]

Cohen's concern is that an emphasis on "culture"—and particularly an emphasis on cultural differences between "China" and "the West—will lead to cultural essentialism—"the radical reduction of a culture to a particular set of values or traits that other cultures are believed incapable of experiencing."[23] Yet throughout his introduction (and throughout his book) he repeatedly speaks of culture unproblematically, and at one point even acknowledges the obvious fact that "there are important differences between the cultural traditions of China and the West."[24]

Why this apparent inconsistency? The reason, I believe, is that we have to generalize in order to make meaning, even as we recognize, at least proverbially, that "all generalizations are false, including this one." What is the difference between generalizing and essentializing? Perhaps it is simply a matter of motive. In "Mapping China's world," I address this issue directly in the light of claims by various scholars that a Parsonian version of culture (i.e., culture as "systems of symbols and meanings") has contributed to various "totalizing" and "essentializing" orientalist projects, including the rise of "academic modernization theory" and "imperialist development policy."

This certainly does not have to be the case, however. Nor should it be. In my opinion, it is possible to posit a close relationship between ideas, values, intentions, ideologies and other forms of consciousness, on the one hand, and human behavior or "action," on the other, without becoming either an orientalist or an imperialist. My particular conception of culture places primary emphasis on the many and varied ways that groups of people, large and small, categorize and interpret experiences and phenomena in order to create a coherent version of reality, and how, in turn, their understanding of this "reality" influences their behavior.[25] This perpetual process of world-making is, it should go without saying, a complex one.

However defined, culture is never static; it constantly undergoes transformations in response to new stimuli, whether generated from within or introduced from without. Ideas, values, and customs change, as do institutions, laws, rituals and language. Seen in this light, "tradition" is not a negative value, nor is it necessarily inimical to "modernization," whatever the latter term might mean. As my colleagues and I put the matter in the Introduction to *Different Worlds of Discourse: Transformations of Gender and Genre in Late Qing and Early Republican China* (2008), "terms such as 'tradition' and 'modernity' can no longer be seen as mutually exclusive concepts with fixed characteristics but rather as fluid categories that existed in a vast crucible of cultural choices."[26]

Another controversy, also addressed in "Mapping China's world," is whether it makes sense to refer to a Chinese "tributary system." A number of scholars, including John Wills Jr., James Hevia and Laura Hostetler, have argued that it does not—largely on the grounds that previous China scholars, John K. Fairbank "and his followers" in particular, have over-generalized its historical significance.[27] The result, however, has been that much of Fairbank's painstaking and valuable research on the structure and function of the tributary system has been ignored.[28] Although Fairbank may well have overestimated the degree to which Chinese assumptions about tribute shaped Qing policy toward foreigners, it seems absurd to suggest that they were of no consequence whatsoever.

Another volume in this Routledge series provides a perspective on the tributary system that avoids the extremes of both Fairbank and his critics. In *China: East Asia and the Global Economy: Regional and Historical Perspectives* (2008), Takeshi Hamashita makes a powerful case for the existence of a long-standing, vibrant, multifaceted and organically evolved "network of [tributary] relations linking the center and its peripheries, including the provinces and dependencies of the [Chinese] Empire, rulers of native tribes and districts, tributary states and even trading partners."[29] In other words, the tributary system provided a framework—both land-based and maritime—for economic and other interactions in which multiple actors played multiple roles. It may have been a "Sinocentric" system, as Hamashita avers, but it was certainly not a static or a monolithic one.

A balanced analysis of the tributary system requires, then, a historically sensitive appreciation of its assumptions and its institutions, its theories and its practices, its goals and its actual outcomes. This kind of understanding compels us to consider, among other things, exactly how the offering and acceptance of "tribute" were conceived (by all parties, not simply the Chinese) and how much flexibility the system allowed.[30] Clearly any conception of the tributary system that suggests a stagnant, "unchanging China" is hopelessly wrong-headed. Yet to ignore or downplay the tributary system as an important historical and cultural frame of reference for Chinese emperors and officials would be equally misguided. Once again the salient point is that cultural constructions of this sort, whether we think of them as administrative "realities" or as cultural "myths," are not unchanging, and they do not predetermine a certain course of action. But they are not irrelevant to policymaking either.

A third controversy, which is either implicit or explicit in all of the chapters contained in this volume, revolves around the elusive and contested notions of "Chineseness" and "otherness." How can "China" be defined and how should "others" be described and understood? Opinions differ, and they are often forcefully expressed.[31] As with the concept of "culture" and the idea of a "tributary system," the category of "China" is a fluid one, historically contingent, with different ideological, political and even spatial connotations over time. There is no timeless China, just as there is no timeless Chinese "tradition." And yet from the Shang dynasty (*c.* 1600–*c.* 1045 BCE) onward, efforts have been made to distinguish "Chinese" people (initially identified primarily with the semi-historical Xia dynasty, *c.* 2000–*c.* 1600 BCE) from "others," variously described.[32] What can we say about this process?

The enduring term *Zhongguo* (Central Kingdom[s]), which appears on Shang dynasty oracle bones, generally referred in these early inscriptions to the "Xia people" rather than to the Shang state, but during the Zhou period (*c.* 1045–256 BCE) this expression acquired three primary connotations: geographical China, political China and cultural China. Of the three, cultural China emerged as the most important conception. Consider, for example, the following quotation from the *Zhanguo ce* (Strategies of the Warring States):

> China [*Zhongguo*] is where intelligent and discerning people dwell, where the myriad creatures and useful implements are gathered together, where the sages and worthies instruct, where benevolence and right behavior are expressed, where the books [later, "classics"] of Poetry, History, Ritual and Music are used, where different ideas and techniques are tried, where distant people come to observe, and where even the [non-Chinese] Man and Yi people exhibit appropriate conduct.[33]

By late imperial times, a common term for China was *Zhonghua* (the "Central Cultural Florescence"), which reflects an emphasis on the cultural foundations of the Chinese state. Consider, for instance, the following fourteenth-century definition of China:

> Central Cultural Florescence is another term for Central Kingdom. When a people subjects itself to the Kingly Teachings [i.e., Confucianism] and subordinates itself to the Central Kingdom; when in clothing it is dignified and decorous, and when its customs are marked by filial respect and brotherly submission; when conduct follows the accepted norms [of ritual] and the principle of righteousness, then one may call it [a part of the] Central Cultural Florescence.[34]

Viewing the matter from the standpoint of the many Chinese reference works dealing with foreigners, Benjamin Schwartz remarks:

> A random perusal of discussions of barbarians in the various encyclopedias and other sources reveals again and again the degree of emphasis on the five relationships, the "three bonds" … and the whole body of *li* [ritual] as providing the absolute criteria dividing barbarians from the men of the Middle Kingdom.[35]

Schwartz's use of the term "barbarian" raises an important question: Is this an appropriate translation for the various Chinese terms that generally denote foreignness (*yi*, *fan*, *hu*, etc.)? On the whole, since Chineseness tended to be viewed primarily in terms of cultural attributes, and since foreign people were usually defined as those who lacked these qualities, "barbarian" in the Greek sense seems apposite.[36] But as with many words in many languages (and especially in the case of Classical Chinese), context is everything. Thus, in the chapters in this book I have used a variety of translations in the hope of capturing the flavor of each term in a certain context.

Issues of "Chineseness" and "foreignness" bear directly on one of the most heated contemporary debates among historians of imperial China: the question of "sinicization" (also known as "sinification")—that is, the adoption of significant aspects of Chinese culture by foreign conquerors. Controversy swirls around every aspect of the question, from the basic definition of terms (Chineseness, sinicization, adoption, absorption, assimilation, etc.) to matters of motive and methodology. The discourse of sinicization (Chinese: *Huahua* or *Hanhua*) originated in the Republic of China in the early twentieth century, prompted by a desire on the part of certain nationalistically inclined Chinese historians to emphasize the continuity of Chinese culture despite long periods of alien rule under dynasties such as the Northern Wei (386–584), the Liao (907–1125), the Jin (1115–1234), the Yuan (1271–1368) and, of course, the Qing. The idea that China had always managed to assimilate (*tonghua*) its conquerors was attractive to such individuals and also alluring to a number of scholars in the West—including Franz Michael, John Fairbank, Mary Wright, Ping-ti Ho and Jing-shen Tao. Their views did not go unchallenged, however. Scholars such as Chia-sheng Feng, Karl Wittfogel, John Dardess and Ruth Dunnell questioned "the simplistic assumption that barbarian rulers in China inevitably got converted to 'Chinese' cultural norms and ways of life on account of the inherent superiority of 'Chinese' civilization."[37]

In 1967, Ho published an article on the Qing in which he claimed that the key to the success of the dynasty was "the adoption of the early Manchu rulers of a policy of systematic sinicization." He went on to state that:

> so effective was the crucible of Chinese culture that by the latter half of the eighteenth century the imperially exhorted Manchu nativism had boiled down to little more than a legal obligation on the part of Manchu princes and Manchu examination candidates to practice horsemanship and archery and to study the Manchu language.

Further, he argued, "There is definite evidence that even for imperial princes, Manchu had become a dead language by the beginning of the nineteenth century at the latest."[38] For nearly two decades this article stood without a direct challenge, but in the mid-1980s and thereafter a new generation of scholars trained in the Manchu language (as well as in Chinese) began to paint a radically different picture of the Qing rulers and their empire. An article by Evelyn Rawski in 1996 summarized the results of their labors, offering a direct and powerful challenge to Ho's sinicization thesis.

In her view, the key to the political success of the Manchus was not their sinicization but rather the opposite, their multiculturalism—specifically, their ability to exploit cultural links with "the non-Han peoples of Inner Asia and to differentiate the administration of the non-Han regions from the administration of the former Ming provinces."[39] Similarly, Joanna Waley-Cohen, in a stimulating and insightful review article on "The New Qing History," opined that the Manchus judiciously and self-consciously combined "Inner Asian and Chinese traditions so as to simultaneously appeal to both constituencies," thus creating "a continuum between the sedentary agricultural world of China proper and the pastoral nomadic world ... where once a strict line had been drawn."[40] What is more, advocates of the so-called New Qing History demonstrated that the Manchus kept far more of their ethnic identity than Ho's account of the sinicized Qing allowed. Indeed, Manchu identity grew stronger during the eighteenth century, particularly under the Qianlong emperor (r. 1735–1796), who, by emphasizing genealogy and not simply cultural affinity, helped to sharpen the distinction between who was Manchu and who was not.[41]

Yet by all accounts the Qianlong emperor, like his accomplished grandfather, the Kangxi emperor (r. 1661–1722), was a great patron of Chinese culture, well versed in the Chinese language, deeply wed to Confucian values (especially filial piety), conversant with Chinese art and literature, and eager to promote Chinese scholars at all levels of his administration.[42] Was he "sinicized" or not? Perhaps this is the wrong question to ask. The important point, underscored by the title of a recent biography—*Emperor Qianlong: Son of Heaven, Man of the World* —is that, for essentially pragmatic reasons, Qianlong, and all subsequent Qing emperors, presented a "Chinese" face to their subjects who lived south of the Great Wall while presenting other ethnic faces to their subjects in other parts of the empire. To be sure, there was much to admire about China's cultural heritage—its philosophies, its religious traditions, its art, its literature and so forth—and all Manchu princes received a substantial education in these areas. But they also received a substantial education in other cultural traditions as well, including, quite naturally, their own.[43]

The sinicization debate continues, of course, fueled in part by P.T. Ho's fierce and somewhat personal attack on Rawski's scholarship.[44] On one side, for example, the Introduction to a book edited by Pamela Crossley, Helen Siu and Donald Sutton, titled *Empire at the Margins: Culture, Ethnicity, and Frontier in Early Modern China* (2006), vigorously assails the "conceptually flawed" notion of sinicization, arguing that the story that needs to be told is not

> [Qing China's] march to the tropics [a play on an old book title] or the taming of steppe nomads (or, as the traditional Chinese terminology put it, the "cooking" of "raw" barbarians), but repeated mutual adaptations of countless groups and individuals across plastic intermediate zones.[45]

On the other side of the debate we have the example of Pei Huang's recently published book, titled *Reorienting the Manchus: A Study of Sinicization, 1583–1795* (2011), which devotes several pages to a spirited critique of the "New Qing" scholars (including Rawski) and argues vehemently that "Chinese norms, mores and values" made deep inroads into the "cultural heritage [of the Manchus] and thus weakened their ethnic solidarity." The Manchus were, Huang claims, "numerically overwhelmed by the conquered and culturally penetrated by Chinese ways of life."[46]

This debate, like the others I have mentioned (and still others, to which I shall refer in the chapters that follow), reflects the notion of culture as I conceive it—not as something rigid, timeless and determinative, but rather as a shared framework for analysis, providing—for the moment—certain discursive categories and a sense of the interpretive possibilities. As Marshall Sahlins has pointed out with respect to the issue of cultural similarities and differences, "In order for categories to be contested there must be a common system of intelligibility, extending to the grounds, means, modes, and issues of disagreement." It would be difficult, he argues,

> to understand how a society could function, let alone how any knowledge of it could be constituted, if there were not some meaningful order in the differences. If in regard to some given event or phenomenon the women of a community say one thing and the men another, is it not because men and women have different positions in, and experience of, the same social universe of discourse?[47]

Let us now see some of the directions that a cultural perspective of this sort might take with respect to China.

1

THE LANGUAGES OF THE *YIJING* AND THE REPRESENTATION OF REALITY

> Writing does not exhaust words, and words do not exhaust ideas ... The Master [Confucius] said: "The sages established images [*xiang*] in order to express their ideas exhaustively. They established the hexagrams [*gua*] in order to treat exhaustively the true innate tendency of things and ... they attached phrases [*ci*] to the hexagrams in order to exhaust what they had to say."
>
> (The "Great Commentary" (*Dazhuan*) of the *Yijing*)

Introduction

In traditional Chinese thought, the sixty-four hexagrams of the *Yijing* (Classic of Changes) represented symbolically the images or structures of change in the universe, and, as such, had enormous explanatory value. Like Chinese characters, these hexagrams were a distinctly visual medium of communication, concrete yet ambiguous, with several possible levels of meaning as well as a great many accumulated allusions and associations. As it developed over time, the *Yijing* reveals with striking clarity one of the most important ways that the Chinese in pre-twentieth-century China organized and explained the world around them. Through an analysis of the symbolism, structure, and cultural uses of the *Changes*, we can gain insights into deeply imbedded and long-standing Chinese patterns of perception, forms of logic, styles of argumentation, and approaches to questions of aesthetic and moral value.[1]

The basic structure of the *Changes*

The *Changes* first took shape about three thousand years ago as a divination manual, comprised of sixty-four six-line symbols known as hexagrams (*gua*). Each hexagram was uniquely constructed, distinguished from all the others by its combination of solid (_____) and/or "broken" (__ __) lines (*yao*). The first two hexagrams in the conventional order are Qian and Kun; the remaining sixty-two hexagrams represent permutations of these two paradigmatic symbols:

Qian Kun

At some point in the Zhou dynasty (*c.* 1045–256 BCE), probably no later than the ninth or eighth century, each hexagram acquired a "hexagram name" (*guaming*) a brief written description known as a "judgment" (*tuan* or *guaci*; aka "decision" or "tag") and a short explanatory text for each of its six lines called a "line statement" (*yaoci*).[2]

The hexagram name, which indicates its basic symbolic significance, refers generally to a thing, an activity, a state, a situation, a quality, an emotion, or a relationship, for example, "Well," "Cauldron," "Marrying Maid," "Treading," "Following," "Viewing," "Juvenile Ignorance," "Peace," "Obstruction," "Waiting," "Contention," "Ills to be Cured," "Modesty," "Elegance," "Great Strength," "Contentment," "Inner Trust," "Joy," "Closeness," "Fellowship," "Reciprocity," etc.[3] These are, however, rather conventional renderings of the terms. Much debate revolves around the earliest meanings of the hexagrams—especially since several alternative versions of the *Changes* have been discovered during the past four decades or so.[4]

The judgment of a hexagram provides certain kinds of advice, expressed in extremely cryptic language. Most judgments originally referred to ancient and now obscure divinatory formulas involving sacrifices and offerings to spirits.[5] Here are a few representative examples: "Primary receipt, favorable to divine;" "The determination is favorable for a great man; no misfortune;" "Step on the tiger's tail; it won't bite the person; a sacrificial offering;" "Gather the people in the open country; a sacrificial offering; favorable for crossing a big river; a favorable determination for a noble person."[6]

Line statements, which vary in length from as few as two characters to as many as thirty, often include records from previous divinations that were either transmitted orally or recorded in early divination manuals. Many of these statements seem to be based directly or indirectly on "omen verses" of the sort that can also be found on Shang dynasty oracle bones. Some examples: "In the hunt there is a catch: advantageous to shackle captives; no misfortune. The elder son leads the troops; the younger son carts the corpses; the determination is ominous;" "In crossing the river at the shallows he gets the top of his head wet: ominous; [but] no misfortune."[7]

Taken together, the six lines of a hexagram represent a situation in time and space, a "field of action with multiple actors or factors," all of which are in constant, dynamic play.[8] These lines, reading from the bottom to the top, represent the evolution of the situation and/or the major players involved. The first, second and third lines comprise a "lower" trigram and the fourth, fifth and sixth lines constitute an "upper" trigram, each having its own set of symbolic attributes. Below, the eight trigrams with what may well have been their earliest primary meanings:

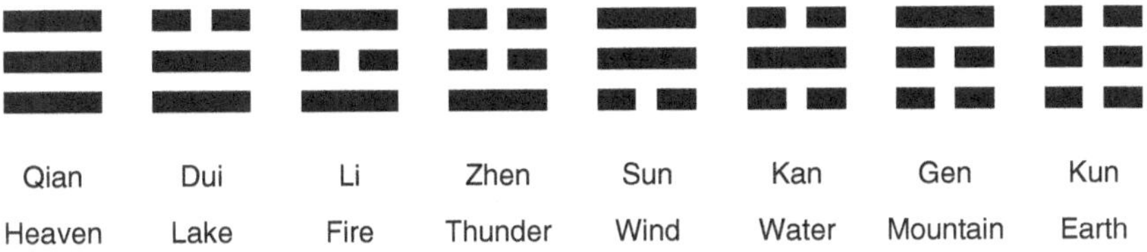

The operating assumption of the *Yijing*, as it developed over time, was that the sixty-four hexagrams represented all of the basic circumstances of change in the universe, and that by selecting a particular hexagram or hexagrams, and correctly interpreting the various symbolic elements of each, a person could gain insight into the patterns of cosmic change and devise a strategy for dealing with problems or uncertainties concerning the present and the future. Interpretation, whether undertaken for inspiration, general guidance, scholarly purposes or in the course of an actual divination, required a deep understanding of the relationship between the lines, the line statements, and the trigrams of the chosen hexagram, and often an appreciation of the way that the selected hexagram might be related to other hexagrams.[9]

To get a sense of the interpretative variables involved, as well as the rich metaphorical possibilities they offer, let us look briefly at the Ge hexagram, as it may have been understood in China around the eighth century BCE.[10]

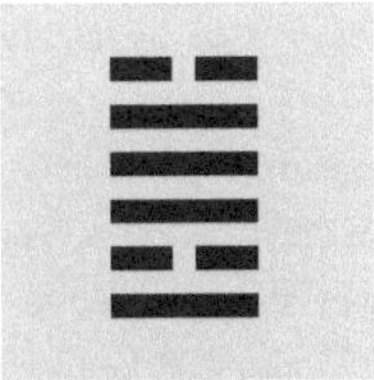

Ge [革]

[Constituent trigrams: below, Li (Fire); above, Dui (Lake)]

革：己日乃孚，元亨利貞，悔亡。

Judgment: On a sacrifice day, take/use captives. Grand offering. A favorable determination. Troubles disappear.

初九：鞏用黃牛之革。

First nine [nine indicates a solid line]: Bind [them?] with the hide [*ge*] of a brown/yellow ox.

六二：己日乃革之，征吉，无咎。

Second six [six indicates a divided line]: On a sacrifice day, make a change [*ge*]; auspicious for an attack; no misfortune.

九三：征凶，貞厲，革言三就，有孚。

Third nine: Ominous for an attack; the determination [of the divination] is threatening. A leather [*ge*] harness with three tassels; there will be captives.

九四：悔亡，有孚改命，吉。

Fourth nine: Trouble disappears; there will be a captive and a change of orders; auspicious.

九五：大人虎變，未占有孚。

Fifth nine: A great man performs a "tiger change;" there will be a capture [or captive(s)] before any divining is done.

上六：君子豹變，小人革面，征凶，居貞吉。

Top six: The noble person performs a "leopard change;" the petty person wears rawhide [*ge*] on his face; ominous for an attack; auspicious in determining a dwelling.

As with most of the other sixty-three hexagrams, very little in the basic text of the Ge hexagram is unambiguously clear—even if we can reasonably assume that the basic theme of the hexagram seems to be warfare. The judgment and several of the line statements contain loan words (always tricky), and there are several possible meanings for the feline "transformations" mentioned in the last two line statements.

Obviously, then, commentaries were needed to make practical, moral and/or metaphysical sense out of cryptic texts of this sort. Over time, literally thousands of such commentaries were written on the *Changes*, reflecting nearly every conceivable political, social, philosophical and religious viewpoint.[11] The most important of these commentaries, at least in the early history of the work, were known collectively as the "Ten Wings" (*Shiyi*). They became attached to the *Yijing* when the *Changes* received imperial sanction in 136 BCE as a major "Confucian" classic. One of the most important reasons for adding the Ten Wings to the basic text at this time was that they were widely (although erroneously) believed to have been written—or at least edited—by Confucius.[12] Had it not been for this close association with the great Sage, we may doubt that Chinese scholars would have scrutinized the document so carefully and searched so relentlessly for its deeper significance over the next two thousand years or so.

The cosmology of the *Changes*

The Ten Wings articulated the *Yijing*'s implicit cosmology and invested the classic with an alluring philosophical flavor and an attractive literary style. This amplified version of the *Changes* emphasized a fundamental unity between Heaven, Earth and Man, and it reflected a distinctive form of correlative logic that emerged in the late Zhou dynasty (*c.* 1050–256 BCE) and became a major feature of Chinese thought throughout the entire imperial era (206 BCE–1912 CE). Metaphysically speaking, this approach was based on the idea that things behave in certain ways not necessarily because of prior actions or the impulsions of other things, but rather because they "resonate" (*ying*) with other entities and forces in a complex network of associations and correspondences.[13] Correlative cosmology of this sort encouraged the idea of mutually implicated "force fields," identified by various specialized terms such as *yin* and *yang*, the "five agents" (*wuxing*; also translated as the five elements, phases, activities, processes, etc.), the eight trigrams, the ten heavenly stems (*shi tiangan*), the twelve earthly branches (*shier dizhi*), the twenty-four solar periods (*ershisi qi*), the twenty-eight lunar lodges (*ershiba xiu*) and, of course, the sixty-four hexagrams. These terms, discussed at greater length below and in two other essays, also came to be linked with specific numerical values.[14]

Qi, literally "breath" or "air," animated everything in the Chinese cosmos. *Qi* is often translated as life breath, energy, pneuma, vital essence, material force, primordial substance, psychophysical stuff, and so forth, but no single rendering serves all philosophical and practical purposes. For now, suffice it to say that in various states of coarseness or refinement, *qi* gave substance and energy to all objects in the world and filled all spaces between them. From late Zhou times onward, everything was assumed to be *qi* in some form, from tangible objects like rocks and trees to more rarefied phenomena such as heat and light. Nathan Sivin has defined *qi* somewhat whimsically, but with considerable point, as "simultaneously 'what makes things happen in stuff' and (depending on context) 'stuff that makes things happen' or 'stuff in which things happen.'"[15]

In its most highly refined form *qi* manifested itself as "spirit" (*shen*). Applied to the world of human beings, "spirit" had a wide range of meanings in Chinese, as it does in contemporary English. But whereas in English the term almost invariably implies a sharp contrast with the material body, in classical Chinese discourse this distinction was never so clear. Spirit was viewed as an entity within the body that was responsible for consciousness, combining what Westerners would generally distinguish as "heart" and "mind." Hence, the Chinese term that I shall be translating as mind (*xin*) in these essays is often rendered heart-and-mind or heart-mind. For well over two thousand years, Chinese of various philosophical persuasions believed that by cultivating their *qi* to the fullest extent, and thus harnessing the highly refined spiritual capabilities of their minds, they could achieve extraordinary things. Daoist-oriented individuals, for instance, could attain immortality; Confucians, for their part, could literally "transform people" and ultimately change the world by means of their ritual rectitude and moral force.[16] Such people could also, according to the *Zhongyong* (conventionally translated as the *Doctrine of the Mean*) "know the future" and be "like a spirit" themselves.[17]

At the heart of late Zhou and early Han systems of correlative cosmology were the ubiquitous concepts of *yin* and *yang*, which came to be conceived in three different but related ways: First, they were viewed as modes of cosmic creativity (female and male, respectively), which not only produced but also animated all natural phenomena through the workings of *qi*. Second, they were used to identify recurrent, cyclical patterns of rise (*yang*) and decline (*yin*), waxing (*yang*) and waning (*yin*). Third, they were employed as comparative categories, representing dualistic relationships that were viewed as inherently unequal but almost invariably complementary. *Yang*, for example, came to be correlated with, among many other things, light, activity, Heaven, the sun, fire, heat, the color red, roundness and odd numbers; *yin*, by contrast, possessed the opposite attributes of darkness, passivity, Earth, the moon, water, coldness, the color black, squareness and even numbers.[18]

Another important feature of Han-style correlative thinking was an emphasis on the five agents, identified with the basic qualities or tendencies of earth (stability), metal (sharpness), fire (heat), water (coolness), and wood (growth). Like *yin* and *yang*, each of the five agents, in various combinations and operating under different temporal and spatial circumstances, had tangible cosmic power embodied in, or exerting influence on, objects of all sorts by virtue of the sympathetic vibration of *qi*. Whether considered as external forces or intrinsic qualities, *yin* and *yang* and the five agents constantly fluctuated and interacted as part of the eternal, cyclical rhythms of nature. Everything depended on timing and the relative strength of the variables involved. It followed, then, that by taking into account these variables, one could predict whether movement on earth or in the heavens would be progressive or retrogressive, fast or slow, auspicious or inauspicious.[19] The same was also true of the eight trigrams and sixty-four hexagrams, as we shall see.

The ten heavenly stems and twelve earthly branches were originally part of a sophisticated sexagenary system of counting hours, days, months and years. For each sixty-unit cycle, one stem was combined with one branch until all combinations had been formed.[20] But by the Han period these stems and branches had come to possess cosmic powers and began to be combined in various ways with *yin* and *yang*, the five agents, the eight trigrams, the sixty-four hexagrams, the twenty-four periods of the solar year, and various heavenly bodies (esp. the twenty-eight lunar lodges) to construct correlative schemes designed to reveal patterns of cosmic change (Figures 1.1, 1.2, 1.3).[21] Throughout the imperial era, scholars and diviners argued incessantly about the coherence and/or predictive value of these schemes, and whether they were even necessary for an understanding of the cosmology of the *Changes*, but the variables persisted in the Chinese cultural imagination across space, time and social class—part of a vast network of correspondences that a great many Chinese thinkers, including scholars of the *Yijing*, relentlessly sought to identify, systematize and quantify.[22]

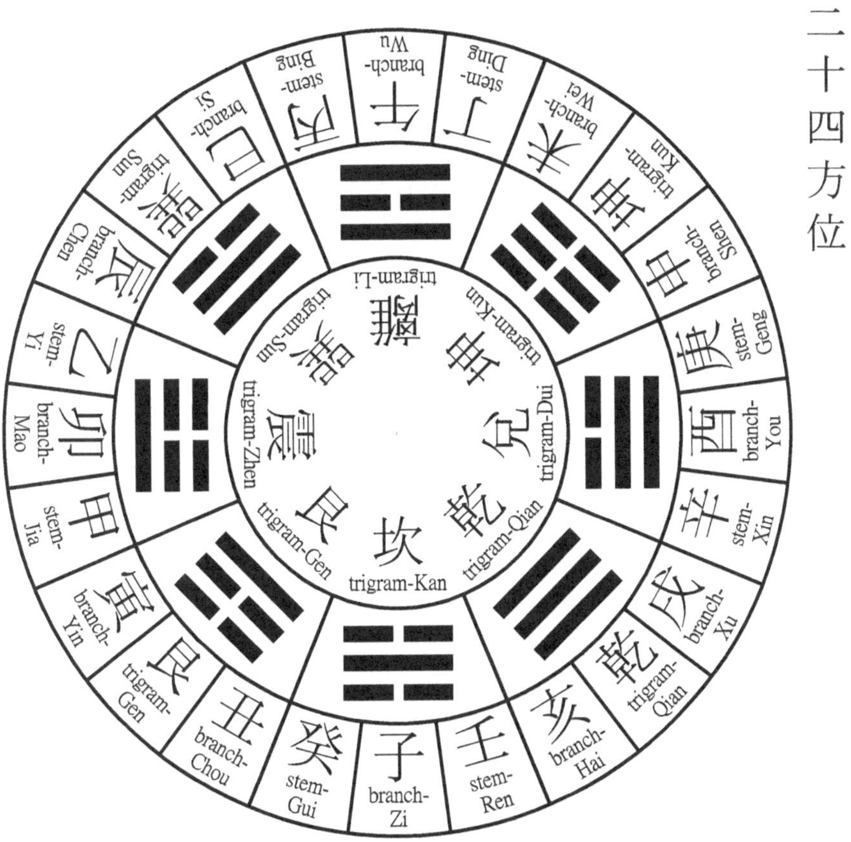

FIGURE 1.1 Correlations between the eight trigrams, the ten heavenly stems, the twelve earthly branches and the twenty-four solar periods

The *Yijing* as a "Confucian" document

With the rise of Confucianism and other moralistic philosophies from about the sixth century BCE onward, a sea change occurred in Chinese thought. Although the oceanography is far too complex to chart here,[23] the salient point is that ethical issues and moral questions increasingly became a standard feature of nearly all Chinese philosophical writing. One significant manifestation in the realm of *Yijing* scholarship was that a number of seemingly value-neutral descriptions of events in the judgments and line statements of the "original" *Changes* came to be seen as prescriptions for proper behavior. For instance, the expression *fu*, which originally seems to have denoted a "capture" or a "captive" in war, came to be seen as a moral quality: "sincerity" or "trustworthiness." Similarly, the term *heng*, which originally had to do with the specific ritual sacrifices and prayers surrounding a divination, came to be glossed as "prevalence," "success," or "penetrating." *Zhen*, originally

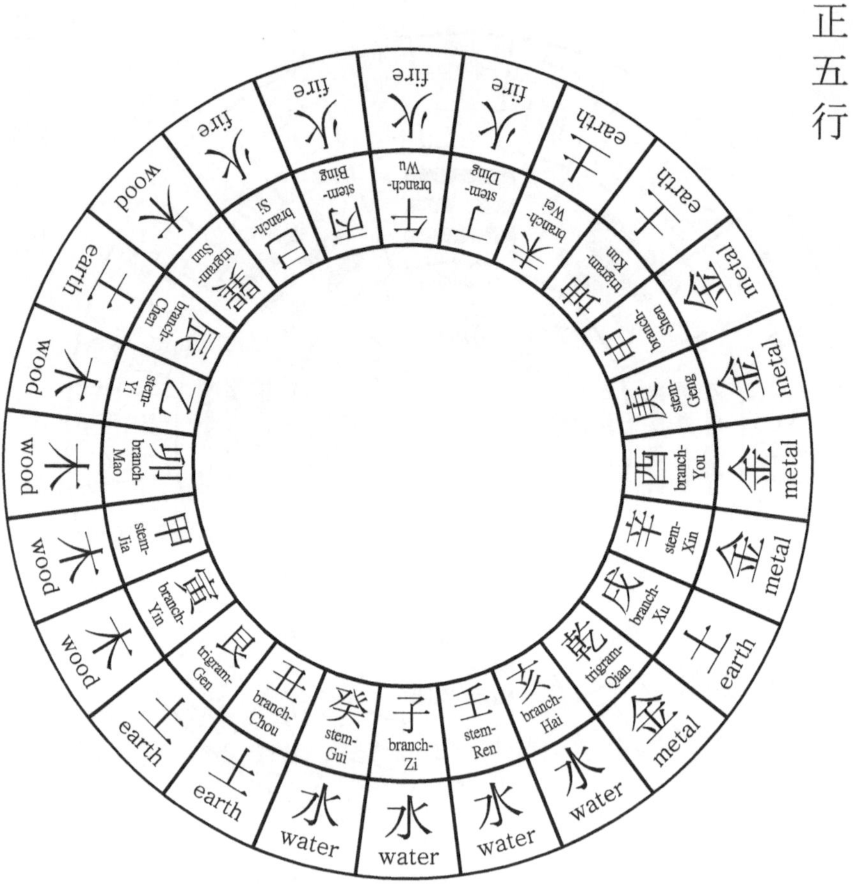

FIGURE 1.2 Correlations between the five agents, the eight trigrams, the ten heavenly stems and the twelve earthly branches

denoting a divinatory "determination," came to be interpreted as "constancy," "perseverance," or "correctness and firmness." By the same token, a term that originally identified a member of the nobility (*junzi*) now described a morally upright "exemplary person." Moreover, words that previously denoted some sort of "trouble" (like *hui* and *lin*) increasingly carried moral connotations of "blame," "remorse," "regret" and even "humiliation."[24]

Let us turn now to the Ten Wings to see how the addition of Confucian values and Han-style correlative cosmology influenced interpretations of the *Changes*. First, a brief description of the "wings" themselves: The first and second, known together as the "Commentary on the Judgments" (*Tuanzhuan*), explain the basic symbolism and/or overall significance of the hexagram. The "Commentary on the Images" consists of two subsections: a "Big Image Commentary," which discusses the "images" (discussed below) associated with the two primary trigrams of each

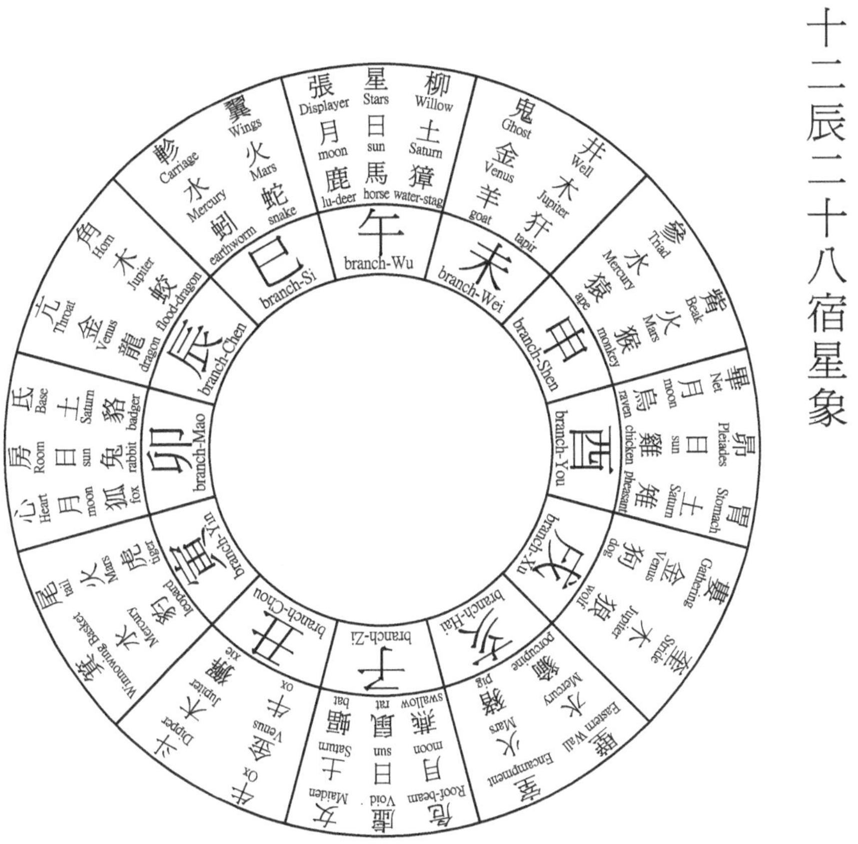

FIGURE 1.3 Correlations between the five agents, the twelve signs of the zodiac and the twenty-eight lunar lodges

hexagram (lines 1–3 and lines 4–6 respectively), and a "Small Image Commentary," which refers to the symbolism of the individual lines. The two parts of the "Great Commentary" (*Dazhuan*), also known as the "Commentary on the Appended Statements" (*Xici zhuan*), are generally described as the fifth and sixth wings. This commentary offers a sophisticated discussion of both the metaphysics and the moral underpinning of the *Changes*, often citing Confucius for authority.

The rest of the Ten Wings lack the divided structure of the first six. The "Commentary on the Words of the Text" (*Wenyan zhuan*) addresses only the first two hexagrams in the received order. The "Explaining the Trigrams" commentary (*Shuogua zhuan*) attaches meanings to each of the eight trigrams that go well beyond the basic significations that they possessed in the early and middle Zhou period. The wing titled "Providing the Sequence of the Hexagrams" (*Xugua zhuan*) aims at justifying the received order of the hexagrams, and the last, the "Hexagrams in Irregular Order" (*Zagua zhuan*) offers definitions of hexagrams that often casts them in terms of contrasting pairs. Different editions of the *Changes* organize this material in different ways.[25]

The Great Commentary is the most philosophically sophisticated of the Ten Wings. It probably assumed something close to its final form around 300 BCE, and from the Han period to the present this document has received far more exegetical attention than any other single wing.[26] The primary goal of the *Dazhuan* was to explain how the hexagrams, trigrams and lines of the document "duplicated the fundamental processes and relationships occurring in nature, enabling those who consult the *Yijing* with sincerity and reverence to partake of a potent, illuminating, activating and transforming spirituality."[27] By participating fully in this spiritual experience, the person using the document could detect the patterns of change in the universe and act appropriately. As the Great Commentary states:

> Looking up, we use it [the *Changes*] to observe the configurations of Heaven, and, looking down, we use it to examine the patterns of Earth. Thus we understand the reasons underlying what is hidden and what is clear. We trace things back to their origins then turn back to their ends. Thus we understand the axiom of life and death … The *Changes* is without consciousness and is without deliberate action. Being utterly still it does not initiate, but when stimulated it is commensurate with all the causes for everything that happens in the world. As such, it has to be the most spiritual thing in the world, for what else could possibly be up to this?[28]

In other words, the *Yijing* showed how human beings could "fill in and pull together the Dao of Heaven and Earth," thus helping to create and maintain cosmic harmony through their spiritual attunement to the patterns and processes of nature. By using the *Changes* responsibly, humans could not only "know fate" (*zhiming*) but also do something about it.[29]

The process of consulting the *Changes* involved careful contemplation of the "images" associated with, and reflected in, the lines, trigrams and hexagrams of the basic text. According to the Great Commentary, sages like the legendary culture hero Fuxi, "had the means to perceive the mysteries of the world and, [by] drawing comparisons to them with analogous things, made images out of those things that seemed appropriate."[30] Initially, then, there were only "hexagram images," "trigram images" and "line images"—"pure signs" unmediated by language. Later on, however, hexagram names, judgments and line statements appeared in written form to help explain these abstract significations. Thus, words came to be used by subsequent sages to identify "images of things" (natural phenomena, such as Heaven and Earth, mountains, rivers, thunder, wind, fire, etc.), "images of affairs" (social and political phenomena, including institutions, war, famine, marriage, divorce, etc.), and "images of ideas" (thoughts, mental pictures, states of mind, emotions, and any other sensory or extra-sensory experiences).[31] The metaphorical possibilities of *Yijing* symbolism proved virtually inexhaustible.

The Great Commentary tells us that the sages of the past appended phrases to the lines of the hexagrams

> in order to clarify whether they signified good fortune [*ji*] or misfortune [*xiong*], and [they] let the hard [*yang*] and the soft [*yin*] lines displace each other so that change [*bian*] and transformation [*hua*] could appear. Therefore, good fortune and misfortune involve images respectively of failure [*shi*] or success [*de*]. "Regret" [*hui*] and "remorse" [*lin*] involve images of sorrow and worry. Change and transformation involve images of advance [*jin*] and retreat [*tui*]. The strong [*gang*] and the weak [*rou*] provide images of day and night ... The judgments address the images [i.e. the concept of the entire hexagram], and the line texts address the states of change. The terms "auspicious" and "inauspicious" address the failure or success involved. The terms "regret" and "remorse" address the small faults involved. The expression "there is no blame" [*wujiu*] indicates success at repairing transgressions ... The distinction between a tendency either to the petty or to the great is an inherent feature of the hexagrams. The differentiation of good fortune and misfortune depends on the phrases [i.e. the line statements].[32]

The *Dazhuan* goes on to point out that the hexagrams "reproduce every action that occurs in the world," and as a result,

> once an exemplary person [*junzi*] finds himself in a situation, he observes its image and ponders the phrases involved, and, once he takes action, he observes the change [of the lines] and ponders the prognostications involved. This is why, since Heaven helps him, "it is auspicious" and "nothing will fail to be advantageous."[33]

The "wing" known as the Explaining the Trigrams commentary vastly expanded the interpretive possibilities of each trigram by attaching meanings to it that went well beyond the basic significations ("trigram qualities" or *guade*) that it possessed prior to the Han. For instance, we discover that among its "new" qualities, Dui, associated in earlier times with the idea of a lake, can now "make things happy;" it "works like the mouth" and represents the "youngest daughter." It can also refer to a female shaman, a concubine, or a sheep, and it can represent the idea of breakage or the deterioration of plant life. Similarly, the Li trigram, initially associated primarily with fire, can dry things out. It can also make things bright and visible, and it can facilitate attachment. It "works like the eye" and represents the middle daughter. In addition, it can refer to the sun, lightning, weapons (the halberd and the sword) and several different animals, including pheasants, turtles, tortoises, crabs, snails and clams.[34] This expansion of interpretive possibilities, as we shall see, opened floodgates that could never be completely closed.

In order to appreciate some of the ways that the Ten Wings influenced interpretations of the *Yijing*, let us return to the Ge hexagram.[35] Here is one prominent and enduring Han dynasty understanding of the text, together with various commentaries from the Ten Wings.

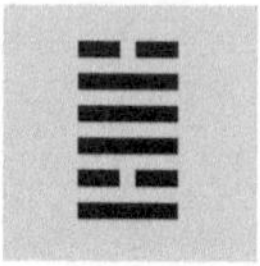

Ge "Radical Change"

Trigrams: below, Li (Fire); above, Dui (Lake)

Judgment: Radical Change is such that only on the day when it comes to an end does one begin to enjoy trust, and then he shall have fundamental prevalence and find it fitting to practice constancy, and his regret will disappear.

Commentary on the Judgment: Radical Change is such that just as Water [exemplified by the Dui trigram] and Fire [exemplified by the Li trigram] try to extinguish each other, so it is when two women live together and find their will at odds. This we call Radical Change. "Only on the day when it comes to an end does one begin to enjoy trust" means that once Radical Change has occurred people trust the one who undertakes it. Such a one brings about joy through the practice of civility and enlightenment, and he shall have great prevalence thanks to his practice of righteousness. If Radical Change were to happen and be right, any regret should consequently disappear. Just as Heaven and Earth make use of Radical Change so that the four seasons come to pass, so did [the great dynastic founders] Tang and Wu bring about Radical Change in compliance with the will of Heaven and in accordance to the wishes of mankind. A Time of Radical Change is indeed great!

Commentary on the [Big] Image: Inside the Lake [the Dui trigram] there is Fire [the Li trigram]. This is the image of Radical Change.

Providing the Sequence of the Hexagrams: The way of wells cannot help but involve radical change. This is why *Jing* [The Well hexagram, number 48] is followed by Ge Radical Change.

The Hexagrams in Irregular Order: Ge Radical Change] means "to get rid of the old."

Line 1: To bind himself tight, this one uses the hide of a brown/yellow cow.

Commentary on the [Small] Image: "To bind himself tight, this one uses the hide of a brown/yellow cow," for he may not attempt anything new on his own.

Line 2: This one should fall in with Radical Change only on the day it comes to an end, and if he were to set forth, it would mean no blame.

Commentary on the [Small] Image: "This one should fall in with Radical Change only on the day it comes to an end," for to set forth then would result in blessings.

Line 3: For this one to go out and attack would mean misfortune, and although he were to practice constancy, he would cause danger. Addressing themselves to Radical Change, three [people] say that they will accede to it; in this he should trust.

Commentary on the [Small] Image: "Addressing themselves to Radical Change, the three say that they will accede to it," so what reason does this one still have to proceed?"

Line 4: Regret disappears, and as this one changes the mandate to rule with sincerity, he has good fortune.

Commentary on the [Small] Image: Changing the mandate to rule results in good fortune because this one trusts that it is so willed.

Line 5: When the great man does a "tiger change," one can trust in the outcome before any divining is done.

Commentary on the [Small] Image: "When the great man does a tiger change," the markings are manifest.

Line 6: Whereas the exemplary person here would do a "leopard change," the petty person should radically change his countenance.

Commentary on the [Small] Image: When the exemplary person does a leopard change, it means that his pattern [*wen*, signifying "culture" or "refinement"] becomes magnificent. When the petty person radically changes his countenance, it means that he will follow his sovereign with obedience.[36]

What we see here is a decidedly moralist interpretation of the judgment and lines of the Ge hexagram, emphasizing values such as trust, steadfastness, sincerity, civility, enlightenment, refinement and righteousness. The trigram relationships suggest some sort of "natural" struggle, and the interests and actions of exemplary people are juxtaposed explicitly to those of petty persons. Although the theme of "attack" still appears in line 3, it is no longer a prominent motif. The emphasis now is on something larger—the possibility of a change in a sovereign's "mandate" (*geming*; the modern Chinese term for "revolution") to rule. This long-standing cosmologically grounded concept, known generally as the Mandate of Heaven (*Tianming*), was based on the idea that a sovereign had the right to rule only as long as he governed in morally responsible ways. If not, Heaven expressed its displeasure with anomalies such as unpredicted earthquakes or eclipses, flood, droughts, and so forth—signs that the people had the "right to rebel." Naturally it was in the interest of all sectors of Chinese society to know what was in store under such circumstances.

Some hermeneutical and divinatory approaches to the *Changes*

From the Han period through the Qing (1644–1912), the *Yijing* was the primary instrument by which Chinese elites sought to understand the present, fathom the future, and devise effective ways of dealing with cosmically-ordained situations. Over this vast span of time, scholars developed a great many different approaches to the classic, all of which were designed to explicate the work and to exploit its enormous interpretive potential. I have described the historical evolution of these various systems elsewhere.[37] My goal here is simply to discuss in a general way some of the most important hermeneutical schemes devised by Chinese scholars to make sense of the *Yijing* in imperial times.[38]

Most of these schemes can be traced to passages in the Ten Wings that refer to: (1) the qualities and capabilities of the eight trigrams, both alone and in various combinations or configurations; and (2) the ways that the individual lines of the sixty-four hexagrams are constantly "changing," "moving," "alternating," being "displaced," "interchanging" (*xiangcuo*) and "intermingling" (*canwu*).[39] Such schemes were also inspired by certain phrases in the Ten Wings that emphasized the importance of both numbers (*shu*) and "calculations" (*zhuan*) in fathoming the "mysteries" of the *Yijing*.[40] The Great Commentary states, for instance, that "The numbers [of the *Changes*] are combined in various ways … [and these combinations] exhaust all aspects of change; in consequence, the hexagrams form the patterns of Heaven and Earth.[41]

Two main schools of *Yijing* exegesis have been identified in imperial China. One is known as the Study of Images and Numbers (*xiangshu xue*); the other is known as the Study of Meanings and Principles (*yili xue*). Exponents of Images and Numbers tended to emphasize numerical calculations and calendrical correlations in their approaches to the *Changes*. For individuals of this persuasion, numbers provided a way of "domesticating" nature, of submitting it to the premodern equivalent of equations. Like the Pythagoreans of ancient Greece, they saw numbers as a way of understanding the workings of the cosmos, thus enabling them to determine how best to situate themselves in a harmonious relationship with their environment.[42] Naturally enough, this approach encouraged the development of a wide range of correlative systems of greater or lesser scope. By contrast, intellectuals who identified themselves with the School of Meanings and Principles paid primary attention to what they saw as the "moral" content of the judgments, line statements and commentaries to the *Changes*.[43] They had little use for systems-builders, feeling that their interpretive schemes were often contrived and overdetermined. The distinctions between these two "schools" have often been far too sharply drawn, obscuring a great deal of overlap and interaction,[44] but they are nonetheless useful in suggesting certain basic orientations or predispositions.

Some integrative systems

Several of the most enduring and influential methods of *Yijing* interpretation arose in the early Han period, inspired by efforts to link the lines, trigrams and hexagrams of the *Changes* with various natural phenomena. Some of these interpretive systems, such as "hexagram breaths" (*guaqi*), "attached initial stems" (*najia*), "attached notes" (*nayin*) and the "eight palaces" (*bagong*) established elaborate systems of correspondence involving one or more of the cosmic variables mentioned earlier in this chapter, including: *yin* and *yang*, the five agents, the eight trigrams, the ten heavenly stems and twelve earthly branches, the twenty-four calendrical periods, the twenty-eight lunar lodges, the sixty-four hexagrams, and so forth.[45] The salient feature of all these systems, whether designed for divination or for more general explanatory purposes, is that they reflected and therefore illumined patterns of cosmic change, enabling individuals to attune themselves to, and harmonize with, the predictable rhythms of nature.[46]

In the post-Han era, devotees of the *Yijing* continued to seek ways of integrating the eight trigrams, the sixty-four hexagrams and other cosmic variables into systems of meaning based on images and numbers. Four such interpretive devices deserve at least brief mention here. References or allusions to all four can be found in the Ten Wings and other late Zhou/early Han writings, but they were not widely employed until the Song period (960–1279). They are: (1) the "Former Heaven Chart" (*Xiantian tu*); (2) the "Later Heaven Chart" (*Houtian tu*); (3) The "Yellow River Chart" (*Hetu*) and (4) The "Luo River Writing" (*Luoshu*).[47] All four illustrations came to be viewed as "maps" of the cosmos, symbolic guides to its inner workings and outward manifestations, and they were often attached directly to editions of the *Yijing* in late imperial times.[48]

The Former Heaven Chart (Figure 1.4), attributed to Fuxi, displays the eight trigrams in four sets, each corresponding to one of the four seasons (usually

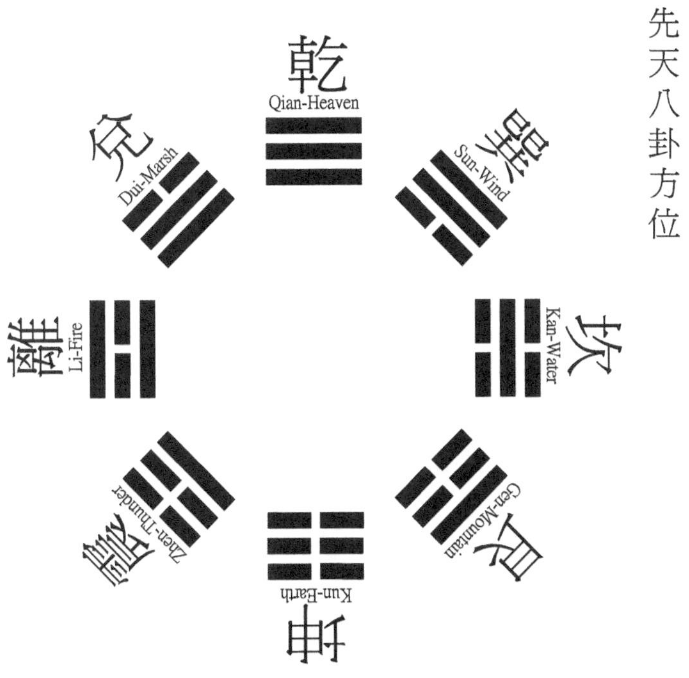

FIGURE 1.4 The Former Heaven sequence of the eight trigrams

depicted in clockwise order, with summer located in the south, at the top of the diagram). The juxtapositions in this configuration are: Qian (South) and Kun (North); Sun (Southwest) and Zhen (Northeast); Kan (West) and Li (East); and Gen (Northwest) and Dui (Southeast).[49] As we discussed in the Explaining the Trigrams commentary, the clockwise movement of the trigrams from Zhen, through Li and Dui to Qian, takes into account what is already existing while the counter-clockwise movement of the trigrams from Sun, through Kan and Gen to Kun, takes into account what has not yet come into existence. In terms of their direct "effects in nature," Zhen (Thunder) causes things to move, Sun (Wind) disperses things, Kan (Water) moistens things, Li (Fire) dries things, Gen (Mountain) causes things to stop, Dui (Lake) pleases things, Qian (Heaven) provides governance, and Kun (Earth) shelters things.[50]

The Later Heaven Chart (Figure 1.5), attributed to King Wen, presents the trigrams in a spatial order that depicts yet another kind of developmental change. In the most common version of this scheme, Zhen (East) marks the beginning of the Later Heaven cycle, followed by Sun (Southeast), Li (South), Kun (Southwest), Dui (West), Qian (Northwest), Kan (North), and Gen (Northeast). In this sequence, all things come forth in Zhen (Thunder), they are set in order in Sun (Wind), they are made visible to one another in Li (Fire), they are nourished by Kun (Earth), they are pleased by Dui (Lake), they contend in Qian (Heaven), they toil in Kan (Water), and they reach maturity in Gen (Mountain).[51]

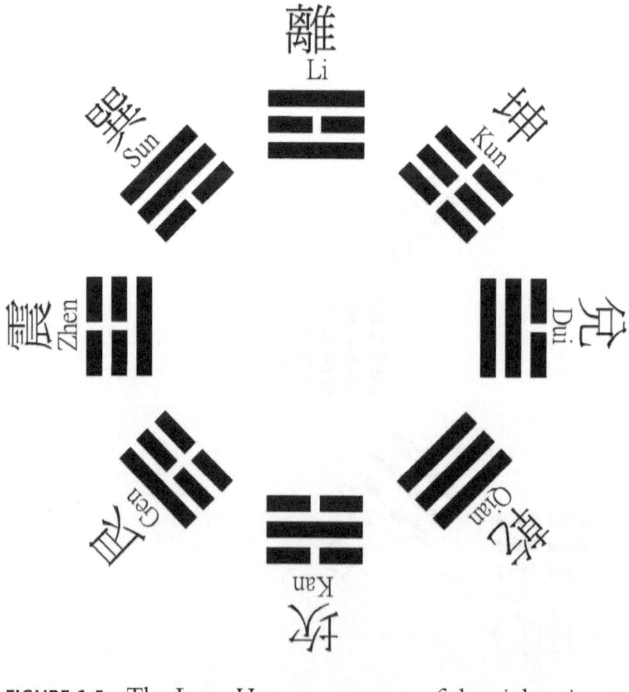

FIGURE 1.5 The Later Heaven sequence of the eight trigrams

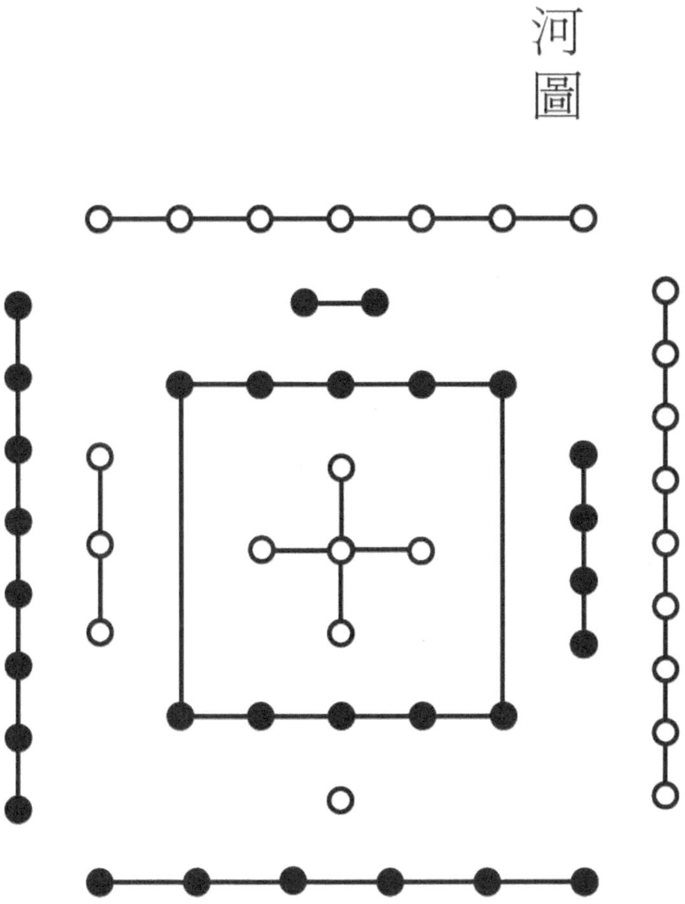

FIGURE 1.6 The Yellow River Chart

The Yellow River Chart (Figure 1.6) arranges its numbers from one to ten in such a way as to pair odd (*yang*) numbers with even (*yin*) ones. These numbers are then correlated with the five directions (and hence, the five agents): two and seven to the south (fire), one and six to the north (water), three and eight to the east (wood), four and nine in the west (metal), and five and ten at the center (earth). This chart, in which all of the odd numbers add up to 25, and all of the even numbers add up to 30, reflects the mutual production sequence of the five agents.[52]

In the Luo River Writing, we find a "magic square," in which the numbers in any row of three, whether perpendicular, horizontal, or diagonal, add up to fifteen (Figure 1.7). Even numbers occupy all four corners, and the five-agents sequence is one of mutual conquest. Thus, for example, wood (three and eight) overcomes earth (five and ten), earth overcomes water (one and six), water overcomes fire (two and seven), fire overcomes metal (four and nine) and metal overcomes wood.[53]

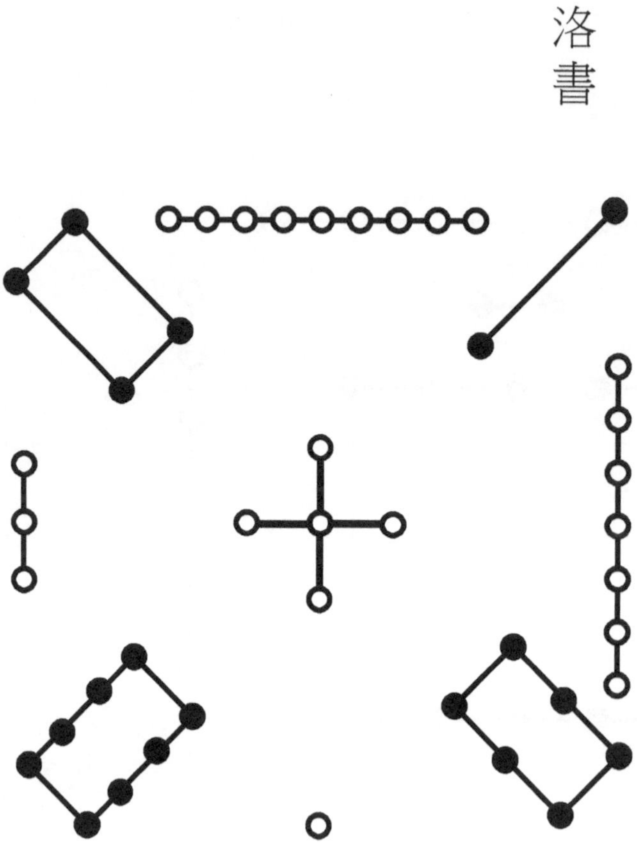

FIGURE 1.7 The Luo River Writing

Predictably, these four illustrations became mutually implicated, and over time the cosmological configurations became increasingly complex. For example, the decimal number system of the Yellow River Chart came to be linked with the eight trigrams, the five agents, the ten heavenly stems and the twelve earthly branches.[54] Meanwhile, expansive thinkers such as Shao Yong (1011–1077) and Lai Zhide (1525–1604), created vast *Changes*-related interpretive schemes of their own. Shao, for example, produced a comprehensive work titled the *Huangji jingshi shu* (Supreme principles that rule the world), based on eight trigrams symbolism and designed to do nothing less than "define the relationship of things, predict the future, and comprehend vast quantities of space and time."[55] He also developed a sequence of the hexagrams that offered a mathematically logical "binary" system of ordering.[56] Similarly, Lai devised a grand theory of hexagram relationships (*guazong*) in which systematic line movement from bottom to top created new hexagrams—as when the first line of Qian (number 1) changes from *yang* to *yin* to produce Gou (number 44), the second line of Gou changes to produce Dun (number 33), and so forth.[57]

In each case, the purpose of these correlative systems was to create a comprehensive vision of reality, one in which number and image, as well as past, present and future, were seamlessly integrated.

Naturally, the wide variety and questionable coherence of these cosmological constructions invited a great deal of criticism from different angles, especially during the seventeenth and eighteenth centuries, when the iconoclastic School of Evidential Research (*kaozheng xue*) flourished. At this time the Yellow River Chart and the Luo River Writing came under particularly fierce fire for being late accretions to the *Changes* and therefore non-canonical.[58] Yet a great many Chinese scholars continued to seek in such illustrations the keys to understanding the universe—among them, such highly regarded and scientifically minded individuals as Fang Yizhi (1611–1671) and Jiang Yong (1681–1762). Although closely associated with *kaozheng* scholarship, both men believed that the eternal principles of astronomy, calendrics, mathematics, music, and medicine could all be found in the Yellow River Chart, the Luo River Writing, hexagrams, trigrams and their individual lines.[59]

The interpretation of individual hexagrams

Although correlative systems of the sort described above could be, and were, used to predict the future, the most common techniques of *Yijing* interpretation throughout the imperial era focused on individual hexagrams—in particular, their lines and constituent trigrams. As the Great Commentary states:

> Good fortune, misfortune, regret, and remorse are all generated from the way lines move ... The lines reproduce how particular things act, and the images [of the trigrams] provide likenesses of particular things.[60]

In other words, by virtue of their position or movement within (and among) the hexagrams, the lines and trigrams of the *Yijing* imitated (*xiao*) the transformations of the universe and thus provided a working model of cosmic change. To fathom the hexagrams was to fathom the cosmos.

Let us look first at some widely shared assumptions about hexagram lines. Each is designated either *yin* (yielding; *rou*) or *yang* (firm, *gang*). Divided lines represent *yin*; solid lines, *yang*. *Yin* lines are even-numbered and inferior; *yang* lines are odd-numbered and superior.[61] Lines are considered "correct" (*zheng*) or "appropriate" (*dang*) if they correspond to the number of their position in the hexagram—that is, if *yang* lines are in odd-numbered places (1, 3 and 5), and *yin* lines are in even-numbered places (2, 4, and 6). From this standpoint, the most "correct" hexagram is Jiji (number 63); the most "incorrect" is Weiji (number 64). From a developmental standpoint, the first line of a hexagram denotes the beginning of a situation, and the sixth, the end, but the first and last lines are generally considered to be relatively unimportant compared to the middle lines, especially in divination. When one or more of the six lines is "in motion" during the process of constructing a hexagram with coins or milfoil stalks, each moving line turns into its opposite, creating a derivative hexagram (*zhigua*) that has to be taken into account along with the original in any prognostication.[62]

Jiji Weiji

Lines may also be viewed hierarchically. In such cases, the places they occupy are differentiated as superior and inferior according to their relative elevation within the hexagram. The sixth place is generally occupied by the sage; the fifth, by the ruler; the fourth, by a minister close to the ruler; the third by a middle-ranking official; the second, a lower official; and the first or bottom, a commoner.[63] When opposite lines occupy analogous *yinyang* positions within a given hexagram, they are said to "correspond" or "resonate" (*ying*), which is generally indicative of a favorable situation. In a majority of cases, the most important resonant lines in a hexagram are those that occupy the second and fifth places, which represent respectively the "proper" relationship between the official and the ruler, the son and the father, or the wife and the husband.[64]

A relationship described as "holding together" (*bi*) may exist between any two adjacent lines of different character. Under these conditions, the upper line is said to "rest upon" the lower line, while the lower line "receives" the upper one. If this occurs between a yielding fourth line and a firm fifth line, the relationship is generally viewed as auspicious, as with the hexagram Bi (number 8). On the other hand, the firm fourth line of Li (number 30) does not complement the weak fifth line; hence its respective line statements speak of abandonment and extreme sorrow.[65]

Bi (8) Li (30) Guai (43)

Hexagrams have two kinds of "ruling" (i.e. dominant) lines—those that "govern" by virtue of their position (usually the fifth place), known as *zhugua zhi zhu*, and those that give the hexagram its overall meaning (*chenggua zhi zhu*), such as the weak top line in Guai (aka Kuai, number 43)—which, as the only *yin* line in the hexagram, indicates a "breakthrough." The ruler(s) of a hexagram can be determined by consulting the Commentary on the Judgment, although other commentaries and calculations may also come into play.[66] When the two kinds of ruling lines are identical, the hexagram has only one ruler; otherwise it may have two or even three. In the case of the Fu hexagram (number 24), since the *yang* line in the first place is the only "firm" one, it becomes the ruler. In the case of Pi (number 12), the yielding second line gives the hexagram its overall meaning ("obstruction" or "standstill") while the firm fifth line rules by virtue of its position.[67]

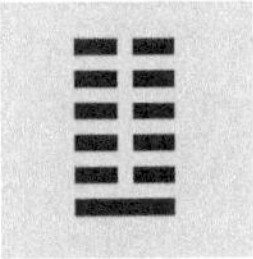

Fu (24)

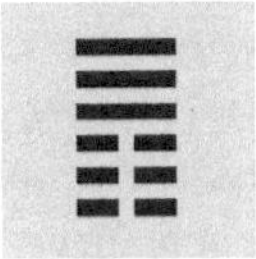

Pi (12)

Another common way of viewing the relationship between the lines in a given hexagram is to see them as representing spatially the "three powers" (*sancai*) of Heaven, Earth and Man. That is, the first two lines symbolize Earth; the next two, Man; and the top two, Heaven. This tripartite structure, emphasized in the "Great Commentary," has its counterpart in the individual lines of each constituent trigram. According to the Explaining the Trigrams commentary, the principles of Heaven are *yin* and *yang*; those of Earth are yielding (*rou*) and firmness (*gang*); and those of Man, humaneness (*ren*) and right behavior (*yi*).[68] Heaven provides the predestined time, Earth the place, and Man the action of any given situation or event.[69]

Relationships may also exist between the individual lines of different hexagrams in which the same or similar phrases appear in the line statements—such as the expression "don't act" (*wuyong*; lit. "don't use [the divinatory information under consideration]), which is shared by Qian (number 1), Shi (number 7), Yi (number 27), Kan (number 29), and Jiji (number 63). Although such relationships were almost invariably taken seriously by both diviners and exegetes, they were sometimes treated as sophisticated riddles. A typical question might be: Why in the hexagram Tongren (number 13) does the subject of the fifth line first cry out and then laugh, while in the sixth line of Lü (number 56) the subject first laughs and then cries out? A similar "riddle" might be: Why is the expression "dense clouds do not rain" (*miyun buyu*), shared by the judgment of Xiaochu (number 9) and one of the line statements of Xiaoguo (number 62)? When the great Qing scholar Jiao Xun (1763–1820) was only 14 years old his father posed this latter question to him and Jiao could not answer it. In apparent reaction, he spent the remainder of his scholarly life devising a sophisticated mathematical theory of analogues (*bili*) to explain such hexagram relationships.[70]

We may now turn briefly to the trigram configurations that exist within each hexagram. As indicated above, the basic symbolism of the trigrams is naturalistic: Qian represents the "creativity" of Heaven; Kun, the "receptiveness" of Earth; Zhen, the "arousing" capacities of thunder; Kan, the "lowly" characteristics of water; Gen, the "stationary" attributes of a mountain; Sun, the "gentleness" of wind; Li, the "cohesion" of fire; Dui, the "joyousness" of a lake.[71] But, as also noted above, each of these primary trigrams possesses a wide range of other qualities or characteristics—attributes that refer not only to intrinsic powers or capabilities but also to natural phenomena, colors, times, seasons, directions, types of people, parts of the body, animals, and so forth. Thus, for example, Qian, in addition to being strong and active, can also represent deep red, metal, early evening, late autumn, south or north-west (depending on the trigram sequence), the sovereign, the head, a horse or a dragon. Kun, in addition to possessing the attributes of docility and receptiveness, can also represent the mother, black, earth, afternoon, late summer/early autumn, north or southwest, the people, the abdomen, a mare or an ox. Zhen not only connotes movement and speed (as with a chariot), but also yellow, wood, morning, spring, northeast or east, a young man, the foot, a galloping horse or a flying dragon. Although these and other such correlations are not always consistently applied in hexagram analysis, they vastly increase the interpretive possibilities of any given trigram configuration.

Two main trigram configurations can be found in any hexagram. The most fundamental of these consists of a primary "inner" trigram (*neigua*, lines 1, 2 and 3) and a primary "outer" trigram (*waigua*, lines 4, 5 and 6). As we have seen, the Big Image Commentary identifies the basic symbolism of each such configuration. Often, however, this symbolism is quite obscure—either because the relationship between the trigrams is not particularly clear or because the implications of the combined imagery do not seem to follow "logically." For instance, the Big Image Commentary for the Sui hexagram (number 17), which has the Zhen trigram below and the Dui trigram above, reads: "Within the Lake there is Thunder; this constitutes the image of Sui ["Following"]. In the same way the exemplary person goes indoors at night for rest and leisure."[72] Neither of these connections seems self-evident, and so, as might well be expected, different commentators explain this relational symbolism in sometimes radically different ways.[73]

The other main trigram configuration, known as *hugua* or *huti*, is based on the assumption that every hexagram has, in addition to its two primary trigrams, a pair of "overlapping" or "intersecting" trigrams, comprised of lines 2, 3 and 4 and lines 3, 4 and 5, respectively.[74] Naturally enough, the presence and presumed activity of these so-called "nuclear" trigrams—each with its own full complement of symbolic associations—provide yet another rich source of interpretive possibilities.[75]

For a sense of how nuclear trigrams might work in conjunction with the lines and primary trigrams of a single hexagram, let us return to Ge, "Radical Change" (number 49).[76] We know from certain comments in the Ten Wings that Ge means "to get rid of the old," and that its primary trigrams, Li (Fire below) and Dui (Lake, above), are bent on extinguishing one another. But which will prevail and why? An analysis of the nuclear trigrams suggests that "Fire," indicating intelligence and clarity, will dominate, because the flames of Li are "fanned" by the lower nuclear trigram Sun (identified with wind), while the fluidity of the "Lake" is stabilized by the "firmness" of the Qian trigram.[77] This indicates a successful enterprise, as, for instance, when the legendary sage Tang established the Shang dynasty and King Wu founded the Zhou.

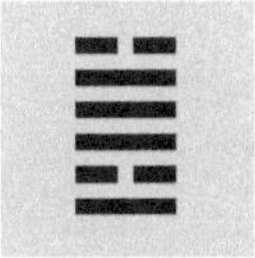

Ge "Radical Change"

Primary trigrams: below, Li (Fire); above, Dui (Lake)
Nuclear trigrams: below, Sun (Wind); above, Qian (Heaven)

From the standpoint of the individual lines, we see that although an initial *yang* occupies its proper place in the hexagram (i.e. an odd-numbered position), it has no corresponding opposite line in the fourth place to offer a response; thus the line statement warns against premature change. The second line, however, is "correct, central and clear." It occupies the appropriate location as a *yin* (even-numbered) line in an even-numbered position, and it has a "responsive" opposite line in the fifth place. Moreover, it lies in the middle of the Li trigram, a symbol of brightness and intelligence. This suggests, then, an appropriate moment to inaugurate action. The third line occupies its proper position in terms of both its place in the hexagram and its correspondence with the sixth line, but the situation remains perilous because the third line has moved out of the center of the lower nuclear hexagram (Sun) and is thus somewhat unstable. Acting prematurely would be dangerous here.

Although the fourth *yang* line of Ge is in an inappropriate *yin* place, its subject is close to the "ruler" of the hexagram (the fifth line), and thus the person planning a righteous uprising may be able to succeed in concert with his superior. Encouraging this strategy is the idea that the fourth line has passed from the lower (Fire) trigram into the upper one (Water), easing the way for cooperation. "Regret disappears and there is trust" (*hui wang you fu*). The fifth (*yang*) line, in correspondence with the *yin* line in the second place as noted, has the clarity of the Li trigram at its disposal. Moreover, it occupies the center of the Dui trigram, which signifies not only the west (the realm of the "White Tiger" in popular astrological lore), but also autumn (a time when animals such as the tiger, which is mentioned explicitly in the fifth line statement, change their coats). The sixth *yang* line, although out of position, "holds together" with the fifth line, remaining steadfast without overt forward movement. Here, too, we have animal symbolism (a leopard), which is linked expressly to the notion that the exemplary person can now stand firm, inspiring subordinates without any further action on his own.

In all, the Ge hexagram seems to provide concrete advice for a person of rectitude desiring to change the existing political order.[78] But this is not the only way that the Ge hexagram could be, or was, understood. For instance, Zhang Zai (1020–1077), one of the towering figures of Song dynasty scholarship on the *Yijing*, chose to relate it directly to moral transformation—a classic example of Meanings and Principles exegesis. Focusing on the often-neglected top line, which refers to a "leopard change" (*baobian*) on the part of an exemplary person, Zhang equated this line explicitly with the moral development of his historical hero, Yan Hui (aka Yan Yuan or Yanzi), one of Confucius's most dedicated disciples. To Zhang, the link between this line and the kind of "drastic" moral change effected by Yan was admirable and undeniable, "even without a divination."[79]

What I have tried to emphasize in the preceding remarks is that the interpretive possibilities of any given hexagram were virtually limitless, constrained only by one's intelligence, learning, imagination and perhaps a commitment to certain philosophical principles and/or exegetical conventions. This same sort of flexibility, combined with the enormous scriptural authority of the *Changes*, made the work a particularly powerful instrument for use in the construction, elaboration and refinement of traditional Chinese culture.

The cultural significance of the *Yijing*

It is no exaggeration to say that the *Yijing* had an importance in traditional Chinese culture comparable in certain significant respects to that of the Bible in the West.[80] Like the Bible, it was a timeless work of enormous scriptural authority, which inspired the elite and served as a source of comfort to the illiterate masses as well (although Buddhist and Daoist religious works probably performed this latter function more effectively). But unlike the Bible, the *Yijing* was the self-conscious product of natural observation by the ancient Chinese sages, not the holy word of a transcendent God; and the order or *Dao* upon which it was based had no Creator or Supreme Ordainer, much less the concept of original sin and evil as an active personal force. The *Yijing* posited neither a purposeful beginning nor an apocalyptic end; and whereas the Bible insists that "man is answerable not to his culture, but to a being that transcends all culture," the *Yi* takes essentially the opposite position. Furthermore, in the Western biblical tradition, God reveals only what he chooses to reveal, while in traditional China, the mind of Heaven—whether viewed impersonally as Tian, or more personally as Shangdi—was considered ultimately knowable and accessible. The "absolute gulf" between God and his creatures in the Western tradition had no counterpart in the Chinese tradition.[81]

We can certainly say, however, that, like the Bible in the West, the *Changes* profoundly affected the development of philosophy, art, literature and social life in imperial China—although not always with the same dramatic impact. A book that lacks such colorful elements as a jealous and angry God, an evil presence like Satan, a savior like Jesus, or the sinfulness of Sodom and Gomorrah, operates at a certain disadvantage, after all. A complete inventory of the *Yijing*'s influences would certainly require several books, but a few general indications of its cultural significance may not be out of place here.[82]

From a philosophical standpoint, the *Changes* exerted enormous influence. It established, for instance, the conceptual underpinnings for much of traditional Chinese cosmology, as well as the point of departure for most philosophical discussions of space and time. Not only did it inspire various early derivative schemes, including Jiao Yanshou's (*c.* 70–10 BCE) *Yilin* (Forest of Changes) and Yang Xiong's highly-refined system of eighty-one tetragrams in the *Taixuan jing* (Classic of Great Mystery),[83] but it also served as the inspiration for alternative cosmologies—notably the worldview of the radical Qing scholar Yan Yuan (1635–1704). Yan, an ardent enemy of Song neo-Confucianism, developed a view of cosmic creativity and change that replaced the conventional "five agents" (*wuxing*) with the "four powers" (*side*), one of which corresponded with each of the four characters comprising the judgment of Qian (1)—translated by Richard Lynn as "fundamentality" (*yuan*), "prevalence" (*heng*), "fitness" (*li*) and "constancy" (*zhen*).[84]

The *Changes* also provided a virtually indispensable philosophical vocabulary for a wide range of Chinese thinkers. It served as the *locus classicus* for such fundamental concepts as *yin* and *yang* and *Taiji* (the "Supreme Ultimate," the sum and source of all the "principles" [*li*] of things), as well as for such crucial ontological distinctions as between "what exists before physical form" (*xing er shang*)—for example "images," principles, and ultimately the Dao—and "what exists after physical form" (*xing er xia*); that is "implements" (*qi*) or more generally, "things" (*wu*). And because so many provocative concepts in the *Yijing* were not well defined in the text itself, they could be used in many different ways. As a result of both the prestige and the ambiguity of the *Changes*, Confucians, Daoists, and Buddhists alike employed its terminology, and argued incessantly over its cosmological, epistemological, ontological and moral implications.[85]

Over time, the *Yijing* became an important intellectual and cultural common denominator of the "Three Teachings" (Confucianism, Daoism and Buddhism)—even though its original messages were sometimes modified beyond recognition. For example, the Ming Buddhist scholar Zhixu Ouyi (also known as Wan Yi) glossed the hexagram Qian (1)—conventionally identified with the assertive idea of a Confucian gentleman who "strengthens himself unceasingly"—as representing "the Buddha-nature and the essence of enlightenment." Similarly, the seventeenth-century Daoist Liu Yiming maintained that Qian referred to creative energy that was "tranquil and unstirring, yet sensitive and effective."[86]

Of course, a number of Confucian scholars claimed that Buddhism and Religious Daoism were themselves derived from the *Yijing*, and at least a few Chinese went so far as to claim affinities between the Bible and the *Changes*.[87] But such efforts by intellectuals to neutralize the effect of "strange doctrines" (*yiduan*) ultimately had the effect of bringing the "Three Teachings" closer together. The Qing scholar Zhang Xuecheng, for example, might hold steadfastly to the view that Buddhism came "originally from the teaching of the *Changes*," but his exposure to, and admiration for, Buddhist texts, impelled him to argue that the abstruse symbolism of the sutras could in fact be understood metaphorically, like the *Yi*, which referred to such things as "dragons with dark and yellow blood."[88]

Whatever its role in harmonizing divergent philosophies in China, the *Yijing* unquestionably contributed in significant ways to the shaping of Chinese discourse. In the earliest strata of the basic text we can already find a sensitivity to rhymes and homophony conducive to puns and double entendres, as well as a tendency to pair words and concepts with opposite or complementary meanings in such a way as to encourage associational or correlative thinking. Although a number of other classical texts exhibit similar tendencies, these linguistic forms are particularly well developed in the Ten Wings—which together display a highly refined if rather diversified system of symbolic logic. In this system, abstract ideas are embodied in concrete instances of things and their relations. The key notion of complementary opposition, for example, is vividly illustrated in the "Commentary on the Judgment" of the hexagram Kui (38), which states in part: "Heaven and Earth may be contrary entities, but their task is the same. Male and female may be contrary entities, but they share the same goal. The myriad things may be contrary entities each to the other, but as functioning entities they are all similar."[89]

The *Yijing*'s numerological and metaphorical symbolism, together with its *yin-yang*-oriented "logic of correlative duality" (to borrow a phrase from Zhang Dongsun), contributed substantially to the Chinese preference for allegory, analogy, and the use of numerical and other forms of relational symbolism in making an argument. The "Great Commentary" describes the *Yi*'s discursive style as one in which the language is indirect: It "twists and turns but hits the mark" (*yan qu er zhong*). Although the names connected to the hexagrams and lines may appear trivial, the commentary states, "the meanings are far-reaching and the phrasing is elegant … The things and events dealt with are obviously set forth, but hidden implications are involved."[90]

Guided by this sort of an intellectual orientation, most Chinese scholars had no particular interest in the ideas of separate qualities or "laws of identity" of the sort that have preoccupied many thinkers in the West, and they saw little need for anything approximating an Aristotelian syllogism as a form of logic.[91] Nor did they deem it necessarily desirable to unpack and systematically elaborate the implicit symbolic connections that might exist within a given text. Put another way, the aphoristic and highly metaphorical nature of the *Yijing* encouraged an intuitive as well as an intellectual approach to understanding. Although Qing scholars distinguished between learning (*xue*) and thinking (*si*), and between erudition (*bo*) and grasping the essence (*yue*), traditional Chinese thought did not on the whole involve a conscious exaltation of "reason" over "intuition."[92]

In the realm of art, the *Yijing*'s brevity and paucity of parables and anecdotes diminished its importance as a direct source of representational inspiration. It is true, of course, that trigrams served as an extremely common decorative motif on craft productions of all types—from ceramics, bronzes, wood carvings and jade objects to architecture, flags and charms—and that certain specific animal symbols from the classic, such as flying dragons, were depicted vividly in Chinese paintings, as well as crafts. One can also find examples of paintings that represent the idea of scholars contemplating the *Changes*, such as Liu Songnian's well-known Song dynasty work titled "Reading the *Yijing* in the Pine Shade," or drawings of friends discussing the work (the Qing bannerman Linqing's illustrated autobiography comes immediately to mind). But the *Yi* did not—and could not, of course—provoke an artistic avalanche of the sort that biblical stories did in the West. On the other hand, it did encourage an artistic preoccupation with nature, and provided an analytic vocabulary that proved as serviceable and enduring in art and literature as it did in philosophy.[93]

As early as the fifth century AD, Chinese art critics had already begun to assert that "Painting should correspond with the *Yijing*." This did not, of course, mean that brushwork should simply imitate trigrams and hexagrams; rather, it meant that the form and symbolism of paintings should reflect the patterns and ever-changing processes of nature that the *Changes* revealed in its own unique way. Thus, for example, the famous Qing dynasty work titled *Jieziyuan huazhuan* (The mustard seed garden painting manual) describes the symbolism of plum tree painting almost exclusively in terms of *Yijing* categories:

> The blossoms are of the *yang* principle, that of Heaven. The wood of its trunk and branches is of the *yin* principle. Its basic number is five, and its various parts and aspects are based on odd and even numbers [like those of the *Yijing*]. [The base of the flower] symbolizes *Taiji* [the Supreme Ultimate] … the branches symbolize the six lines [of the hexagrams] … and the tips of the branches have eight knots or forks, symbolizing the Eight Trigrams.[94]

Yijing symbolism also informed literary and musical criticism. The hexagram Bi (22), for example, stood for beauty, grace and simplicity of form, while Yu (16) indicated energy, enthusiasm and emotion. Guai (43) symbolized resolute, critical judgment, and Li (30), logical clarity. Qian (1) generally denoted creativity and spirituality, while Kun (2) suggested passive intelligence. Thus we find the poet Yuan Mei justifying his preoccupation with landscape gardens by reference to "the grace of hills and gardens;" and Zhang Xuecheng, in a much-admired critique of Han dynasty historical scholarship, describing the writing of Sima Qian as "round and spiritual," and that of his successor, Ban Gu, as "square and sagacious."[95] In both cases, the critical vocabulary employed by these scholars is drawn verbatim from the *Changes*.

Liu Xie, the enormously influential literary critic of fifth and sixth century China, tells us in his *Wenxin diaolong* (The literary mind and the carving of dragons), that "the images of the *Yi* first brought to light the spiritual presences (*shenming*) that previously had been concealed," thus illuminating "human patterns (*renwen*)." Throughout his work, Liu repeatedly and reverentially cites the *Changes*—not only as the foundation of Chinese aesthetics, but also as a concrete model for composition. In discussing linguistic parallelism, for example, he states that the "Commentary on the Words" and the "Great Commentary" of the *Changes*

> embody the profound thought of the Sage. In narrating the four virtues of the hexagram Qian [1], the sentences are matched in couplets, and in describing the kinds of responses evoked by the dragon and the tiger, the words are all paired and paralleled.

He goes on to say:

> When describing the hexagrams of Qian and Kun [2] as easy and simple respectively, the passage winds and turns, with lines smoothly woven into one another; and in depicting the going and coming of the sun and the moon, the alternate lines form couplets. Occasionally there may be some variation in the structure of the sentence, or some change in word order, but parallelism is always the aim.[96]

The *Yijing* also affected the narrative tradition in China, although in very different ways than the Bible influenced Western literature. In the first place, the *Changes* deliberately avoids recourse to elaborated myth in presenting patterns of temporal flux, while the Bible, decidedly diachronic, revels in it. And whereas the latter work presents us with the image of a self-conscious alienation from civilization, an ever-present awareness of discontinuity, and the idea of "sacred discontent"—that is, joy over "breaking free from redundant patterns," in the words of Herbert Schneidau—we find in the *Yijing* precisely such recurrent, cyclical patterns.[97] These patterns, in turn, were the foundation of Chinese aesthetics, whether expressed in art, music, landscape gardens or literature. Andrew Plaks has argued, for example, that the *Yijing*'s principles of bipolarity, ceaseless alternation, presence within absence, and infinite overlapping, are not only the key to the structure of China's greatest novel, *Honglou meng* (Dream of the Red Chamber), but also "basic to a major portion of the Chinese literary tradition."[98]

The *Changes* provided not only structural elements but also particular concepts that informed Chinese narrative works, both consciously and unconsciously. Wen Tong, a Manchu bannerman and divination specialist in Qing times, developed the unusual but at least defensible theory that several of the main characters in the Ming novel, *Shuihu zhuan* (Water margin), as well as a few more minor figures, were directly related to images derived from the *Changes*. Thus, for example, he argued that events in the text foreshadowed Song Jiang's identification with the hexagram Song (6), Li Kui's association with Sheng (46) and Liu Tang's affinity with Ding (50).[99] In the same spirit, Wang Mingxiong, a contemporary scholar/diviner in Taiwan, adduces evidence from *Honglou meng* to suggest that the author, Cao Xueqin, used images from the *Changes* as well as other divinatory systems, in constructing his narrative of events in "Prospect Garden."[100] Plaks suggests that the name itself (*Daguan yuan*; lit. Garden of Great View) may have been inspired by the "Commentary on the Judgment" for Guan (20), which refers to "a great view (*daguan*) from above."[101]

Chinese scholars wrote literally thousands of essays on the *Yijing* from the Han period to the Qing—many of which, along with various inscriptions, memorials, eulogies, and works of rhyme-prose (*fu*) focusing on the classic, found a place in the massive Qing encyclopedia *Gujin tushu jicheng* (Collection of illustrations and writings, past and present) under the heading "Literature" in the section on the *Yi*.[102] In addition, the editors brought together in this work thirty-four poems on the *Changes*—the earliest by Fu Xian in the third century. Most of the verses included in the encyclopedia are relatively undistinguished, but the authors in the collection include several well-known *Yijing* specialists such as Shao Yong, Ouyang Xiu (1007–1072), Zhu Xi (1130–1200), Hu Bingwen (1250–1333), and Lai Zhide, in addition to a number of other famous scholars.[103]

Needless to say, we do not find in this collection anything like Milton's epic *Paradise Lost*; nor, despite the theme of prophecy, mystic poetry akin to that of, say, William Blake. Specific references to the *Changes* abound, however. Many poems focus on the idea of *yinyang* interaction, and a number employ trigram and hexagram names and images. Some include lines from the *Yijing*, usually judgments; and one poem, by Shao Yong, reads like a miniature essay on hexagram relationships. Two works refer to Zhou Dunyi's "Diagram of the Supreme Ultimate" and another to the Yellow River Chart and the Luo River Writing. Approximately one-fifth of the *Yi*-related poems in the *Tushu jicheng* mention the supposed inventor of the trigrams, Fu Xi (alternatively Bao Xi), and about the same number refer to Confucius. Several poems allude to the famous story about the Sage breaking the bindings of his copy of the *Changes* three times, and several more refer to Shao Yong, either by elliptical reference to plum blossoms or to the concept of "Former Heaven."[104]

In the realm of Chinese social and political life, the *Yijing* exerted enormous influence. Quite apart from its use as a divination text and as a repository of moral knowledge, it provided a cosmologically-grounded justification for the political and social hierarchies of imperial China from Han times through the Qing. For just as the eight trigrams symbolized unequal family relationships, various hexagrams legitimated other relations of subordination in terms of *yinyang* inequality. The hexagram Guimei (54), for example, casts the role of women solely in terms of their subordination to men in marriage, concubinage, or slavery. Jiaren (37) indicates that the woman of a household should submit totally to her husband's authority, attending only to her domestic chores and neither following her whims nor dallying and laughing.[105]

By definition, all hexagrams had potential application to human affairs. Among those most relevant to Chinese social and political life, in addition to those mentioned above, were Qian (1), signifying male control; Kun (2), female compliance; Song (6), litigation; Shi (7), military affairs; Bi (8), union and accord; Li (10), circumspect behavior; Qian (15), modesty; Yu (16), comfort or satisfaction; Gu (18), decay; Shihe (21), criminal law; Fu (24), return; Wuwang (25), absence of falsehood; Daguo (28), excess; Heng (32), perseverance; Dun (33), retreat; Jin (35), advance in rank; Mingyi (36), failure to be appreciated; Kui (38), separation or alienation; Guai (43), breakthrough; Gou (44), social intercourse; Cui (45), people gathered around a good ruler; Sheng (46), the career of a good official; Kun (47), difficulty; Ding (50), nourishment of talents; Jian (53), slow and steady advance; Feng (55), prosperity; Lü (56), travel and strangers; Huan (59), dispersion; Jie (60), restraint; Zhongfu (61), kingly sway; Jiji (63), accomplishment; and Weiji (64), something not yet completed.[106]

The influential neo-Confucian compilation known as the *Jinsi lu* (Reflections on things at hand) employs about fifty different hexagrams, including most of the above-mentioned, to illustrate various social and political themes. In the chapter on governing, for example, Zhou Dunyi is cited:

> It is difficult to govern a family whereas it is easy to govern the world, for the family is near while the world is distant. If members of the family are separated, the cause surely lies with women. This is why the hexagram Kui [to part; 38] follows the hexagram Jiaren [family; 37], for "When two women live together, their wills move in different directions." This is why [the sage emperor] Yao, having put the empire in order, gave his two daughters in marriage to Shun in order to test him and see whether the throne should be given to him. Thus it is that, in order to see how he governs his empire, we observe the government of his family.[107]

The official eighteenth-century assessment of the *Yijing* by the editors of the *Siku quanshu* tells us: "The way of the *Changes* is broad and great. It encompasses everything, including astronomy, geography, music, military methods, phonetics, numerical calculations, and even alchemy."[108] The scientific implications of this outlook were significant. For if, as the Great Commentary claimed, the trigrams and hexagrams of the *Yijing* truly expressed the principles and patterns of all phenomena in Heaven and on Earth through images and numbers, then everything was equally susceptible to interpretation and classification in terms of the classic—not only social and political life, art, literature and music, but also, as the *Siku quanshu* confidently asserts, science, technology, and mathematics. And if the great sages of the past who devised these images did so on the basis of their direct observation of nature, further investigation might be—or at least it might seem to be—unnecessary.[109]

Explanations of natural phenomena by reference to the *Yijing* took a variety of forms, from grand elaborations of time and space to specific examples of scientific inspiration.[110] The index to almost any volume of Joseph Needham's monumental *Science and Civilisation in China* under the subject heading "*I Ching*" will reveal that there were very few realms in the natural world for which the *Changes* did not provide some sort of meaningful explanation: The color and flow of blood, the anatomy of crustaceans, the physical constitution of people from different areas of China, the movements of the eye and jaw, acupuncture and pulse points, chemical and alchemical reactions, the nature of earthquakes, musical tonality and even male and female sexual responses could all be explained by reference to trigrams and/or hexagrams.[111] As one concrete example, the constituent trigrams of the hexagram Jiji (63), Li (Fire) and Kan (Water) came to represent the respective sexual propensities of men (quick to heat; quick to cool) and women (slow to heat; slow to cool).[112]

But hexagrams did more than simply represent correlative functions. They also "controlled" time, phenomena, and situations in a concrete way, rather like "force fields" as I have suggested above. Some were considered donators, others receptors;

some involved movement; others, immobility; some encouraged aggregation, others disaggregation; some entailed progression, others retrogression. Trigrams possessed similar attributes, whether paired as opposites in the "Former Heaven" sequence, or arranged developmentally in the "Later Heaven" sequence. The important point is that in exerting this sort of cosmic control, trigrams and hexagrams acted much the same way bureaucrats did in the Chinese administrative order, and spirits did in the supernatural world. Joseph Needham has observed in this regard that "The *Book of Changes* might almost be said to have constituted an organization for 'routing ideas through the right channels to the right departments.'"[113]

In Needham's view, the *Yijing* served as a "universal filing system"—a device for "pigeon-holing novelty and then doing nothing about it." It "tempted those who were interested in Nature to rest in explanations that were no real explanation at all."[114] This judgment is unduly harsh, for Chinese science achieved remarkable things well after the *Changes* had attained the exalted status of a "classic." And although China's scientific accomplishments from the Song period may appear to be relatively unimpressive when measured by the standard of modern Western science, a number of scholars have questioned whether the West's unique historical trajectory should serve as the model for all (or any) other non-Western societies.[115]

Furthermore, the seventeenth and eighteenth centuries in China witnessed a burst of critical and creative scholarship by individuals who applied rigorous scientific methods to the analysis of ancient texts, and who used new methodologies to explore the scientific realms of astronomy, mathematics, geography and archaeology.[116]

Nathan Sivin describes these developments—undertaken in the spirit of "evidential research" (*kaozheng xue*), and marked by an emphasis on "solid learning" and "seeking truth from facts"—as constituting a genuine "scientific revolution" in China.[117]

Most twentieth-century Chinese intellectuals have had a less sanguine interpretation of their own past, however. On the whole, their view has been that even the most iconoclastic exponents of "evidential research" in the late Ming and early Qing periods were concerned more with narrow philological issues and a text-based quest for moral knowledge than with the investigation of nature itself.[118] Science in premodern China never witnessed—at least not on a significant scale—a Newtonian phase of classical mechanics, with an emphasis on direct observation and the mathematization of hypothesis. Thus, with relatively few exceptions, the number symbolism of the *Yijing* remained numerological and never truly mathematical.[119]

Even after the traumatic impact of Western imperialism on China during the latter half of the nineteenth century—a development that brought a host of new ideas to the Middle Kingdom—many Chinese claimed, as they had done since the time of the Jesuits in the seventeenth century—that foreign science and mathematics were all based on the principles of the *Yijing*.[120] A few progressive scholars referred to the concept of change in the *Yijing* to argue for fundamental reform, and the abortive "self-strengthening movement" of the period from 1860 to 1895 in fact derived its name from the Commentary on the Images for Qian (1). But when radical change finally came after the Sino-Japanese War—first with the abolition of the civil service examinations in 1905, and then with the revolution of 1911—the *Yijing* as a classic of now-discredited Confucianism lost a great deal of its scriptural authority.[121]

Yet today the *Yijing* remains a vital document—studied, consulted, and often revered by Chinese (and others) throughout the world. Hong Kong, Taiwan, Japan, South Korea and Singapore have remained havens for twentieth-century devotees of the *Changes*, not to mention the source of much valuable and stimulating scholarship. And even on the Chinese Mainland, where investigation and especially use of the *Yijing* have long been stigmatized as a reflection of "feudal" superstition, there is evidence of a revival of interest in the one-time classic. An historic 1984 conference in Wuhan, for example, produced 120 papers and five specialized volumes, generating much enthusiasm and material for further research by Chinese and foreign scholars alike.[122]

During the late 1980s, "*Yijing* fever" (*Yijing re*) swept over the Mainland. Even the *People's Daily* got into the act, publishing three successive articles (on November 18, 19, and 21, 1988), which were designed to introduce readers to new developments in *Yijing* research. A year later, Professor Liu Zheng, writing in the journal *Zhexue yanjiu* (Philosophical Research), pointed out that

> Within the last decade, our country's publishers, large and small, have published and republished thirty to forty works on the study of the *Yijing*, and two to three hundred articles on the subject have . . . [appeared] in our scholarly journals.[123]

Part of the reason for a revival in *Yijing* studies on the Mainland has been the "Open Policy" of the Chinese government, initiated in 1978. This deliberate effort to expand the parameters of scholarly debate in China, although inconsistently applied over the past few decades, has given Chinese intellectuals far more latitude in their studies than previously. Furthermore, the "spiritual" crisis that has seized many people in China recently, together with the continued search for a "modern" but still "Chinese" identity, has revitalized interest in all aspects of China's ancient past, including, of course, the *Changes*.[124]

Given this quest for a unique "modern" identity, we should not be surprised to find that a number of Chinese scholars have tried to find in the *Yijing* elements of modern mathematics and science.[125] Feng Youlan, for example, maintains that the *Changes* contains an incipient "algebra of the universe;" and Xie Qiucheng goes so far as to claim that the hexagrams of the classic were originally designed as a high efficiency information transfer system analogous to contemporary computer coding based on optimal units of two (the number of basic trigrams in each hexagram) and three (the number of lines in each trigram). Tang Mingbang, drawing on the writings of Xie and other contemporary Chinese scholars, asserts that the forms of atomic structure in nuclear physics, the genetic code in molecular biology, the eight-tier matrix in linear algebra, all seem to be related to the logic of the *Yijing*.[126]

In my view, such nationalistically inspired attempts to link the *Changes* with modern scientific thought are not likely to yield many useful insights, but the effort is probably worth making anyway. After all, the basic challenge of the *Yijing* is to make creative connections, to establish meaningful relationships. Why arbitrarily limit the possibilities?

On an individual level, the contemporary attraction of the *Yijing* may be either aesthetic or pragmatic—perhaps both. Certainly the language and imagery of the document remain entrancing, even for those not fully conversant with its rich and elusive symbolism. Moreover, the philosophy of the *Changes*, with its subtle intermingling of Confucian and Daoist themes, is both profound and culturally satisfying. Psychologically speaking, the *Yijing* encourages introspection and self-knowledge by challenging the reader to make unanticipated connections between events and conditions. Chinese scholars have long employed the document in this way, even if only unconsciously, as they attempted to "resolve doubts" (to borrow a cliché from the *Changes*). The psychologist Carl Jung has emphasized that despite "the multitude of problems that arise when we seek to harmonize the oracle ... with our accepted [Western] scientific canons," the book has genuine value as an admonition to "careful scrutiny of one's own character, attitude, and motives."[127]

And, of course, the classic is still used as an explicitly "spiritual" work—a divination manual designed to offer practical guidance. Fortune-telling books are big business on the Chinese mainland these days. The trend began with an enormously popular work by the husband-and-wife team of Hong Pimo and Jiang Yuzhen, titled *Zhongguo gudai suanming shu* (The divinatory arts of ancient China), published in 1990.[128] Although ostensibly critical of divination as a form of "feudal superstition," the authors devote a great amount of space to specific fortune-telling techniques, including those linked with the *Yijing*. In the eyes of many contemporary scholars, however, this popularization of the *Changes* has only succeeded in debasing it. For instance, Liu Zheng wrote at about the same time that Hong and Jiang were writing:

> [I]n many bookstores and bookselling stalls ... throughout the country, we can easily find books relating to the study of the *Yijing* sitting side by side with books that deal with such subjects as sexual physiology and psychology, or even "literary" magazines and periodicals that are highly and explicitly sexual or "erotic" in their contents and descriptions. This is a fact that those of us who are used to discussing the *Yijing* as some lofty and highbrow philosophy ... cannot deny or dismiss out of hand.[129]

In short, the hermeneutical and divinatory traditions of the *Changes*, both past and present, have always allowed, and even encouraged, an extraordinarily wide, almost unimaginable, range of interpretive techniques, and the expression of a great many different philosophical viewpoints. There is little doubt that, for better or worse, they will continue to do so.

2

MAPPING CHINA'S WORLD

Cultural cartography in late imperial China

For encapsulating a worldview there is nothing quite like a world map. As with other forms of cartography, *mappaemundi*—whether medieval or modern, Asian or Western—tell us about values and attitudes, aims and aspirations, hopes and fears; but they express them on a particularly grand, indeed global, scale.[1] To the extent that such productions in any given society share affinities across space and time, they reveal significant features of that culture's self-image (and, of course, its conceptions of the "other"); and to the degree that they do not, they suggest changes, ruptures, tensions, and conflicts within the larger cultural system. With these considerations in mind, I would like to look at the evolution of Chinese maps of the world during late imperial times—from the twelfth to the twentieth centuries—focusing on two basic questions: How did changing conceptions of "the world" shape the contours of Chinese cartography?, and how did changing (as well as enduring) cartographic practices affect Chinese conceptions of the world?

Significant methodological and practical problems attend such questions. In the first place, it is often difficult to determine where a map of "China" ends and a map of "the world" begins. Large-scale cartographic representations of space in late imperial times present us with a number of overlapping political, cultural, and geographical images, identified either by dynastic names (Song, Yuan, Ming, and Qing) or by more enduring, but still fluid, designations such as the Central [Cultural] Florescence (*Zhonghua*), the Spiritual Region (*Shenzhou*), the Nine Regions (*Jiuzhou*), the Central Kingdom (*Zhongguo*), the Central Land (*Zhongtu*), and All under Heaven (*Tianxia*). The relationship—as well as the distinction—between these time-honored concepts is by no means always clear in traditional Chinese maps.[2]

Another difficulty has to do with access. Not surprisingly, a number of politically sensitive maps that deal with disputed territory are still not generally available to foreign scholars doing cartographic research in the People's Republic of China.

Moreover, as far as I know, there is no single, comprehensive collection of Chinese "world maps" anywhere. Rather, they are scattered all over the globe—not only in Asian archives (primarily China, Taiwan, Japan, and Korea), but also in Russia, various European countries (including Italy, Spain, France, Germany, England, and Holland), and North America.[3]

In the course of my research, I have been able to examine well over a hundred Chinese manuscript maps and woodblock prints in various Asian, European and American collections.[4] I have also consulted a number of other valuable primary materials in Chinese, dating from the Song through the Qing dynasties, and have investigated much of the large and growing body of secondary literature on Chinese cartography in both Asian and Western languages.[5] Still, this chapter must be seen as only a preliminary investigation, an initial voyage of exploration into a vast and potentially very productive field of study.

Some methodological issues

A number of scholars, both within and outside the field of Chinese studies, have vehemently criticized what is generally described as a Parsonian version of culture (i.e., culture as "systems of symbols and meanings") for contributing to various "totalizing" and "essentializing" orientalist projects, including the rise of "academic modernization theory" and "imperialist development policy." It has been blamed for creating a "neat divide between 'Oriental' culture and 'Western' reason," and for providing "the most convenient" explanation for the "willful backwardness and irrationality [of so-called traditional societies] in the face of rapid global modernization." In the view of critics such as Judith Farquhar and James Hevia, the reification of ideas and values encouraged by Parsons and his disciples has led to a "static and stagnant" conception of culture that justifies Western aggression and represents imperialism as "a salvation project."[6]

It is not clear to me that all, or even most, of those who have employed some sort of Parsonian notion of culture in their academic writing (including John Fairbank within the China field and Clifford Geertz without) are guilty of such crimes. Nor am I convinced that the long-posited relationship between ideas, values, intentions, ideologies and other forms of consciousness on the one hand and human behavior or "action" on the other is wrong-headed. One can argue, I think, for positioning culture "in the materiality and (messiness) of everyday life"[7] without disengaging it entirely from the realm of thought.

My own work on Chinese culture has tended more toward the generalizing than the particularizing side of the interpretive spectrum. I would hope, however, that it is not viewed as either a "totalizing" or an "imperialist" enterprise, and certainly not one that valorizes or privileges Western culture in any way.[8] As I have tried to indicate in numerous writings, my interest remains in sustaining a dialectic between holistic and particularistic studies of China. Pamela Crossley is assuredly correct in maintaining that the geographical and cultural entity of China is "a totality of convergently and divergently related localisms," and that Chinese culture is a product of the "challenging and differentiating effects of aboriginal, border and heterodox cultures."[9] But simply to speak of aboriginal, border and heterodox cultures is to acknowledge implicitly a hegemonic, "central," and "orthodox" one in constant tension with them. What, one may legitimately ask, were the features of that larger culture and how were versions of it transmitted, transformed and reinforced over time?

My interest is in the idea of culture as "classification"—that is, the way groups of Chinese (whether they see themselves primarily as "the Han people," "Guangdong people," or whatever) name and arrange things and ideas into coherent systems of meaning. In this respect I identify with the interpretive outlook of Marshall Sahlins.[10] I am perfectly willing to acknowledge, however, that the "Chinese" cultures under consideration are neither static nor monolithic. They vary across time and space and according to class and gender; they are constantly changing, and are always situated in particular social and political contexts. I would also grant that these cultures are invariably the product of some sort of "invention," and that the cultural meanings produced are constantly contested by different groups and individuals.

Yet there remains a sense in which people share an identity that not only can be encapsulated by one or more self-referential terms (as opposed to designations imposed from without) but also described (again from within) as a constellation of commonly accepted attributes, attitudes, and concerns. As Sahlins points out, "In order for categories to be contested . . . there must be a common system of intelligibility, extending to the grounds, means, modes, and issues of disagreement." It would be difficult, he argues,

> to understand how a society could function, let alone how any knowledge of it could be constituted, if there were not some meaningful order in the differences. If in regard to some given event or phenomenon the women of a community say one thing and the men another, is it not because men and women have different positions in, and experience of, the same social universe of discourse?[11]

There would seem to be no point in asserting a collective identity unless there is an identifiable "other" that stands in opposition to it—"us" versus "them." I am interested in what the "us" consists of, and how it may influence the way individuals within that self-identified group operate. Maps—"world maps" in particular—seem to be a revealing yet rather neglected way to reveal conceptions of the other, thus revealing something important about the collective self. As Sahlins remarks, "Divinities or enemies, ancestors or affines, the Others are in various ways the necessary conditions of a society's existence." A more or less "self-conscious fabrication of culture," constructed in response to "imperious outside 'pressures'" is thus a "normal" historical process.[12]

Robert Rundstrom has observed that mapping "is fundamental to the process of lending order to the world."[13] Yet quite clearly there are many ways of worldmaking.

In Denis Wood's vivid formulation: "Every map shows *this* . . . but not *that*, and every map shows what it shows this *way* . . . but not *the other*."[14] In other words, cartographers construct the world, they do not reproduce it. Places are where they are, but maps represent them where the mapmakers want them (or need them, or think them) to be. Every map, then, has an author, a subject and a theme (or themes). No map is a neutral document. All reflect efforts of one kind or another to impose oneself (or one's culture) on physical space. A map is an interpretation that needs, in turn, to be interpreted.

How should one go about doing this? William Boelhower emphasizes the need to take special notice of the map's hybrid nature, its "dissimulating cleverness" and "the complexity of its conventions." He urges us to feel the map's "political muscle," salute its "military potential," and delight in its "aesthetic seductiveness." In Boelhower's view, to understand the peculiar "generative logic" of maps requires a perspective that gives to the viewer a sense of the map's "spatial dynamism, its temporal narrativity, and its unfailing subjectivity."[15] Such an interpretive stance requires the "reader" to understand how images, words and lines produce symbolic information together, and how the world is transformed into these three different but related ways of "encoding space."[16]

G. N. G. Clarke alerts us to the importance of inscriptions and other forms of "decorative" art on maps. To view cartouches as merely a source of visual pleasure, or to see them as fundamentally a distraction from the main content of a map, is to "deny the complex textuality held within the look of the map." Such a perspective, he says,

> not only fails to give the map its necessary cultural status; it ignores the subtle relationship between the scientific and decorative; it fails to see them ... as a series of interrelated indexes which bind the map within a series of ideological assumptions as to the way the land is viewed.[17]

Thus, inscriptions, colors, the presence or absence of overt symbols, even the thickness of lines, may provide clues as to the cultural purposes of maps.[18]

Many theorists emphasize the use of cartography as a means of asserting political and social control. J. B. Harley writes, for instance:

> Both in the selectivity of their content and in their signs and styles of representation, maps are a way of conceiving, articulating and structuring the human world which is biased towards, promoted by, and exerts influence upon particular sets of social relations. By accepting such premises it becomes easier to see how appropriate they are to manipulation by the powerful in society.[19]

David Harvey states more succinctly, "command over space is a fundamental and all-pervasive source of social power."[20] As products and symbols of various kinds of authority (moral, "scientific," etc.), maps make distinctions that favor certain interests, "culturalizing the natural"[21] through the process of identifying and naming, categorizing and containing.

Although maps are usually viewed as representations of space, they can also be taken as spaces of representation—fields of opportunity, waiting to be cultivated by acts of physical or intellectual appropriation or both.[22] Indeed, as Boelhower points out in "Inventing America: A Model of Cartographic Semiosis," the map, as a "cultural sign," provides an "ideal text for studying the way Indian land was transformed into Euro-American territory and settlers from various nations into a homogeneous ethnos, as the ideological boast goes."[23] According to Boelhower, it was not so much the "discovery" of the New World that mattered as the particular way that it was *seen*—the sense of possibility that maps opened up. How, we might now ask, did Chinese cartographers view their craft and their world? What sort of possibilities did Chinese maps present?

Some general features of Chinese cartography

From ancient times, maps have served a variety of purposes in China.[24] Many were designed as practical educational tools for scholar-officials, to guide, instruct and edify in times of both peace and war. They were also employed as a concrete means of asserting the emperor's territorial claims, whether local, empire-wide, or world-wide. Maps became symbolic tokens of exchange in China's domestic and foreign relations, and were even used to depict a perceived link between the realms of Heaven and Earth. Significantly, they also provided a means by which viewers could take "spiritual" journeys to distant lands—the cartographic equivalent of "travelling [through a landscape painting] while remaining at rest [*woyou*]."[25]

Chinese mapmakers tended to be broadly gauged scholars and artists rather than narrow technicians. Until the late nineteenth century there were no professional or specialist cartographers as such in China.[26] The scholars who created maps saw their productions as part of a larger intellectual and cultural enterprise—one that embraced not only science (especially astronomy and geography) but also history, philosophy, religion, art, literature, and religion (including divination). "History" was an especially prominent value in Chinese maps. Many cartographic collections, and even individual maps, bear titles indicating that they are concerned with the relationship between the "past and present" (*gujin*), or between successive dynastic periods (*lidai*). In other words, time and space remained closely connected in imperial China.[27]

On the whole, explicitly religious maps seem to have been less popular in the Central Kingdom than in other parts of Asia, such as Burma, Korea, Japan and Tibet. We do, however, find Chinese cartographic works in both the Religious Daoist and Buddhist traditions. One such work, known as the *Sihai Huayi zongtu* (General map of Chinese and barbarian [lands] within the four seas) (Figure 2.1), purports to show the Buddhist continent of Jambudvipa (Chinese: Nanshanbuzhou), but replaces India as the principal geographical focus with China. The Chinese landscape, with its provinces, major rivers, mountains, and the Great Wall, is

FIGURE 2.1 General Map of Chinese and Barbarian Lands within the Four Seas, undated

Source: Zhang Huang, 1613, in SKQS.

depicted in considerable detail, while India recedes to comparative insignificance in the southwest.[28] A distinctive feature of this sort of map is its strong affinity with Korean "wheel maps."[29]

A number of Chinese maps indicate, sometimes explicitly, a concern with the principles of "siting" or "geomancy" (*kanyu*, *fengshui*, etc.). A central feature of this cosmological system is the belief that certain geographical forms and/or spaces will bring good fortune.[30] Softly undulating rectangular shapes, for instance, are generally considered auspicious, as are lines of protective hills and mountains.[31] Land configurations that envelope important spaces (in the fashion of the flanks of well-positioned gravesites), and waterways that nourish these areas, are also esteemed. As Philippe Foret and others have pointed out, Chinese mapmakers were not above adding such topographical features to their cartographic productions in order to depict (create) a more favorable geomantic environment.[32] And where hills and mountains already existed but were separated by flat expanses of land that seemed to diminish their collective power, mapmakers might edit their rendering of the scene to give it greater geomantic strength. Sometimes places would simply be relocated in maps to give them a more favorable geomantic position, or altered in appearance for similar reasons.[33]

Overall, Chinese cartographers treated large-scale space, including the world itself, as essentially flat. Although mathematical astronomers used ecliptic as well as equatorial coordinates in their celestial mapping, cartographers saw no need to project them on the earth. As a result, they "simply acted as if they were transferring points from a very large flat surface to a smaller one."[34] At the same time, however, Chinese mapmakers often employed variable perspective and variable scale. Thus, for example, mountains might be drawn in elevation while rivers would appear in plane. Moreover, the size of objects relative to one another, as well as their distance from one another, were usually dictated not by their actual dimensions or by geometrical perspective but rather by the specific purposes for which the map was produced. Heavy annotation provided valuable information that might otherwise have been expressed by graphic images of scale.[35]

Chinese maps often devote more space to the written text than to the actual geographical image. Although the tendency for historians of cartography has been to denigrate heavily annotated maps in favor of more "representational" ones, there is no intrinsic reason for doing so. It was not, after all, lack of skill or "backwardness" that determined the nature of traditional Chinese cartography. In China, for cultural reasons, the written word, rather than visual images, remained the primary source of representational authority (Figure 2.2).[36]

Cartographic texts in China commonly provided technical data concerning roads, waterways, landmarks, distances, and so forth. But they also supplied important

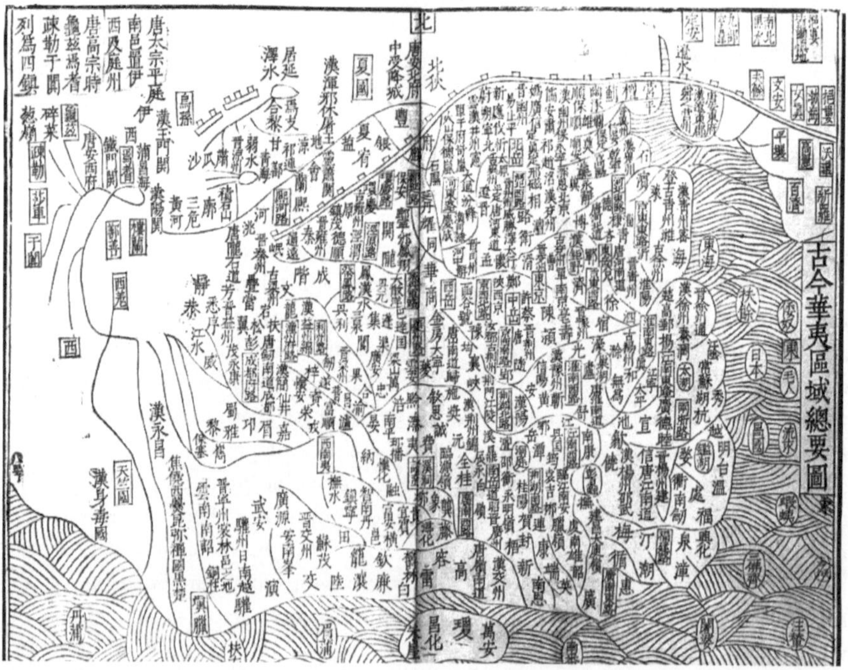

FIGURE 2.2 General Map of Chinese and Barbarian Territories, Past and Present

Source: http://geog.hkbu.edu.hk/GEOG1150/Chinese/catalog/Catalog.htm.

political and cultural information. An excellent illustration can be found in a "geographic map [of China]" (*Zhuili [Dili] tu*), created by a scholar named Huang Chang in the 1190s—several decades after the fall of the Northern Song capital of Kaifeng to the invading Ruzhen people. The map was intended as an illustration for the future Song emperor (Ningzong, r. 1194–1224) of how much land had been lost to the northern invaders, and as a reminder of the sovereign's responsibility to reunite the empire. The commentary to the map addresses the perennial problem of keeping China together, observing that "only one out of every ten [rulers] has been able to bring unity to all under Heaven."[37]

This discussion—which constantly emphasizes morality as the key to administrative success—is full of historical allusions to events such as the invasion of China by the Qidan people in the early tenth century and the rebellion of An Lushan in the mid-eighth century, as well as to the noteworthy unifying accomplishments of the sage-rulers Tang and Wen, who, despite having only modest territories to begin with (like the Southern Song), founded the great Shang and Zhou dynasties, respectively. The commentary naturally includes references to northern landmarks that had recently fallen under "barbarian" control—including the Yellow River, the Great Wall, and "a vast forest stretching several thousands of *li*."[38]

The textual emphasis of traditional Chinese cartography did not in any way undermine the aesthetic appeal of maps. On the contrary, inscriptions often enhanced it. In contrast to the development of cartography in Europe, where manuscript maps became rather rare following the spread of copper engraving in the late fifteenth century, manuscript maps continued to be produced in large numbers in China throughout the late imperial era.[39] These documents, like landscape paintings, were tastefully shaded and often complemented by substantial amounts of calligraphy—sometimes even poetry. Printed maps could also be extraordinarily beautiful, with handsome, well-cut cartouches, and carefully colored natural features. Neither type of map could be considered true art, however, for both lacked the qualities of "life force" (*qi*) and "kinesthetic power" (*shi*) that distinguished artistic creativity from mere craftsmanship.[40]

Yet another distinctive feature of Chinese cartography is what Cordell Yee describes as its tendency toward introspection—a self-conscious preoccupation with concrete administrative concerns.[41] Buildings and walls, for example, tend to loom large, quite literally, in many kinds of maps.

Paradoxically, Chinese "introspection" included looking outward. That is, one of the emperor's traditional "domestic" concerns as the ruler of "all under Heaven" was the management of foreign peoples—whether on the periphery of his realm or beyond. These so-called "barbarians" (*yi*, *fan*, *hu*, etc.), although by definition not fully Chinese, were all at least theoretically the emperor's "subjects."[42] Many of them periodically sent him local products, designated "tribute" (*gong*), and, in return, expected the Son of Heaven to protect and nurture them. From a Chinese standpoint, this highly refined system of "guest ritual" (*binli*), which allowed foreigners the opportunity to demonstrate their loyalty to the Chinese emperor, was the logical extension of an ancient "feudal" structure of lord–vassal relationships, to

FIGURE 2.3 Map of China and the Barbarians, 1185

Source: Library of Congress, Geography and Map Division.

be discussed more fully below. It was this Sinocentric assumption of universalistic overlordship that often blurred the distinction between maps of "China" and Chinese maps of "the world."

Images of All under Heaven

The earliest extant "world maps" in China date from the Song dynasty (960–1279).[43] One example is the *Gujin Huayi quyu zongyao tu* (General map of Chinese and barbarian territories, past and present), which dates from about 1100.[44] Another closely related and far more famous example is the *Huayi tu* (Map of China and the barbarians; 1136, Figure 2.3).[45] This latter work, about 3-foot square and carved in stone, supplies hundreds of place names and identifies a dozen or so

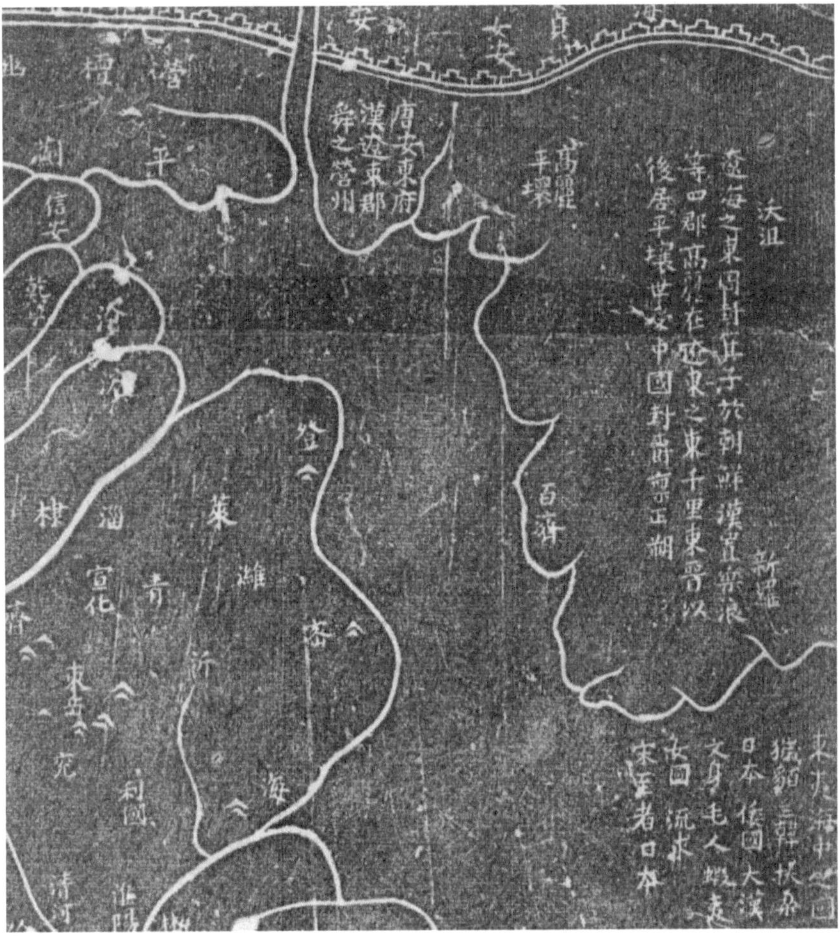

FIGURE 2.4 Detail from the Map of China and the Barbarians, showing the Great Wall and including descriptions of the Koreans, the Japanese and other "barbarians"

Source: Library of Congress, Geography and Map Division.

rivers and tributaries in China. A few foreign lands are represented visually in the map—notably, Korea and India—but a great many more groups of "barbarian" peoples are indicated by written notes on the margins of the map (Figure 2.4). Near the top, on the northwestern side we learn, for instance, that the area of the formerly enfeoffed Qidan people "is now called 'the Great Liao Country'." Several such notes refer specifically to tributary relationships, past and present.[46]

Not all Song dynasty renderings of space arose from the same source, however. Indeed, inscribed on the reverse side of the *Huayi tu* is an astonishingly "modern-looking" version of an ancient work called the *Yuji tu* (Map of the tracks of Yu; 1136, Figure 2.5), probably created about 1080.[47] It marks the earliest extant

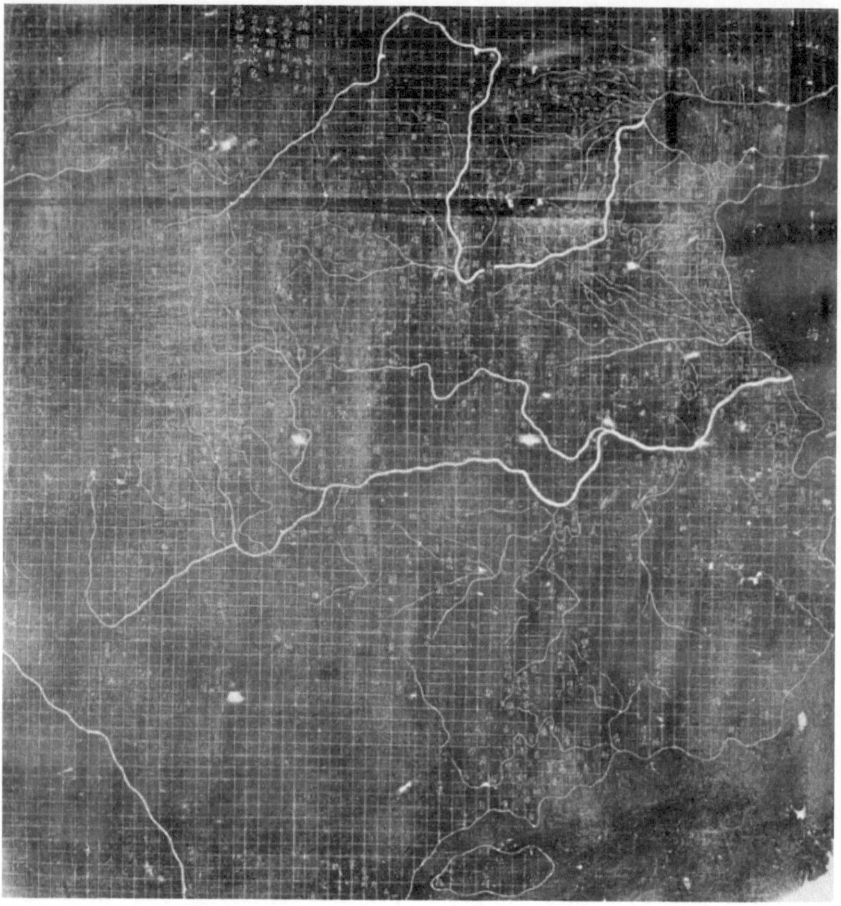

FIGURE 2.5 Map of the Tracks of Yu, 1136

Source: Library of Congress, Geography and Map Division.

example of the so-called "latticework" cartographic grid in China. Each side of each square represents 100 *li* (*c.* 33 miles), yielding a scale of about 1:1,500,000. The outstanding feature of this map, in addition to the near total absence of written commentary, is its remarkably accurate depiction of major landforms, including the Chinese coastline and the Yellow and Yangzi Rivers.[48]

The grid system of the *Yuji tu* apparently provided the model for Zhu Siben's influential *Yutu* (Terrestrial map; *c.* 1320). The only extant version of this work is Luo Hongxian's *Guang Yutu* (Enlargement of the terrestrial map), first published in 1579 (Figure 2.6). Luo's production takes the form of an atlas, with more than forty separate maps—including a "General map [of China]" (*Yudi zongtu*) and a "General map of China and the barbarians" (*Huayi zongtu*). Like the *Yutu*, the *Guang Yutu* employs a grid system, but unlike Zhu's map, Luo includes a number of cartographic

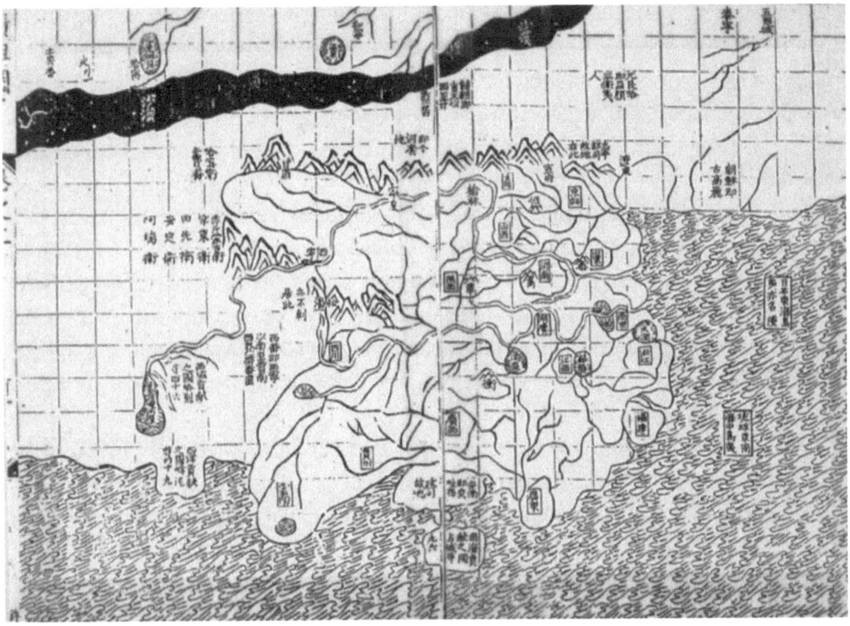

FIGURE 2.6 Luo Hongxian's General Map of China, 1579

Source: Luo, 1579.

legends—twenty-four in all—for mountains, rivers, boundaries, roads, and other landmarks.[49]

Luo's atlas obviously reflects, not least in its abundant written texts, the expansion of Chinese knowledge about the rest of the world gained during the Mongol ascendancy and then in the course of the eunuch-admiral Zheng He's extensive naval expeditions during the early fifteenth century—voyages that took him as far west as the shores of East Africa.[50] As one measure of its comprehensive scope, the *Guang Yutu* includes an elaborate chart that distinguishes the residents of over 120 foreign countries by area: Eastern Barbarians (Koreans and Japanese), Southeastern Barbarians (Liuqiu Islanders), Southern Barbarians (Southeast Asians), Southwestern Barbarians (Filipinos, Indians, Westerners, etc.), Barbarians of the "Western Regions" (including various Turkic peoples) and Northwestern Barbarians (Mongols and other such tribes). Many of these peoples are designated "tributaries," not only in the chart but also on some of the maps themselves.

Luo's work spawned a number of imitations, including the *Da Ming guangyu kao* (An examination of the enlarged terrestrial [map] of the Great Ming dynasty; 1610) and Chen Zushou's *Huang Ming zhifang ditu* (An administrative map of the Ming dynasty; 1636), banned during the Qing period.[51] It is important to remember, however, that most large-scale Chinese maps of the late imperial era continued to conform to the gridless *Huayi tu* cartographic model.

The most striking and expansive example from the early Ming period is the magnificent, multicolored *Da Ming hunyi tu* (Amalgamated map of the great Ming empire; *c.* 1390, Figure 2.7). Drawn on a horizontal scale of 1:820,000 and a vertical scale of 1:1,060,000, it covers an area extending all the way from Japan to the Atlantic Ocean (including both Europe and Africa), and from Mongolia to Java. Although the section on China seems to be derived primarily from Zhu Siben's *Yutu*, the renderings of Africa, Europe and Southeast Asia appear to have been based at least in part on Li Zemin's *Shengjiao guangbei tu* (Map of the vast reach of [China's moral] teaching; *c.* 1330), no longer extant.[52] A Korean version of the *Da Ming hunyi tu*, titled *Yoktae chewang honil kangnido* (Amalgamated map of the borders and terrain of historical countries and capitals; aka Kangnido; 1402), is noteworthy for depicting Korea as nearly twice the size, proportionally speaking, as it appears in the *Da Ming hunyi tu*.[53]

Subsequent Chinese maps based on the *Da Ming hunyi tu* model tended to be somewhat more restricted in geographical scope, but still impressive in their coverage. The best sixteenth-century example is Yu Shi's *Gujin xingsheng zhi tu* (Map of

FIGURE 2.7 Amalgamated Map of the Great Ming Empire, *c.* 1390

Source: http://geog.hkbu.edu.hk/GEOG1150/Chinese/catalog/Catalog.htm.

advantageous terrain, past and present; 1555).[54] This beautifully colored and heavily annotated work—representing an expanse of territory stretching from Samarkand, India and Arabia in the west to Japan in the east, and from present-day Mongolia in the north to Java and Sumatra in the south—lacks any sort of grid. It elongates Korea, treats the Shandong peninsula as if it were an island, and, like the *Huang Ming yitong dili zhi tu*, refers to the existence of several mythical places derived from the ancient *Shanhai jing* (Classic of mountains and seas). This work—the earliest known illustrated account of aliens in China—describes a great number of foreign lands with all sorts of exotic inhabitants: societies consisting of women, or giants, or dwarfs; people with multiple heads or bodies; creatures with the heads of humans and the bodies of snakes; and so forth.[55]

A horizontally oriented version of Yu's map, titled *Gujin tianxia xingsheng zhi tu* (Map of the advantageous terrain under Heaven, past and present; n.d., Figure 2.8).

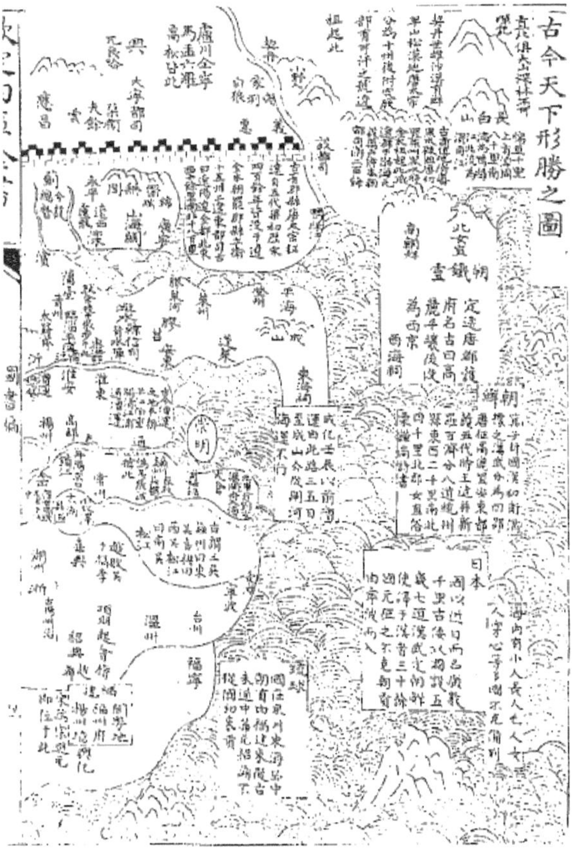

FIGURE 2.8 Detail of the Map of the Advantageous Terrain under Heaven, Past and Present in Zhang Huang, 1613

Source: http://geog.hkbu.edu.hk/GEOG1150/Chinese/catalog/Catalog.htm.

appears in Zhang Huang's *Tushu bian* (Compilation of illustrations and writings; 1613). Like Yu's production, it is full of historical references, including information on the activities of China's "barbarian" neighbors, the development of the Chinese tributary system, and various administrative changes within the Chinese empire. To an even greater extent than Yu's map, the *Gujin tianxia xingsheng zhi tu* identifies the homes and/or exploits of China's great cultural heroes, ranging from Confucius and his followers, to the Tang poet, Li Bai, to the founding emperors of the Tang, Song and Ming dynasties. It also refers to the activities of several prominent Chinese loyalists, including Zhuge Liang of the Three Kingdoms period and both Yue Fei and Wen Tianxiang of the Song.[56]

Joseph Needham argues that there was a general "advance" in Chinese mathematical cartography from the Song period into the seventeenth century.[57] In fact, however, the evolution of map-making in China cannot be characterized as simply a linear process of "progressive" improvement. Rather, Chinese cartographers continued to produce two distinctly different types of maps—one based on relatively precise mathematical measurements, and one based primarily on cultural data—without explicitly recognizing the existence of two competing traditions. If a characterization is required, it would have to be that maps of the latter sort greatly outnumbered those based on more mathematical models—not only up to the seventeenth century but well beyond. On the other hand, as we shall see, a number of cartographic documents of the late Ming and Qing periods placed the two types of maps together, in the spirit, one suspects, of the complementary maps engraved on the two sides of the Song stele of 1136.

Jesuit cartography and its limits

Many modern scholars, both Western and Chinese, have seen the arrival of the Jesuits in China during the late sixteenth century as a landmark in the history of Chinese map-making.[58] In fact, however, their influence was rather limited. To be sure, Jesuit scientific methods, including sophisticated surveying techniques, enabled the Qing dynasty (1644–1912) to create a far more mathematically "accurate" map of the Chinese empire than had ever been produced before—the *Huangyu quanlan tu* (Map of a comprehensive view of imperial territory; 1718).[59] This massive work, the product of many years of dedication by both the Jesuits and Qing scholars, provided China's Manchu rulers with an important instrument of political and military control, and it remained the most authoritative atlas of the realm for nearly two centuries.[60]

But from the standpoint of world maps, Jesuit *mappaemundi*—including Matteo Ricci's *Yudi shanhai quantu* (Complete map of the earth's mountains and seas; 1584, Figure 2.9; Japanese version, Figure 2.10), his *Kunyu wanguo quantu* (A complete map of the myriad countries of the world; 1602), Giulio Aleni's *Zhifang waiji* (Notes on [world] geography, 1623), and Ferdinand Verbiest's *Kunyu tushuo* (Illustrated discussion of the geography of the earth; 1674)—had little long-term influence.[61] Whereas precise maps of the empire had obvious strategic value, especially for the

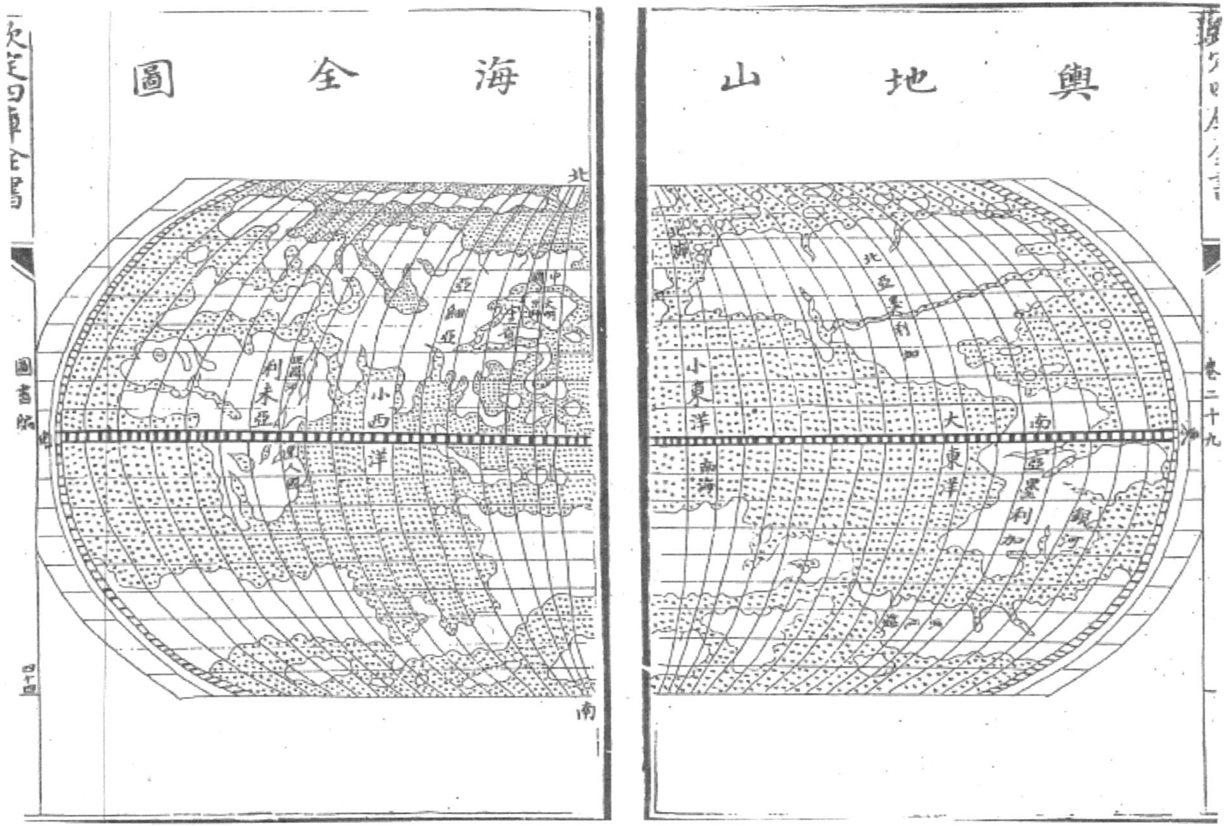

FIGURE 2.9 Complete Map of the Earth's Mountains and Seas, 1584, reproduced in Zhang Huang, 1613

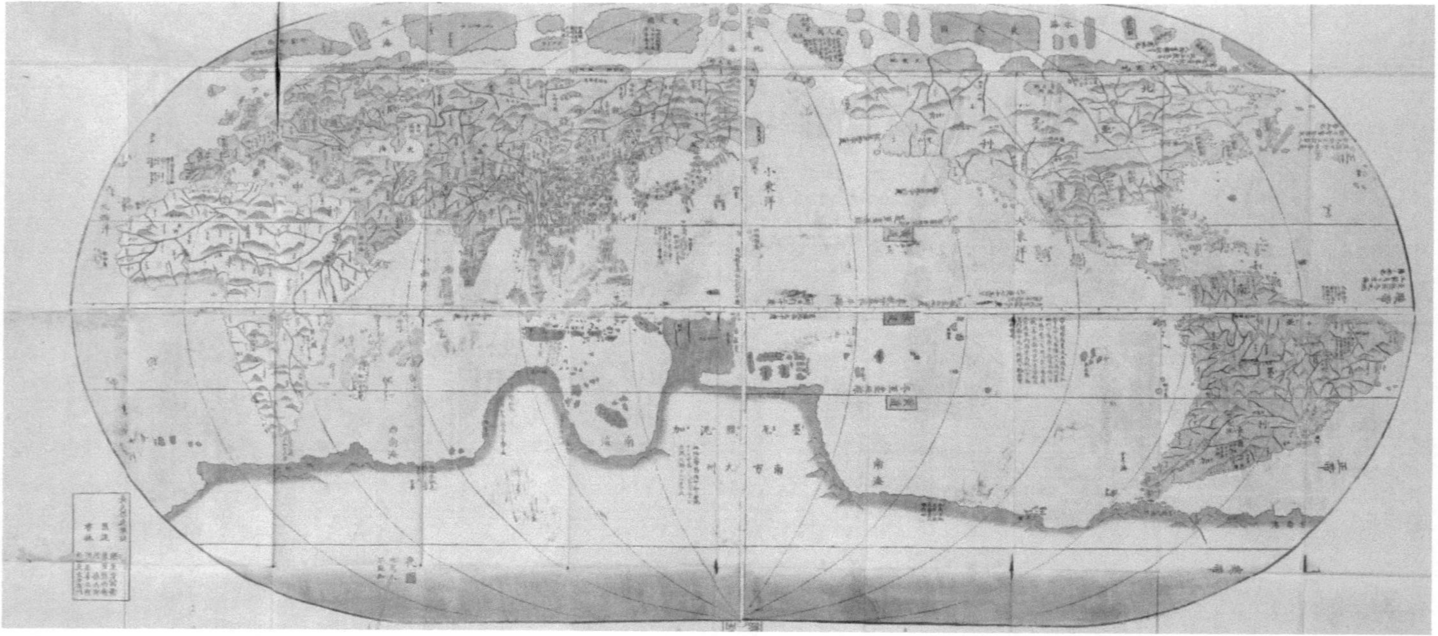

FIGURE 2.10 A Japanese version of Ricci's Complete Map of the Earth's Mountains and Seas (*Sankai yochi zenzu*; *c.* 1785)

Source: Library of Congress, Geography and Map Division.

expansive but alien and somewhat insecure Manchus, world maps had a different function altogether. They were designed primarily as visual statements about a great and glorious culture, a universal order focused squarely on the Chinese tributary system.[62] Indeed, one gains the impression that many Chinese world maps were constructed as if they were to be seen by the emperor himself.

It is not surprising, then, to find that a number of Chinese scholars bitterly attacked the Jesuits for misrepresenting the world and China's place in it. According to one Ming scholar, Wei Jun, Ricci's map not only contained "fabulous and mysterious" information that could not be verified, but in locating China to the west of center and inclined to the north, it dislodged the "Central Kingdom" from its rightful position at "the center of the world." How, Wei asked, "can China be treated like a small unimportant country?"[63] Similarly, the *Huangchao wenxian tongkao* (The imperial dynasty's comprehensive examination of source materials; 1787) denounced Ricci's account of the world as full of contradictions, misguided statements and "boastful lies" (*dankuang*). It accused him of belittling China, aggrandizing his own culture, and spreading misinformation in the course of his cartographic work.[64]

Even individuals who claimed to have been directly inspired by the Jesuits often borrowed little of cartographic substance from them.[65] One noteworthy example is a map by the scholar-official Liang Zhou, titled *Qiankun wanguo quantu gujin renwu shiji* (Universal map of the myriad countries of the world, with traces of human events, past and present; *c.* 1600).[66] This work—which appears to have been created more out of defiance than admiration—bears no trace of meridians and arranges foreign locations topologically rather than topographically. About eighty transoceanic lands outside of China appear in this form, in addition to a hundred or so additional foreign places to the north and west. Locations such as North America (on the upper right-hand side of the map) and South America (on the lower right-hand side)—like the Land of Tall People, the Land of Small People, the Land of Women, and many other places drawn directly from the pages of the *Shanhai jing*—are shown as inconsequential islands surrounding the large nucleus of the Chinese Empire.[67]

One might well think that scholars of "empirical research" (*kaozheng xue*) in the seventeenth and eighteenth centuries would have appreciated Jesuit learning, and indeed many did; but most Chinese intellectuals drew quite selectively from the available pool of Western scientific knowledge. Ironically, a deep distrust of symmetry and regularity on the part of *kaozheng* scholars hostile to traditional cosmography led them to reject the notion of a lawful, uniform, and mathematically predictable universe. Thus, for instance, the great Qing intellectual, Wang Fuzhi, dismissed the round-earth concept of the Jesuits out of hand.[68]

Meanwhile, a turn inward in Chinese thought after 1644 diverted attention away from Jesuit-style conceptions of the external world. Gu Yanwu, a towering figure in early Qing scholarship, makes no mention of Jesuit world maps in his otherwise comprehensive *Tianxia junguo libing shu* (Treatise on the advantages and disadvantages of the commandaries and states of the empire; 1662). This lack of a

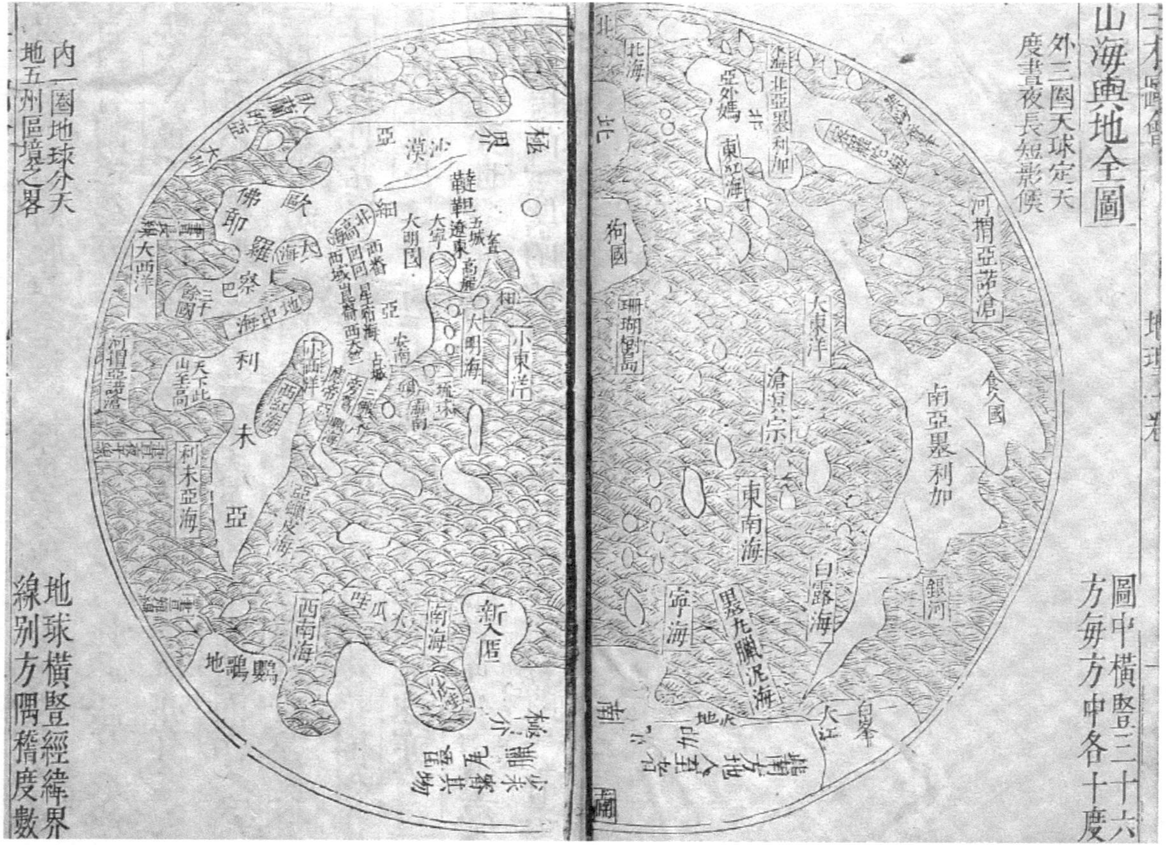

FIGURE 2.11 A world map purportedly based on one of Ricci's maps of the world

Source: The *Sancai tuhui* (Illustrated compendium of the Three Powers; *c.* 1607).

serious interest in the Western world encouraged Gu to describe Portugal (*Fulangji*) as simply a one-time tributary state, located "south of Java," whose early contact with China was for the purpose of studying trade routes and "buying small children to cook and eat."[69]

Of course, cartographic decisions do not necessarily involve either/or choices. In fact, a spirit of compromise animated a number of Chinese mapmakers in late imperial times. Beginning in the waning years of the Ming dynasty, Chinese scholars tried various techniques designed to unite Jesuit-inspired knowledge and more traditional Chinese cartographic renderings of space. An excellent example can be found in Cao Junyi's ambitiously titled *Tianxia jiubian fenye renji lucheng quantu* (A complete map of allotted fields, human events and travel routes [within and without] the nine borders under Heaven; 1644, Figure 2.12).[70] This handsome

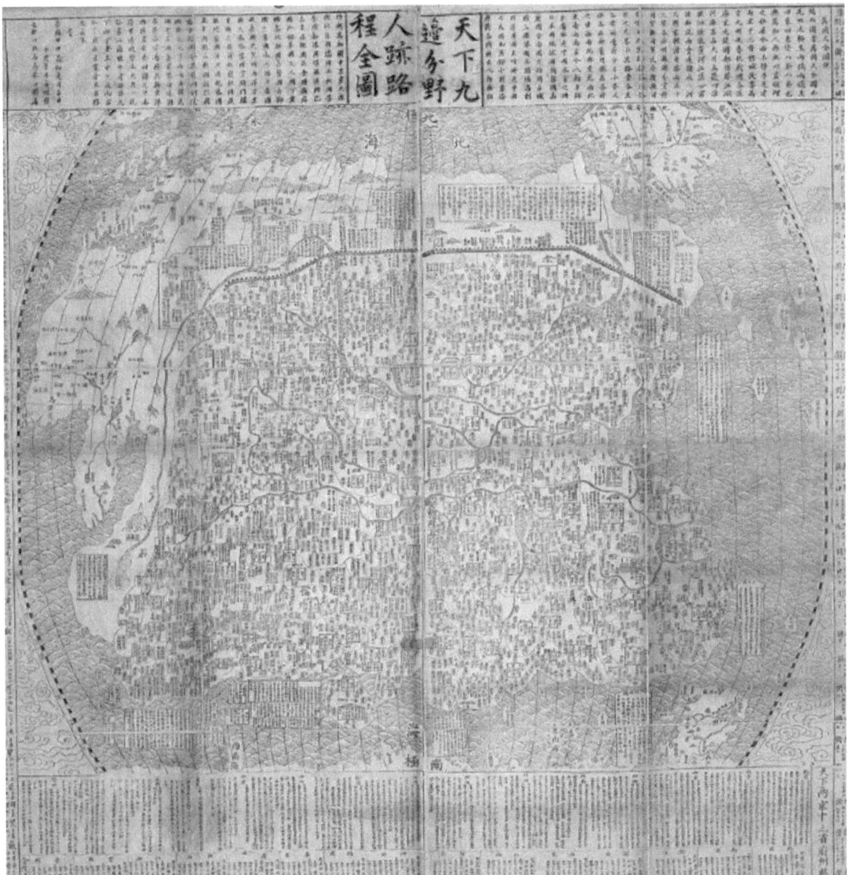

FIGURE 2.12 A Complete Map of Allotted Fields, Human Events and Travel Routes [within and without] the Nine Borders under Heaven, *c.* 1644

Source: By permission of the British Library Board.

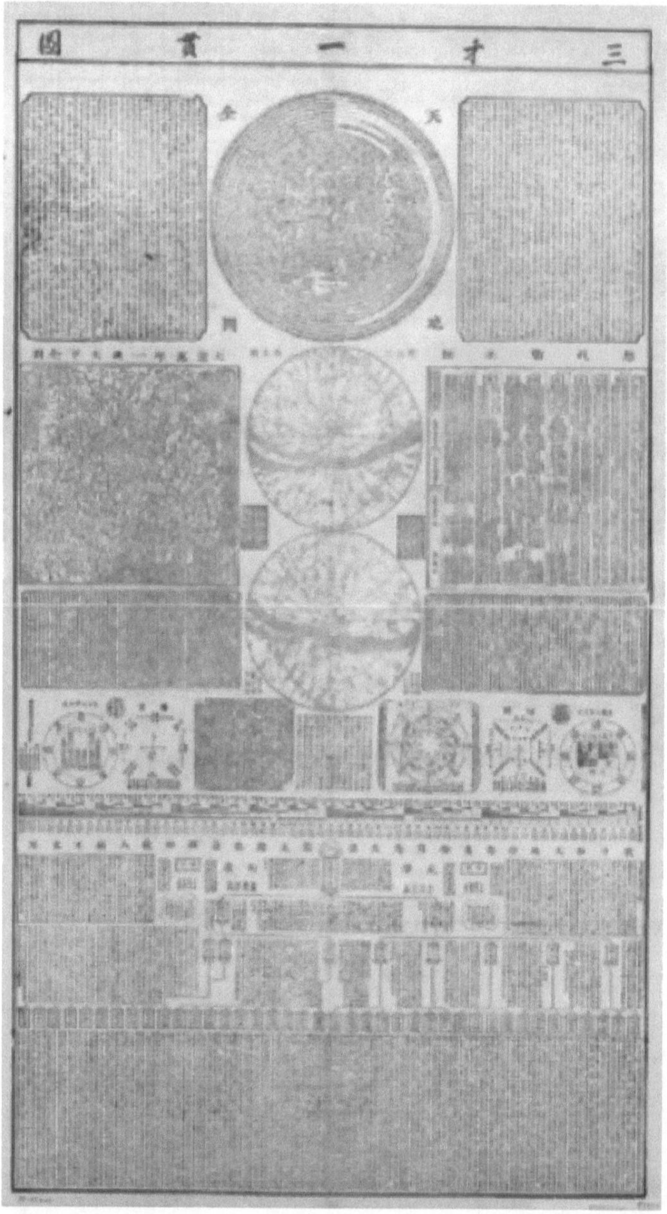

FIGURE 2.13 Illustrations of the Unity of the Three Powers [Heaven, Earth and Man], 1722

Source: Library of Congress, Geography and Map Division.

cartographic document, which continued to serve as a model for cartographers during the Qing period, acknowledges the existence of Europe, Africa, the Middle

East and India, but the two latter areas are represented primarily by cartouches, and Africa—which appears only about one-tenth the size of China—hangs down on the west side of Cao's map as if it were little more than a protective flank. Europe, tiny and even more marginal, is barely recognizable in the upper north-west portion of the map. Most of the place names in these distant areas have been derived from Jesuit sources; but in the south-eastern seas there are a number of mythical countries taken directly from the *Shanhai jing*.

The map gestures toward mathematical accuracy by providing longitudinal lines and degrees, and by supplying the estimated distances of various "barbarian" countries from the southern Ming capital (modern-day Nanjing). Moreover, in his extensive written commentaries, Cao provides a great deal of solid administrative data and historically grounded information on China's strategic rivers, lakes, mountains and seas. At the same time, however, he is pains to locate his discussions of world geography within the traditional confines of both the Chinese tributary system and Chinese cosmology. Furthermore, in his discussions of aliens, he does not differentiate clearly between actual foreign countries and the lands and peoples described in the *Shanhai jing*. The general Sinocentric spirit of Cao's map is captured in the remarks of his contemporary, the cartographer Chen Zushou: "All the barbarian people within the Four Seas should come to pay tribute to the Chinese emperor. Although they [the Jesuits] might describe the world as comprising Five Continents, yet four of them should surround the nucleus of China."[71]

Another kind of cartographic compromise appears in the form of a large anonymous scroll known as the *Sancai yiguan tu* (Illustrations of the unity of the Three Powers [Heaven, Earth, and Man]; 1722, Figure 2.13), archived in the Library of Congress. Although this document consists primarily of a written text dealing with history, morality, cosmology and military affairs, it includes two red planispheres, a "Comprehensive map of Heaven and Earth" (Figure 2.14) and a "Perpetual map of the unified Qing empire" (Figure 2.15). The former map is quite clearly based on Jesuit cartography, while the latter seems to represent an unusual amalgamation of the Song *Huayi tu* and *Yuji tu* traditions.[72]

A similar approach appears in a map produced by Ma Junliang, a 1761 *jinshi* degree-holder who was well known for his skill as a mapmaker. In the 1780s or early 1790s, Ma produced a large and widely distributed woodblock print titled *Jingban tianwen quantu* (Capital edition of a complete map [based on] astronomy, Figure 2.16), which featured a traditional-style rendering of "the world" based more or less on the time-honored model of Liang Zhou. But Ma also offered on the same sheet of paper a pair of global maps—one derived from a loose rendering of Matteo Ricci's *mappamundi* that appears in the Ming encyclopedia *Sancai tuhui* (Illustrated compilation of the Three Powers; *c*. 1607) (Figure 2.17) and one borrowed from a similarly structured Chinese map of the eastern hemisphere, first published by Chen Lunjiong in his *Haiguo wenjian lu* (Record of things heard and seen in the maritime countries; 1730, Figure 2.18).[73]

FIGURE 2.14 Detail from the Illustrations of the Unity of the Three Powers, showing a Comprehensive Map of Heaven and Earth

Source: Library of Congress, Geography and Map Division.

Variations on a theme

Ma's production may be considered an innovative offshoot of a genre of "complete maps of all under Heaven" (*Tianxia quantu*) that arose in the late seventeenth century and seems to have dominated Chinese visual representations of "the world" until at least the mid-nineteenth century.[74] Although these maps all conformed to the basic Liang Zhou model, they also included cartographic symbols of the sort found in the *Guang Yutu*, and easily accommodated variations such as those in Ma's *Jingban tianwen quantu*. Different editions of these attractive, often delicately tinted maps were produced by a series of Qing scholars, including Ma, at least two predecessors—Yan Yong (fl. *c*. 1710) and Huang Qianren (fl. *c*. 1770)—and at least one

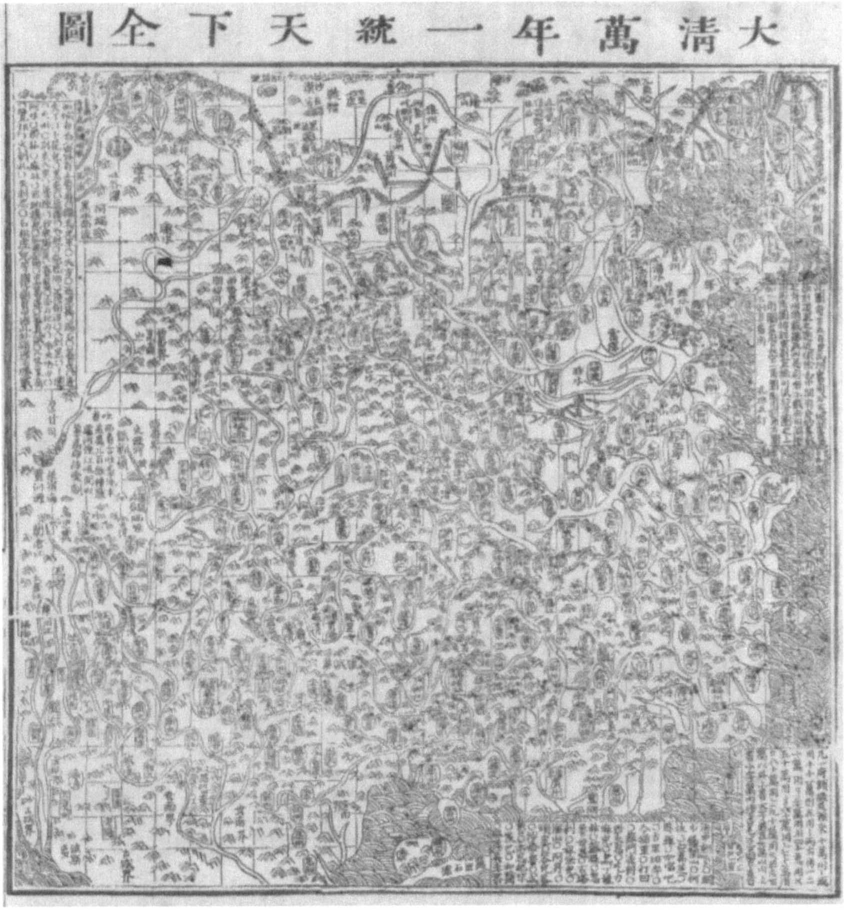

FIGURE 2.15 Detail from Illustrations of the Unity of the Three Powers, showing a Perpetual Map of the Unified Qing Empire

Source: Library of Congress, Geography and Map Division.

successor, Zhu Xiling (fl. *c.* 1820). Of these individuals, only Yan is known to have employed a grid system, following the lead of the great Qing scholar, Huang Zongxi (1610–1695), who produced a rather stark "map of China" (*Zhongguo ditu*) in 1673.[75] Later versions of the *Tianxia quantu* genre—at least the ones that I have seen—lack any traces of a grid, although they generally acknowledge a direct cartographic debt to Huang Zongxi (Qianren's grandfather).

Most maps of this variety are known by the title *Da Qing yitong tianxia quantu* (Complete map of the comprehensive great Qing empire) or a close equivalent. Some, however, bear significantly different names as a way of highlighting certain additions to the basic cartographic format. For example, there is an anonymous work titled *Jingban tiandi quantu* (Capital edition of a complete map of Heaven and Earth) that

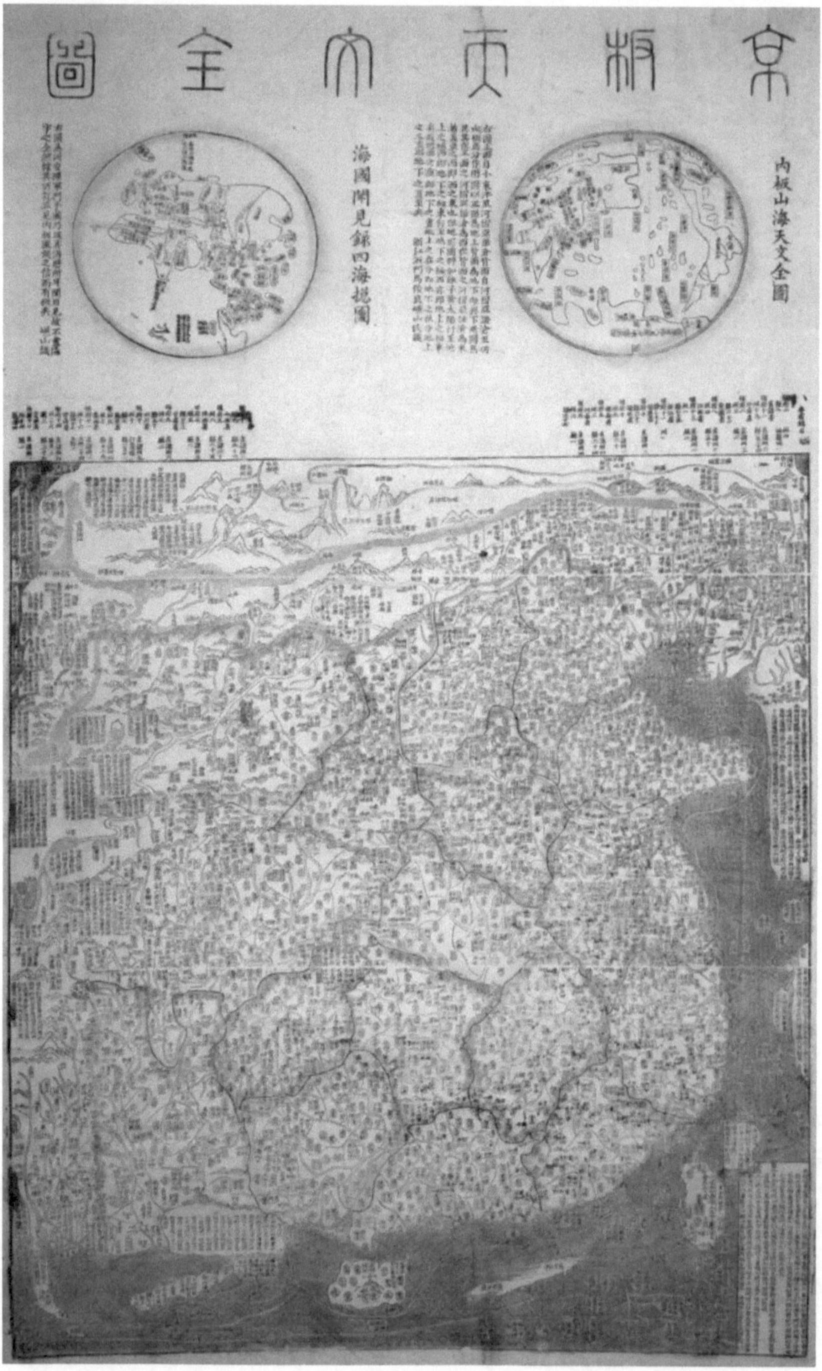

FIGURE 2.16 Capital Edition of a Complete Map [Based on] Astronomy

Source: Rice University Fondren Library, Woodson Collection.

FIGURE 2.17 Detail from a Capital Edition of a Complete Map [Based on] Astronomy, showing a late Ming rendering of Matteo Ricci's *mappamundi*

Source: Library of Congress, Geography and Map Division.

includes a round star chart above the standard terrestrial image.[76] There are also certain minor differences in quality, color, commentaries and a few place names in *Tianxia*-style maps. Some discrepancies appear to be simple scribal mistakes, such as writing "ten thousand *li*" instead of "twenty thousand *li*" (the usual figure) for the extent of the Russian empire; or confusing Zhu Siben's family name with another, similar-looking character. Others involve the expansion or contraction of information—the inclusion or omission of a certain source of authority, or varying degrees of specificity regarding time periods and other minor details. Once in a while there is a major discrepancy. For instance, on some maps, the characters "Small Western Ocean" appear where the characters "Great Western Ocean" would be expected. Occasionally, delicate coloring gives way to much darker and less attractive tones. On the whole, however, the similarities are far more striking than the differences.

Like virtually all large-scale Chinese maps, works of the *Tianxia quantu* genre convey a vivid sense of China's vast and varied landscape: its mountain ranges, overland travel routes, river systems, lakes, coastal communications, and deserts (particularly the Gobi), as well as other prominent landmarks—notably the Great Wall and the "Sea of Constellations" (*Xingxu hai*)—the legendary source of the Yellow River. An intriguing feature of every *Tianxia* map I have seen is a prominent stone tablet (*bei*) erected to Zhuge Liang in the far southwest, presumably out of appreciation for his role in pacifying China's borders during the Three Kingdoms period (222–265). This is the only example of an individual so honored in these maps; even the birthplace of Confucius at Qufu has only a general reference to the Kong family graveyard (*Konglin*), not to the sage himself.[77]

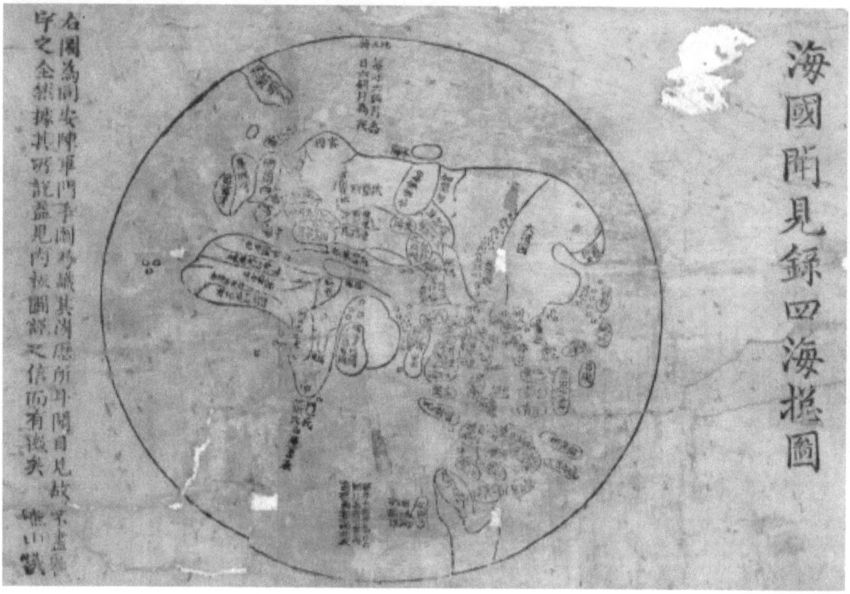

FIGURE 2.18 Detail from a Capital Edition of a Complete Map [Based on] Astronomy, showing Chen Lunjiong's map of the eastern hemisphere

Source: Library of Congress, Geography and Map Division.

As with works of the Yu Shi tradition and those undertaken with Jesuit assistance in the early eighteenth century, maps of the *Tianxia quantu* variety pay close and careful attention to administrative changes within China's provincial boundaries—changes resulting from divisions, consolidations and other adjustments. All maps of this sort employ sets of eight or so cartographic symbols to indicate important administrative centers such as provincial capitals (*sheng*), prefectures (*fu*), departments (*zhou*), districts (*xian*), passes (*guan*), garrison towns (*yingzhen*), and so forth. They also mark the presence of local headmen (*tusi*)—members of ethnic minorities responsible for supervising their own people and for reporting periodically on them to regular Qing officials.

Of particular importance to the authors of *Tianxia* maps are changes in the size and shape of the Chinese empire occasioned by Qing military conquests during the seventeenth and eighteenth centuries. The introduction to every new edition of these maps begins with exactly the same proud phrase: "The land ruled by the present dynasty is unprecedented in its extent." Each cartographer then proceeds to describe the specific political and military developments that made a new map necessary. For the most part, these developments had to do with campaigns by either the Kangxi or the Qianlong emperor, which brought large areas of the Inner Asian frontier under direct Qing control. But they also involved negotiations between the Chinese and representatives of various "outer" peoples, from Tibetans to Europeans.

Significantly, "outer" areas are not at all well defined in maps of the *Tianxia quantu* genre. Representationally speaking, there are no obvious borders separating China from Russia or clearly delineating the individual kingdoms and territories of, say, Central Asia, India, or mainland Southeast Asia. In a few cases boundaries are suggested by written inscriptions; but only the oceans and seas allow certain countries to appear fully detached from the Central Kingdom. On three sides of the Mainland such places are represented as islands, with written descriptions of varying lengths. There is no attempt to show the relative size of foreign realms pictorially, however. Rather, following a pattern established by the *Huayi tu* and continued by such cartographers as Yu Shi and Liang Zhou, the size of foreign territories often appears to be a function of the amount of text deemed necessary to tell the viewer what needs to be known about them. (Unlike these works, however, *Tianxi* maps do not refer to mythical lands from the *Shanhai jing*.) Most inscriptions provide useful historical background on the "barbarians" in question, including place name changes and changes in their relationship to China over time. Sometimes they also supply data on distances, including travel routes and the length of an occasional border.

We should not think, however, that cartographers working in the *Tianxia quantu* tradition were unconcerned with representing foreign lands and peoples as accurately as possible. In a lengthy introduction to his map of 1714, Yan Yong candidly admits that limitations of both information and cartographic space prevent him from showing the actual locations of faraway places. Nonetheless, he has tried to indicate their relative positions and to include textual information on their approximate distance from China.[78] Although most later maps of the *Tianxia quantu* genre do not bother to make this point or to offer systematic data on distances, Ma Junliang's *Jingban tianwen quantu* offers an interesting solution to the problem, as we have seen. By combining two radically different types of maps in one document, he gives his viewers an epistemological choice. Rather than trying to reconcile the two versions, Ma leaves the issue open—a cartographic compromise reminiscent of the Song dynasty stele bearing the two radically different maps of 1136.

The tributary system and Qing cartography

From the Song dynasty onward, virtually all large-scale maps that deal with foreigners of any sort refer to the Chinese tributary system. Works of the *Tianxia quantu* genre are certainly no exception. Invariably they include textual information on the background and evolution of important Sino-foreign tributary relationships, the frequency of certain missions, and major tributary routes. In so doing they reveal a rich lexicon of tributary terminology. The preface to each map refers explicitly to the process by which barbarian envoys come to China and offer themselves as vassals of the Qing dynasty. This process of symbolic submission is often described as an arduous one, involving "the scaling [of mountains], the sailing [of seas], and several stages of translation [*ti hang chongyi*]."

Such cartographic clichés reflect deep-seated attitudes expressed in a number of official documents, including the *Huang Qing zhigong tu* (Illustrations of the tribute-bearing people of the imperial Qing; 1761), the *Huang Qing fanbu yaolüe* (Essentials of the vassal [tribes] of the imperial Qing; 1845), and the *Chouban yiwu shimo* (Management of barbarian affairs from beginning to end; 1880). The prefaces to each of these compilations display the same condescending tone. The first emphasizes how "within and without the empire united under our dynasty, the barbarian tribes have submitted their allegiance and turned toward [Chinese] civilization [*xianghua*]."[79] The second, by the great Qing geographer, Li Zhaoluo, refers to the way the emperor "nourishes [his dependencies] like their father and their mother," and "illuminates them like the sun and the moon."[80] And the third, using much the same language as the first, describes the historic process by which foreigners gravitate to China, become "cultivated" and learn "elegance and etiquette."[81]

The ten volumes of the *Huang Qing zhigong tu* provide a detailed picture of the Qing tributary system in its theoretical heyday. Most of these volumes deal with the peoples of Inner Asia and the ethnic minorities of southwest China. The first, however, focuses on China's "overseas" tributaries, listed in the standard order: Korea, the Liuqiu Islands, Annam, Siam, Sulu, Laos, Burma, and the Great Western Ocean (Da Xiyang). These discussions are followed by sections on the Small Western Ocean (Xiao Xiyang), England, France, Sweden, Holland, Russia, and the Philippines.

Here, without benefit of cartographic representation, the so-called Great Western Ocean Country is located vaguely in the Atlantic region and identified both with Italy and Portugal. Other Western nations, including England, France, Sweden, Holland, and Russia are lumped together indiscriminately with Asian countries such as Japan, Borneo, Cambodia, Java, and Sumatra. Modern France is confused with Ming dynasty Portugal; and England and Sweden are recorded as countries dependent on Holland. According to the *Huang Qing zhigong tu*, Italy presented tribute in 1667 (it was actually Holland that did so) and the Pope himself is reported to have once brought tribute to China. In religious matters the *Huang Qing zhigong tu* informs us that the Portuguese/French were Buddhist countries before they accepted Catholicism.[82]

The same kind of misinformation can be found in the section on "tributary states" in various editions of the *Da Qing huidian* (Collected statutes of the great Qing dynasty). Thus we read in the *Collected Statutes* of the Jiaqing reign (1796–1820), 31: 3–4, that "Portugal [*Gansila*] is in the northwestern sea near England," and that "France [*Falanxi*], also called *Fulangxi*, is the same as Portugal [here, *Folangji*]." After absorbing the Philippines (*Lüsong*), this account goes on to say, "they [the French/Portuguese] divided their people and lived there, still governing it at a distance … The sea route from this country to China is more than 50,000 *li* [*c.* 17,000 miles]." Sweden (*Ruikuo*), we are told, is in the northwestern sea; the distance by sea is calculated to be over 60,000 *li* [*c.* 20,000 miles]. "Denmark [*Lianguo*] is [also] in the northwestern sea, and its route to Guangdong province is the same as that for Sweden."[83] Small wonder, then, that Chinese mapmakers found it difficult to acquire accurate data on foreigners. In fact, some of the same misinformation cited above is repeated in maps of the *Tianxia quantu* variety—for example, that the Portuguese (*Gansila*) "absorbed" the Philippines.[84]

More reliable information was, however, available, as we can see from a large, hand-colored cartographic scroll produced by a scholar named Zhuang Tingfu in 1794. The title of Zhuang's production is: *Da Qing tongshu zhigong wanguo jingwei diqiu shi* (Model of the myriad tributary states of the Great Qing dynasty from around the globe). Although this work borrows certain cartographic elements from Ma Junliang's *Jingban tianwen quantu*—specifically, the *Sancai tuhui* version of Ricci's map and Chen Lunjiong's depiction of the eastern hemisphere—it pointedly ignores the lower part of Ma's map. Instead, it provides two extremely "modern-looking" renderings of the Eastern and Western hemispheres, both produced by Zhuang himself. These latter two maps were reprinted by Western-oriented Korean exponents of "practical learning" during the 1830s.[85]

A pair of long written inscriptions, totaling about five thousand characters, illustrate Zhuang's two major themes: one, the transmission to China of new Western scientific knowledge by the Jesuits; the other, the historic process by which foreigners came to be ruled (*laiwang*) as vassals; that is, they "knocked on [China's] gates," "sincerely offered tribute," and asked to become "attached" (*shu*) to the Central Kingdom. The interesting feature of Zhuang's document is the way it accommodates simultaneously the idea of embracing new knowledge from the West and the notion of enrolling Westerners as traditional-style tributaries.

From a scientific standpoint, Zhuang seeks to show that he has learned a great deal from the Jesuits about geography, cartography and astronomy, which, indeed, he did. He waxes at length about latitude and longitude, time and seasonal change, the circumference of the earth (90,000 *li*), the North and South Poles, and so forth. He also writes knowledgeably about how different cartographic projections yield different pictures of the world. According to Zhuang, previous maps, including those offered by the Jesuit fathers, Matteo Ricci and Ferdinand Verbiest, distorted China's size by placing it too far north, thus compressing it (making China appear too small and the foreign countries, too big). His map, drawing upon the work of the famous Qing scientist Mei Wending, provides, he says, a more accurate picture. Significantly, Zhuang cannot resist remarking on how, cosmologically speaking, the Chinese are fortunate to have been born in the Central Land (*Zhongtu*), where the radiance of the sun nourishes them like a sovereign or a father—unlike those people whose misfortune it is to be in far northern or southern regions, where beneficial *qi* is less direct and therefore not very helpful.[86]

Although Zhuang devotes a great deal of attention to science, his primary concern is a cultural one: the Chinese tributary system. The *Da Qing tongshu zhigong wanguo jingwei diqiu shi* commemorates the well-known Macartney embassy of 1793, which, in turn, marked what Zhuang considers to be the high-water mark in the development of China's age-old system of hierarchical foreign relations. This system, he notes, expanded significantly during the Kangxi and Qianlong reigns to include many new parts of the "Western Regions." The peoples of these areas, Zhuang goes on to say, have been registered as part of the Chinese empire [*ru banji*], and have offered tribute to the Qing dynasty along with the British, who had engaged in no official communication with China prior to 1793.

Earlier maps, Zhuang tells us, did not include all of China's tributaries; but the Macartney mission, together with the "coming to court" of other tributaries, and the "return" (*laigui*) of various tributary peoples from the "Western regions" during the eighteenth century, offers a fitting moment to celebrate the transformative effect of the throne's glory (*shenghua*) with a set of maps. His renderings, then, are respectfully offered on this magnificent occasion. Significantly, but not at all surprisingly, Zhuang's remarks about the civilizing role of the Chinese emperor (*shengjiao*) correspond closely to those provided in the major cartouche of Ma Junliang's *Jingban tianwen quantu*.

The late eighteenth-century maps by Ma, Zhuang, and others bring into sharp focus the issue of how best to characterize the Qing tributary system. Was it, in fact, a "system" and, if so, what role did it play in China's foreign relations? James Hevia's book, *Cherishing Men from Afar: Qing Guest Ritual and the Macartney Embassy of 1793* (1995), provides a fascinating case study. Hevia considers the Sino-foreign negotiations surrounding the Macartney mission to represent a confrontation involving two great empires, the Qing and Great Britain, rather than one involving two different "cultures." In his view, instead of interpreting the Macartney episode as "an encounter between civilizations," we should consider it as "one between two imperial formations, each with universalistic pretensions and complex metaphysical systems to buttress such claims."[87] There is much of value in this approach, and Hevia performs a useful role in describing at length the ritual beliefs and practices that seem to have been at the center of the Qing approach to tributary relations. Ironically, however, he fails to link these beliefs and practices to any sort of tributary system itself, steadfastly refusing to see Chinese "guest ritual" as a manifestation of "culture."

The reason for this reluctance, as Hevia states explicitly in his Introduction, is that certain influential historians of China's foreign relations—John Fairbank and John Wills in particular—have treated the tributary system as a reflection of Chinese "culturalism," which he suggests these scholars equate with inflexibility and irrationality. According to Hevia, "Fairbank and his followers" have persistently held a dualistic vision of China, "maintained through the use of binary oppositions such as tribute and trade, ritual and diplomacy, ideology and pragmatism, culture and practical reason, … [and] appearances and political realities."[88] In this view, culture is automatically juxtaposed to practicality and rationality. These are, however, false dichotomies, which neither Fairbank nor Wills would accept as an accurate representation of their approach. In fact, the Qing tributary system was, in their view as well as Hevia's, an extremely sophisticated mechanism for dealing with outsiders; it was flexible, multi-purposed, rational and flexible. It also changed substantially over time in response to varying conditions.[89]

To be sure, tributary rhetoric reflected a long-standing worldview based on the notion of China's cultural superiority to all other peoples and the idea that foreigners would gravitate to China out of admiration and "turn toward civilization" (*xianghua*).[90] This was the theory, even though it was not always the practice. For instance, the Qing rulers, themselves "barbarians" in the eyes of many Japanese and Koreans,[91] recognized that foreigners who came to China were not always motivated by admiration for

Chinese culture. Moreover, as a number of Asian and Western scholars have clearly shown, the Qing emperors, like their predecessors for hundreds of years, developed a wide variety of institutions and practices for "managing" foreigners that had little if anything to do with the formal features of the tributary system.[92] Most of these studies also indicate a great deal of flexibility within the general framework of the tributary system, especially in times of China's military weakness. Nonetheless, it was always important for the ruling dynasty to preserve as much of the rhetoric and as many of the formal features of the tributary system as possible, not least as a reflection of their claim to rule "all under Heaven" (*Tianxia*). Understandably, this was especially the case when tributary representatives from other countries were present (as they were during the crucial month of September, 1793).

The tributary system may thus be seen a political "myth"—not in the sense that it was "false" or "irrational," but rather in the sense that it asserted and sustained uniquely Chinese claims to political legitimacy; it was part of the hoary notion, dating back to the Zhou dynasty, of a "Heavenly mandate (*Tianming*)," an expression of the emperor's morally grounded right to rule "all under Heaven."[93] No impartial reading of the historical record could fail to reveal that the rituals and rhetoric associated with the tributary system were designed to affirm China's claims to cultural superiority and moral suasion, to emphasize the submission and subordination of all foreign peoples, and to bring harmony and solidarity to the world. Hevia may believe that such characterizations of Chinese policies toward tributary states are "orientalist" and somehow serve to "constitute the 'West' as a privileged area of intellectual, political and economic activity,"[94] but I see them in no such light. Nor do most other scholars of the Qing period.

In fact, as Hevia's own analysis of "guest ritual" makes clear, adherence to the ceremonies mandated in the *Da Qing tungli* (Comprehensive rites of the Great Qing dynasty; 1756) created the conditions under which a harmonious world order could literally be performed into being.[95] The closer the British could be brought to accept Chinese ceremonial practices, the more efficacious the ritual. This is precisely why the Qing authorities tried so hard to resist British efforts to alter the terms of their diplomatic interaction in favor of a European understanding of international relations. To accept this sort of understanding was anathema to China's self-image as the superior party responsible for harmonizing the lands and peoples represented by "a multitude of lords."[96]

As it turned out, during the Macartney visit the Qing authorities preserved most major elements of Chinese tributary ritual and refused British requests for regular diplomatic representation and the opening of trade. Macartney's apparent failure to abide by some of China's ceremonial stipulations, notably his refusal to kowtow (*ketou*; lit., to kneel three times and knock the head three times after each kneeling) to the Qianlong emperor, although a cause of some consternation on the part of the Qing authorities, did not undermine the overall ritual significance of the event. (The Qianlong court had, after all, already accepted a modified kowtow in receiving a Portuguese embassy in 1753.) Chinese officials and the throne had long understood that all rituals were susceptible to modification under certain specific conditions.[97] As the Qing scholar, Zhao Yi (1727–1814) put the matter, in speaking specifically of China's foreign relations:

> The teachings of true principle cannot always be reconciled with the circumstances of the times. If one cannot entirely maintain the demands of true principle, then true principle must be adjusted to the circumstances of the time, and only then do we have the practice of true principle.[98]

Throughout the entire period of the Macartney visit, the Qing government continued to view the embassy as a tributary mission and treated it as such. It was also described as a tributary mission in official documents designed for the edification of future generations.[99] The fact is that the Chinese had no "illusions" about the event. The rituals performed and the policies decided upon accorded perfectly with Chinese assumptions about both the theory and the practice of the tributary system. From a Chinese standpoint, the outcome was entirely satisfactory, although the negotiations were certainly frustrating at times.

In short, the Macartney mission involved much more than an encounter between "two imperial formations." It was also a cultural clash, representing fundamentally different understandings of how the world in general, and how foreign relations in particular, should be understood and effectively managed. Hevia argues that Qing guest ritual "does not appear to deal in crude distinctions between civilization and barbarism."[100] Surely the word "crude" is not an apt characterization of the Qing dynasty's approach to foreign relations; yet it seems clear to me that Hevia vastly underestimates the problem of cultural difference. What distinguished the "Chinese" from so-called "barbarians" was precisely the difference in their levels of "civilization"—specifically, differences in their ritual behavior.

There is another problem with Hevia's approach. Although his stated aim is to understand events "through their multiple recountings," his analysis is marked by a curious asymmetry. In his zeal to expose the "orientalizing" tendencies of both Westerners and post-Qing Chinese scholars (who have, according to Hevia, appropriated "the intellectual framework of the colonizer"), he virtually ignores similar "occidentalizing" gestures on the part of the Qing intelligentsia—essentializing and condescending moves that are abundantly evident not only in the Chinese documents that Hevia has quite obviously studied, but also in Chinese cartographic materials, which he apparently has not. The result is an account of historiographic "distortions" that is itself "distorted" by Hevia's inclination to view precolonial China through a postcolonial lens.[101]

Let us now return briefly to the specific issue of whether, all things considered, it is possible and productive to speak of a Chinese tributary system. John Wills, in his introduction to a recent set of excellent essays titled *China and Maritime Europe, 1500–1800: Trade, Settlement, Diplomacy, and Missions* (2011), refers specifically to a "mode of management of [Chinese] foreign relations in which everything was governed by a single set of hierarchical concepts and bureaucratic precedents, a 'tribute system' in the full sense of the word." He hastens to point out, however, that "many important facets of Chinese relations with maritime Europeans … had little or nothing to do with the tribute system." And he goes on to say that if we take a "longer view" of the history of Chinese foreign relations, we find that the tribute system of the Ming and early Qing period—which reflected, in his words, "persistent tendencies toward the unilateral and bureaucratized control of relations with foreigners and toward the ceremonial superiority of the Son of Heaven over all other sovereigns"—was limited in duration. His argument is that "the years from about 1425 to 1550 were the *only* time when a unified tributary system embodying these tendencies was the matrix for policy decisions concerning all foreigners." Thus, he believes that the term "tribute system" should apply only to the Ming system and not be used to refer to the "less systematic and more varied diplomatic practices of other times."[102]

While there is a certain logic to this position, I maintain that as long as tributary institutions, regulations, policies and rhetoric were a significant part of the Chinese worldview and self-image, indicated at least suggestively in maps, encyclopedias and almanacs, and described at length in a great many Qing documents—including the dynasty's Collected Statutes (*huidian*), its Comprehensive Rituals (*tongli*) and its Veritable Records (*shilu*)[103]—until the very end of the nineteenth century, the term "tributary system" is apposite. The fact that many changes took place in the system over time, and that many other forms of Sino-foreign diplomatic interaction occurred outside the formal framework of the system, does not negate the concept. We know, for example, that in contemporary American political life not all negotiations and interactions with foreign countries take place within the formal framework of the State Department, and yet its institutions, regulations and rhetoric obviously remain important to an understanding of American foreign policy.

Wills offers the rather vague idea of pragmatic "defensiveness" as a more satisfactory "master concept" than the tributary system for the interpretation of Sino-foreign relations.[104] I would argue, however, that we do not need master concepts of this or any other sort. We do, however, need to employ certain terms from the Chinese lexicon that, when properly nuanced and time-sensitive, have analytical value. "Tribute" is one of those terms.

Continuity and change

Qing documents from the last Dutch embassy to China in 1794–1795, a year or so after the Macartney episode, help to explain why the Chinese tributary viewpoint proved to be so tenacious, and why the Qing authorities were so confused and dismayed by Great Britain's failure to meet Chinese expectations in 1793. Although the Dutch mission, designed to celebrate the Qianlong emperor's sixtieth birthday (and, of course, to promote trade), was considered an irregular event, it conformed in every respect to the basic requirements of Chinese tributary ritual. Holland's preliminary letter, sent from its fictitious "king" to the Qianlong emperor in the summer of 1794, captures the distinctive flavor of a vassal's petition to his feudal superior. The Chinese version reads in part:

> From the time of the Kangxi emperor's reign [1662–1722] onward … [we foreigners] have all been transformed by China's civilizing influence [*xianghua*]. Throughout history there has never been a monarch with such a peerless reputation as you possess, my exalted emperor.[105]

Small wonder China had such a well-developed sense of its superior status. In response to the Dutch mission, the Qianlong emperor replied:

> I have now reigned for sixty years, so that the four seas are forever pure and all the regions of the world have been transformed by Chinese culture. My virtuous reputation has spread far and wide … and I have [always] treated Chinese and foreigners as one family … Now [representatives of] the myriad countries, scaling mountains and sailing seas, have come, one after another, to offer birthday congratulations … [This] heavenly dynasty views all [people of the world] with equal benevolence, and although some may come to China with only meager [tributary presents], all will leave amply rewarded … [Since you admire Chinese culture [*muhua*] and will be receiving valuable tokens of imperial favor with this edict,] may these gifts strengthen your bonds of loyalty and integrity, preserving good government in your kingdom and making you forever worthy of my esteem.[106]

Well into the nineteenth century, Qing documents on foreign relations, including maps, continued to reflect this condescending tone, using the same basic vocabulary.[107]

It simply will not do to dismiss these statements as empty rhetoric. One can acknowledge the flexibility and sophistication of traditional China's approach to foreign relations—and even accept James Polachek's (1992) argument that the "inertia of the [Chinese] central-government political system … [was] the chief obstacle to foreign policy change"—without denying, as Polachek himself puts the matter (somewhat indelicately), that "the pompous 'Celestialism' of the late Ch'ing [Qing] court posed a very real problem for those would have brought China more speedily into the modern world."[108]

This much we know: On the eve of the first Anglo-Chinese War of 1839–1842, world maps of the sort produced by individuals such as Cao Junyi and Zhuang Tingfu were at best a dim memory for most Chinese scholars. From the late seventeenth century into the nineteenth, the vast majority of Chinese mapmakers ignored Jesuit constructions of the world almost entirely. Most did not even choose to pattern their cartographic productions after Luo Guangxian's grid-oriented *Guang Yutu*. Far more popular were maps of the *Tianxia quantu* variety, or those based on the rhythmic and colorful cartography of Yu Shi. A striking example of the latter type is an untitled and anonymous "world map" of 1743 that is now housed in the Oriental Manuscripts Division of the British Library.[109]

Yet at least a few indigenous mapmakers carried on the cartographic traditions established by the Jesuits—assisted now by the efforts of a new breed of Western missionaries, primarily Protestants. For instance, Li Mingche, a well-known Daoist priest and scientist with foreign contacts, included two relatively "modern" illustrations of the Eastern and Western hemispheres—complete with lines of latitude and longitude—in his *Huantian tushuo* (Illustrations of encompassing Heaven; 1819).[110] Moreover, every late eighteenth- and early nineteenth-century example I have seen of the defensively oriented scrolls known generically as *Haijiang yangjie xingshi quantu* (Complete map of [China's] coastal configurations) begins with a colorful and quite faithful line-drawn rendering of the eastern hemisphere based on Chen Lunjiong's *Haiguo wenjian lu*—the same basic model that appears on Ma Junliang's maps in the *Tianxia quantu* tradition. Although the Chinese empire usually appears in such renderings to be as large as Africa (which embraces one-fifth of the Earth's total land area), the salient feature of the map is its unmistakably "modern" appearance.[111]

After China's defeat at the hands of the British in 1842, editions of the *Haijiang yangjie xingshi quantu* began to reflect a new awareness of the Western presence in treaty port areas—significantly, by means of textual remarks to the effect that "during the Daoguang period [1820–1850] Western countries [began to] trade at this place." Such cartographic changes were part of a growing sense on the part of at least some Chinese scholar-officials that China had to acquire more up-to-date knowledge about the outside world in order to survive. The two most important books to provide this information, both based substantially on Western maps, were Wei Yuan's *Haiguo tuzhi* (Illustrated gazetteer of the maritime countries; 1843), ably analyzed by Jane Kate Leonard (1984), and Xu Jiyu's *Yinghuan zhilüe* (A short account of the maritime circuit; 1849), cogently discussed by Fred Drake (1975). What these two pioneering works had in common was a desire, in Wei's words, "[to] describe the West as it appears to Westerners."[112]

From the 1860s to the 1890s, as part of China's Self-Strengthening Movement (1862–1895), study associations, books, and journals devoted to geographic and cartographic issues began to proliferate in China. The publication of Wang Xiqi's massive *Xiaofanghu zhai yudi congchao* (Collected texts on geography from the Small Square Vessel Studio; 1877–1897), which brought together several hundred individual Qing dynasty titles, marked a watershed in China's geographical awareness. Meanwhile, Chinese cartographers began to produce their own colorful, modern-looking maps. The Beijing Library has collected several of these works, with titles such as *Diqiu wu tazhou quantu* (Complete map of the five great continents of the globe; 1874); *Diqiu quantu* (Complete map of the globe; 1883) and *Diqiu wanguo quantu* (Complete map of the myriad countries of the globe; 1895). Certain mapmakers, including Yao Wentong, Gong Zhai, Chen Zhaotong and Wen Shao, even managed to achieve a certain limited celebrity. In the 1880s and 1890s, the Qing court itself attempted to update and standardize its geographic and cartographic practices.[113]

But it was the Sino-Japanese War of 1894–1895 that sounded the death knell of traditional Chinese cartography. From this time onward, in elite journals and even popular almanacs and encyclopedias, Chinese readers sought ever more accurate knowledge about other parts of the world, including once-despised Japan. The rise of Chinese nationalism—generated by China's humiliating defeat at the hands of the so-called "dwarf-bandits" (*wokou*)—brought a heightened awareness of the

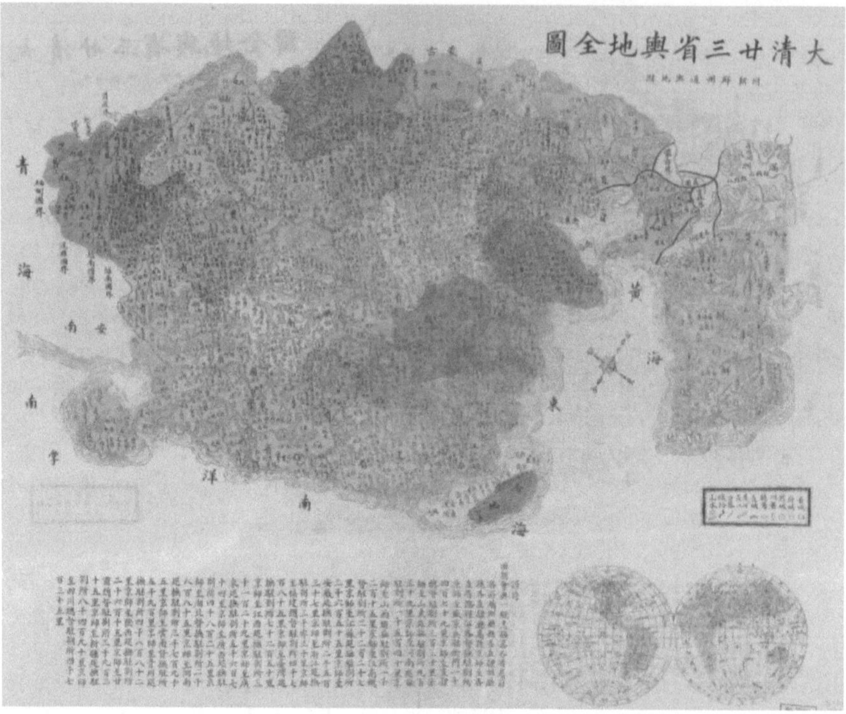

FIGURE 2.19 A strategically-oriented Chinese map of the "Twenty-Three Provinces of China," produced between 1885 and 1894, which depicts Korea, a problem area at the time, larger than scale. Although the map of China itself reflects the traditional *Tianxia* style, the cartographer has also included "modern" maps of the eastern and southern hemispheres at the bottom right.

Source: Library of Congress, Geography and Map Division.

wages of foreign imperialism (Figure 2.19). Chinese cartography, like many other areas of Chinese life after 1895, became suffused with the spirit of patriotism and political action. One revealing illustration can be found in a map contained in a 1912 almanac (*Zhonghua minguo yuannian lishu*), issued in the name of the newly established Republic of China. Although not particularly sophisticated in terms of mathematical cartography, the map is fascinating because its commentaries strikingly identify the various places taken from China by foreign imperialism, including the province of Taiwan and the tributary states of Korea, the Liuqiu Islands and Annam.[114] From this time onward, Chinese nationalism affected in fundamental ways the rendering of geographical space by cartographers in China.

Concluding remarks

Viewing matters from one angle, we might say that Chinese cartography in late imperial times impeded a more "realistic" understanding of foreign lands and peoples. Certainly it both expressed and reinforced a tributary-based perspective on Sino-foreign relations—one that probably over-estimated the submissiveness and dependency of aliens. Moreover, the emphasis in so many Chinese "world maps" on the great military conquests and unprecedented territorial expansion of the Qing dynasty may well have contributed to an exaggerated sense of self-confidence on the eve of the Western intrusion.

The overwhelming majority of Chinese *mappaemundi*—including works produced after the Jesuit interlude—depicted "the world" as if the foreigners inhabiting it existed precariously on the fringes of the Chinese Empire. Whole continents appeared either as tiny offshore islands or as inconsequential appendages to China's land mass—terrestrial afterthoughts, so to speak. It is not at all clear, however, that a "more "realistic" depiction of foreign lands would have produced a greater sense of military threat. Indeed, some Chinese scholars argued that Jesuit-style maps were designed quite deliberately to mislead their countrymen into thinking that the aggressive, avaricious people from "the Great Western Ocean" were further away than they actually were.[115]

Moreover, we should remember that foreigners were not always marginalized in Chinese "world maps"—even those with certain "traditional" features. Scholars such as Cao Junyi, Yan Yong, Chen Lunjiong, Ma Junliang, and Zhuang Tingfu, for example, made concerted and largely successful efforts to depict foreign territories accurately; and the Manchus, for their own political reasons, produced excellent maps of the Qing Empire with Jesuit assistance. In fact, it seems clear that reliable cartographic information existed for those scholars who wanted it, despite Manchu efforts to keep certain types of knowledge to themselves, and notwithstanding the understandable confusion produced by so many different renderings of foreign space. But the incentive to seek this knowledge out, like the incentive to disseminate it widely, does not seem to have been particularly powerful—at least not until the rise of Western imperialism in the mid-nineteenth century. Quite the contrary, there were political incentives to support the status quo.

In any case, this line of analysis assumes a certain pragmatic approach to cartography that obscures other ways of thinking about maps.[116] For many Chinese scholars, maps—world maps in particular—were designed to be appreciated, not simply employed. Although such works had a certain (limited) practical value, they had a much greater emotional appeal. As Liang Zhou put the matter in the introductory remarks to his highly influential map of 1593: "[This work] deals with the grandeur of China's mountains and rivers as well as the excellence of its people, past and present." Emperors and officials may have required certain kinds of finely wrought maps for specific military and administrative purposes, but they also needed large-scale maps as a means taking "spiritual journeys [*shenyou*] across vast space," in the poetic words of Zhuang Tingfu. Indeed, we know that certain *mappaemundi*, such as the *Liangyi xuanlan* (Map for the profound observation of Heaven and Earth; 1603) were designed and used expressly for such purposes.[117]

By combining aesthetics, cosmology, history and culture in particularly creative and compelling ways, the makers of Chinese world maps often sought to blur the conventional distinction between actual, lived space and imaginary, idealized space.

To put the matter somewhat differently, many if not most traditional Chinese *mappaemundi* were not as much renderings of the world as they were cultural statements about the nature of world. Their purpose, at least in part, was to reinforce certain abiding cultural myths which, in turn, sustained China's self-image—stories about the Central Kingdom's advantageous geographical and cosmological location, its glorious conquests; its impressive explorations; its heroes, its famous landmarks, and its powerful influence on other lands and peoples. Many of the places depicted or referred to in Chinese maps of the world provoked powerful reactions—regardless of whether they were actual locations or purely mythological sites.

This was true not only for points to be found in China Proper, such as the legendary "Sea of Constellations," but also for points beyond. Indeed, one reason for including place names from the *Shanhai jing* in so many Chinese *mappaemundi*, was not only to create a sense of comprehensiveness, but also to trigger certain poetic and other literary associations. Every member of the Chinese elite—and probably a number of literate commoners as well—knew of Tao Qian's famous poem, inspired by looking at the illustrations in the *Shanhai jing* ("Du Shanhai jing"). Two lines suggest its value as a source of imaginary inspiration:

> I view the pictures in the *Classic of Mountains and Seas*.
> A drifting glance [*liuguan*] encompasses the ends of the universe.[118]

Quite possibly Tao's poem refers to a map (or maps) no longer extant. In any case, dozens of his lines mention specific places in the *Shanhai jing*—each connected with colorful and well-known images that circulated freely in Chinese folklore. Thus we read:

> Far away, the Locust River Range;
> At hand, the famed Hanging Gardens Hill.
> To the south and west you see the Kunlun Mountains;
> The shining atmosphere is hard to match.[119]

The *Shanhai jing* tells us what every Chinese literatus already knew: that the Locust River Range is topped with the finest gold, silver and jade; that the Hanging Gardens Hill has a tutelary deity with the body of a horse, a human face, tiger stripes and the wings of a bird; and that—as everyone was surely aware—the Kunlun Mountains are the legendary source of the Yellow River, not to mention the abode of the Queen Mother of the West (*Xiwang mu*). References to even more remote places in Chinese maps—regions inhabited by strange beings such as "The People with Perforated Chests," "The People with Crossed Legs," and "The People with Three Heads" (Figure 2.20)—naturally conjured up a different set of exotic associations, images shared by elites and common people through media such as encyclopedias and almanacs.[120]

In short, Chinese world maps in late imperial times had several purposes—not all of them either pragmatic or scientific—and specific claims of territorial jurisdiction in

FIGURE 2.20 Depictions of "barbarians" based on descriptions in the *Shanhai jing* (Classic of mountains and seas) from a Qing almanac, undated

Source: Author's collection.

Chinese *mappaemundi* often took a back seat to more general claims of feudal overlordship. Until the late nineteenth and early twentieth centuries, mathematical precision was never considered a cartographic end in itself. To be sure, Chinese mapmakers understood the utility and appeal of accurate measurement, and their colleagues in astronomy developed sophisticated instruments that made possible the projections and coordinate systems that Westerners associate with Ptolemaic cartography. But throughout most of the imperial era, they found no compelling reason to conceive of the world as spherical, nor did they see any special merit in drawing all maps "to scale." (After all, the eunuch-admiral Zheng He made his way to the coast of Africa in the early fifteenth century without much difficulty.) Besides, cartographers knew that commentaries could always provide precise geographic details, if they should prove necessary.

Despite a long tradition of sophisticated geographical and cartographic scholarship, an equally long history of foreign exploration (and conquest), and the systematic acquisition of information on barbarians of various kinds, the "outer" world as a whole remained relatively unimportant to the vast majority of Chinese—elites and commoners alike—until the late nineteenth century.[121] In the West, by contrast, the great voyages of discovery from the late fifteenth century onward ignited interest in "capturing the world as a single ordered image," which eventually occurred after much rivalry and much scientific trial and error.[122] But Zheng He's earlier—and in some ways far more impressive—sea voyages had no such effect in China;

in fact, they were a source of embarrassment. And whereas the possession and display of a world map or globe from the Renaissance onward in Europe signified that the owner was "a knowledgeable and worldwise citizen,"[123] it meant no such thing in late imperial China. Thus, until forced to reconsider their craft by new political and cultural priorities, Chinese mapmakers generally made the choice to depict the world not so much in terms of how it "actually" was, but rather in terms of how their various constituents wanted it to appear.

3

RITUAL IN QING CULTURE

> Through ritual, [the positions of] honorable and lowly are fixed, intimate and distant are separated, Heaven is served above, Earth is served below, ancestors are respected, and sovereigns and teachers are glorified ... In the end, ritual is what distinguishes men from beasts.
>
> (*Jiali huitong* (1734))

The culture of late imperial China is distinguished, on the one hand, by uniformity and consensus and, on the other hand, by diversity and dissidence. This chapter surveys both aspects of traditional Chinese culture from the often-neglected perspective of ritual. It focuses on the Qing dynasty (1644–1912), a period of material prosperity and expanding literacy. During the Qing period, Confucian values penetrated more deeply into Chinese society, political, social, and economic institutions attained greater maturity, and China's economy and society achieved a greater degree of interregional integration than at any previous time in Chinese history. What was the place of ritual in Qing culture, and how does an examination of ritual shed light on the relationship between cultural unity and cultural diversity in China?

It is well known that the dominant self-image of the traditional Chinese elite was one of cultural uniformity, characterized by the assumption that "all strains of thought, all institutions, [and] all forms of behavior should embody and express a common set of values."[1] This view naturally tended to minimize regional differences, "folk" culture, foreign influences, and intellectual or religious dissent, imposing a symmetry and coherence on Chinese culture that "neither past nor present reality, in its fullness and variety, would confirm."[2] Yet it would be a mistake to assume that the literati had no conception or appreciation of the problem of cultural diversity in China. One version of the Yongzheng emperor's *Shengyu guangxun* (Amplification of the [Kangxi emperor's] Sacred Edict) which was designed to be read aloud to commoners by local officials or scholars throughout the empire on the first and fifteenth days of each month, cites the well-known proverb "Customs differ every ten *li*" (*shi li butong feng*); and another version indicates: "Popular customs [*fengsu*] vary greatly. In some places people are kindly, in others, reserved; in some places they are extravagant and pompous, in others, frugal and simple."[3]

It was through orthodox ritual (*li*)—as well as the closely related systems of law and education—that the Qing state sought to encourage cultural uniformity and minimize both regional diversity and social deviance.[4] In the words of the *Shengyu guangxun* just cited:

> Because the customs of every place differed, the ancient sages created ceremonial practices [*li*] in order to standardize conduct. The sage [Confucius] said that to secure the ease of superiors and bring order to the people [*an shang zhi min*] nothing is better than ritual.[5]

Although under normal circumstances popular customs were assumed to be conducive to order and harmony, they could be subverted by beliefs and practices that were considered "heterodox" (*buzheng*, *buduan*, etc.) or "heretical" (*xie*, *yin*, etc.).[6] Ritual facilitated the rectification (*zheng*), solidification (*hou*), and standardization (*tong*) of customs, and hence the restoration or maintenance of social harmony.[7]

The degree to which this goal of social integration was actually achieved is a matter of substantial debate (see "Some concluding remarks" below), but I am persuaded that overall—and especially in times of peace and prosperity—there was a broad consensus in traditional Chinese society on the importance of certain ethical values and ritual practices, as well as a shared vocabulary of secular and religious symbolism and a common set of basic cosmological assumptions. As we shall see, the particular manifestations of these beliefs and practices naturally differed somewhat over time and across space. It could not be otherwise, given the vastness and diversity of the Qing empire and the limited reach of the traditional Chinese state. Nonetheless, all of the people south of the Great Wall who considered themselves to be "Chinese"—as opposed to ethnic "others"—had a well-developed sense of what the fundamental ritual requirements were for their particular social status.

There are, of course, many kinds and definitions of ritual, but in its broadest sense, the term may be said to include all forms of structured social behavior, from the etiquette of daily greetings to solemn ceremonies and religious sacrifices.[8] Such a broad definition accords well with traditional Chinese usage. Although the term *li* never completely lost either its original religious and mystical connotations or its close association with music as a source of moral cultivation, by late imperial times, *li* had come to embrace all forms of sacred and secular ritual as well as the entire body of social institutions, rules and regulations, conventions, and norms that governed human relations in China.[9]

For hundreds of years the Chinese have referred to themselves as "the land of ritual and right behavior" (*li yi zhi bang*), equating the values of *li* and *yi* with civilization itself.[10] Some scholars, such as Etienne Balazs, have argued that late imperial China witnessed a decline in the importance of ritual as part of a growing "secularization, rationalization, [and] bureaucratization" in Chinese society as a whole.[11] But if ritual did in fact decline in significance, we can only marvel at its earlier strength, for there is every indication that in Ming and Qing times ritual continued to be a matter of extraordinary importance. Charles Hucker writes, for instance: "Anyone who works extensively with Ming documents ... cannot avoid the conclusion that proper government in the Ming view was largely a matter of performing the proper rituals."[12]

The Qing dynasty placed tremendous emphasis on *li*, copying the Ming ritual system down to the smallest detail and often exceeding its dynastic predecessor in ceremonial exuberance.[13] We know, for example, that the Qing emperors paid unprecedented homage to Confucius, including two kneelings and six prostrations in Beijing and the full kowtow (*ketou*) in Qufu, birthplace of the Sage. Zhi Chaozi's account of life at Beijing tells us that the Xianfeng emperor once plowed four furrows instead of the customary three in the annual ritual of Spring Plowing, destroying the symmetry of the rite in his overzealous effort to "show added reverence."[14] And, of course, in the waning years of the dynasty, the Qing government elevated the worship of Confucius to the first level of state sacrifice, presumably in a last-ditch effort to enhance the reputation of the Sage (and the Manchus) after the literary examinations that had reinforced his teachings for many centuries had been abolished.[15] But by this time, the Qing government could no longer maintain its rigid orthodoxy in either Confucian education or the related realm of ritual, as the revival of the dynasty's School of Ritual Studies (*Lixue guan*) in 1907–1908 indicates all too clearly.[16]

Orthodox ritual: the theory

In the Qing period, the imperially endorsed rituals were in theory compatible with the Cheng-Zhu Neo-Confucian tradition. Among the illustrious products of Qing scholarship were a number of scholarly commentaries on the three major Chinese ritual classics—the *Zhouli* (Rites of Zhou), the *Yili* (Etiquette and ritual) and the *Liji* (Record of ritual)—as well as several later works on *li*.[17] The basic ritual institutions and practices of the dynasty were detailed in the Qing collected statutes (*Da Qing huidian*), first issued in 1690 and revised several times. The Qing legal code (*Da Qing lüli*) and its supplements, also revised numerous times after the mid-seventeenth century, established the foundations for the dynasty's ritual law (*lilü*), a closely guarded sphere of imperial control.[18]

Until the Qianlong emperor's reign, there were few major works that gave a distinctive cast to the Qing ritual system. During that period, however, in response to requests dating back at least to the 1680s, two important and distinctive ritual compilations were completed under imperial patronage. The most impressive of these was the *Da Qing tongli* (Comprehensive guide to Qing dynasty rituals), commissioned in 1736, completed in 1759, and printed in 1818. This compilation, together with the collected statutes, became the standard of ritual orthodoxy for the remainder of the

Qing period.[19] The other work, titled *Huangchao liqi tushi* (An illustrated description of ritual implements), was begun in 1759 and published in 1763.[20] At about the same time, Qin Huitian (1702–1764) finished his monumental *Wuli tongkao* (A comprehensive analysis of the five categories of ritual), a private undertaking of great and lasting scholarly value.[21] Other privately sponsored compilations of particular importance for an understanding of Qing ritual are the famous compendium on statecraft *Huangchao jingshi wenbian* (A collection of Qing dynasty writings on statecraft; 1826) and its supplements, especially those published in 1888 and 1897; the *Huangchao xu wenxian tongkao* (Supplement to the encyclopedia of the literary and historical records of the Qing dynasty, 1921); and Wu Rongguang's (1773–1843) extremely valuable study titled *Wuxue lu* (Record of my studies; 1832).[22]

Ritual handbooks of various sorts circulated widely throughout the Qing period. Some of these were written for officials and scholars; others for literate commoners, including ritual specialists.[23] The scope of such handbooks might be comprehensive or narrow, but all were deemed necessary for the proper conduct of everyday affairs in traditional China.[24] Popular proverbs reinforced the sense of a need to be fully conversant with appropriate ritual forms: "A breach of etiquette results in confusion" (*yi shi ze hun*), "There is nothing better than propriety [for gaining ends]" (*mo shan yu li*), "With propriety one can travel throughout the world; without propriety it is difficult to go forward an inch" (*youli zoubian tianxia; wuli cunbu nanxing*).[25] Family instructions (*jiaxun*), genealogies (*jiapu* or *zupu*) and "clan rules" (*zonggui* or *zugui*) all gave considerable attention to questions of ceremony. Many such sources cited with reverence the ritual writings of Sima Guang (1019–1086) and especially Zhu Xi (1130–1200), and some lineages required that formal lectures be given on these and other standard ritual works.[26] Hui-chen Wang Liu points out that among the numerous ancient classics cited in traditional clan rules, the most prominent was the *Liji*, followed by the *Zhouli* and the *Yili*. She adds:

> [W]hen the clan rules cite other classics such as the *Classic of Changes*, the *Classic of Poetry*, and the *Spring and Autumn Annals*, they are concerned mainly with bringing metaphysical interpretations, aesthetic sanction, and historical precedents to the support of the approved conventions and principles called *li*.[27]

The assumptions that lay behind ritual in traditional China may be stated briefly. The first is that ritual was inextricably related to cosmology. The *Liji*, for example, links ritual with the forces of *yin and yang*, the five agents (*wuxing*), and the spiritual world of *shen* and *gui*. It further remarks: "Ritual has its origin in Heaven and its movement reaches Earth. Its distribution extends to all human affairs. It changes with the seasons and accords with various allotments and conditions." Similarly, the *Wuli tongkao* acknowledges the cosmological foundations of *li* and declares: "Everything between Heaven and Earth is bound up [*panji*] in ritual."[28] Not surprisingly, the early classical texts on ritual, and later Chinese writings as well, give explicit attention to the place of divination in ritual activity.[29] I shall return to this important point.

One of the most basic functions of ritual was as a device for ordering government. The *Liji* informs us that ceremonies [*li*] form a great instrument in the hands of the ruler. They provide the means by which to resolve what is doubtful, clarify "what is abstruse, receive the spirits [*bin guishen*], examine regulations, and distinguish humaneness [*ren*] from righteousness [*yi*] ... To govern a state without ritual would be like plowing a field without a plowshare."[30] The seventeenth-century censor Chen Zizhi spoke for many Qing officials in asserting that "for managing the world [*jingshi*] and pacifying the people [*anmin*] there is nothing greater than ritual."[31] Wei Xiangshu advised the early Manchu rulers: "The moral transformation of the people [*jiaohua*] is the dynasty's first task, and the regulations of ritual [*lizhi*] constitute the great item [*daduan*] of moral transformation."[32]

Through orthodox ritual the emperor not only affirmed his position as the Son of Heaven and ruler of all earthly domains, but he and his officials also promoted order, harmony, and stability throughout the realm. State ritual was thus seen as a means of both legitimizing political power and exerting administrative control. In part because official sacrifices bolstered the authority of the bureaucracy, the Qing legal code specified heavy penalties for breaches of etiquette in state-sponsored ceremonies (improper conduct, poor preparations, etc.) and even for failure to honor rules of precedence and seniority in village festivals.[33]

Ritual contributed to political and social order by encouraging submissiveness to authority and by delineating status distinctions and obligations. "The teachings of ritual" (*lijiao*) were, in effect, "the teachings of status" (*mingjiao*)—the very essence of Confucianism.[34] Qing period writings of all sorts—from penal statutes, edicts, and other official documents to ritual handbooks and private essays—place a heavy emphasis on status distinctions.[35] By late imperial times, the "Five Relations" (*wulun*) of Confucianism occupied a central position in Chinese social thought. Several important essays in the *Huangchao jingshi wenbian* reflect this emphasis, including the opening piece on ritual by the mid-Qing scholar Ling Tingkan. Citing Mencius on the virtues inherent in each of the Five Relations, Ling goes on to show how they are also associated with certain specific types of ritual: father and son with capping (*shiguan*), ruler and subject with audiences (*pinjin*), husband and wife with marriage (*shihun*), elder and younger with the community drinking ritual (*xiang yinjiu*), and friend and friend with visitation (*shi xiangjian*).[36]

Also central to Chinese conceptions of social life and ritual was the principle of recompense or reciprocity (*bao*). *Bao* covered all facets of Chinese social intercourse, from entertaining to gift-giving, and was implicit in the pervasive notion of "human feelings" (*renqing*).[37] It was also linked in the popular mind with both Buddhist and Daoist notions of divine retribution. The late Qing scholar Liang Qichao provides us with a fascinating rationalist explanation of ritual recompense involving both social and metaphysical elements:

> Sacrifices are offered to parents because they give birth to us and bring us up; to Heaven and Earth because they give us numerous facilities ... Extending this principle to persons, we want to sacrifice to all those who

> have warded off disasters or promoted the welfare of the state or the locality. Extending this principle to things, even the deities [i.e., spiritual representations] of cats, dogs, cattle and horses should be sacrificed to. This single conception of *bao* penetrates the whole [institution] of sacrificial offerings.[38]

In the conventional view of the *Liji*, ritual was not to extend down to the commoners, just as penal statutes were not to extend upward to the great officials. But in fact, as the *Shengyu guangxun* and other Qing sources indicate, all members of Chinese society were expected to adhere to at least the basic elements of ceremonial behavior. Dedication to ritual, however, was never supposed to be mere blind devotion to outward form, for ritual and morality were inseparable in the orthodox Confucian view. As the Sage once indicated: "If a man is not humane [*ren*], what has he to do with ritual?"[39] To borrow a distinction made by Tu Wei-ming, *li* as sterile "ritualism" was clearly inadequate; rather, the performance of ritual demanded full "ritualization," that is, the process of "intentional appropriation of recognized values."[40] *Li* was, in other words, the manifestation of *ren* in concrete social circumstances.

Together with music, poetry, and other forms of refinement (*wen*), ritual contributed to self-cultivation and the establishment of individual character. The *Liji* states: "To cultivate one's person and fulfill one's words is called good conduct. When conduct is ordered and words are in accordance with the Dao, we have the substance of the rules of propriety."[41] Ling Tingkan emphasizes that ritual controls the emotions and contributes to learning, which in turn restores the goodness of man's "natural" endowment from Heaven. Drawing heavily from both the *Liji* and the famous chain syllogism of the *Daxue*, Ling goes on to state that the foundation for the "pacification of all under Heaven" is self-cultivation (*xiushen*) and that "the foundation of self-cultivation is ritual."[42]

We should not neglect the aesthetic dimension of Chinese ritual, for in both its symbolism and its dramatic action we find the expression of enduring cultural concerns. A. S. Cua points out that by integrating form, emotions, and moral attitude, a Chinese ritual performance, like a work of art, possesses a kind of gestalt, or "organic unity." It represents a fusion of ethical and aesthetic values in which neither aspect can be separated from the other.[43] The same may be said of much of Chinese painting (including temple art) and literature. As Andrew Plaks has observed, it is no accident that Chinese ritual shares with the Chinese literary (and dramatic) tradition common symbols and aesthetic patterns, including an abiding concern with *yinyang* balance, periodic rhythms, and cyclical sequences.[44]

Ritual was always closely associated with music in both its moral and its aesthetic dimensions. In the words of the *Liji*:

> Music contributes to unity, ritual to differentiation. From unity comes mutual affection; from differentiation comes mutual respect … From the establishment of right behavior [*yi*] through ritual comes the distinction between noble and mean, and from the unification of culture [*wen*] through music comes the harmony between high and low.[45]

Furthermore, ritual and music reflected a single shared cosmology. "In the greatest music," states the *Liji*, "there is the same harmony as between Heaven and Earth, and in the greatest ritual there is the same rhythm [*jie*] as between Heaven and Earth."[46]

In short, the general purpose of ritual in Qing China was to demonstrate attunement with the natural order, or Dao, and in so doing help to achieve "mutual respect and love within the four seas."[47] Specifically, *li* was designed to promote social cohesion, preserve status distinctions, sanctify ethical norms, and transmit Chinese tradition. Timeless, yet constantly changing,[48] ritual performed the function Clifford Geertz assigns to "sacred symbols" in synthesizing ethics, aesthetics, and worldview. In the case of China, this meant expressing a cultural outlook stressing particularist Confucian values (sometimes in Buddhist or Daoist dress) and the pervasive notion of "harmony based on hierarchical difference."[49]

Orthodox ritual: practice

By design, and in accordance with the theory outlined above, no two people were completely equal in Chinese society. The standard criteria for social differentiation included generation, age, sex, and education, but wealth remained an important factor in many areas of Chinese social life, including ritual.[50] In the words of a popular proverb (of classical inspiration): "Ritual and right behavior are born of wealth" (*liyi sheng yu fuzu*).[51] Clan rules and other such works often gave explicit recognition to the hard fact of economic inequality: "Capping and wedding ceremonies should be carried out according to one's means. Funerals and burials, being [especially] important matters, should be more elaborate, but one should still be mindful of financial considerations."[52]

As a general rule, the higher the social class in China, the more rigid the adherence to orthodox ritual as a matter of both Confucian responsibility and public prestige or "face" (*mianzi*). But one striking and significant feature of Chinese ritual life was the effort on the part of all classes of society to put on the most impressive ceremonial displays possible, even at great cost and financial hardship. One mid-nineteenth-century Western account of rural life in South China tells us:

> Poverty and death are haunting spectres of the poor. They roam through the village and inspire fear that is not physical but social. It is not that the villager fears death; his belief in Fate relieves him of that worry. But to think of his parent drawing near to the time of departure without funds for proper rites and burial—this is a real fear. To fail in the provision of rites, feasts, coffin and funeral would be conduct most unfilial and condemned by social opinion.[53]

Other works, both Western and Chinese, confirm this view.[54] Domestic rituals that had a public dimension—weddings and funerals in particular—provided opportunities for families to assert their status within their respective communities.[55]

There were several levels of ritual activity in late imperial China, each related to specific spheres of political and social organization. At the top of this structure was state ceremony (*sidian*), which encompassed all five major types of orthodox ritual: auspicious sacrifices (*jili*), ceremonies of celebration (*jiali*), rites of visitation (*binli*), military ritual (*junli*), and ceremonies of misfortune (*xiongli*).[56] In each of these areas, the emperor assumed a leading role as "theocratic ritualist," for it was through his innumerable ceremonial responsibilities that he legitimized his position, promoted confidence among his family members and within the bureaucracy, and inspired awe among the common people.

Everything he did had a sacred flavor to it; the greater the ritual significance, the more regulations and prohibitions surrounding the act.[57] Contemporary Qing sources describe in detail the elaborate dress codes and other stipulations that safeguarded and expressed the emperor's exalted status at every turn, from his performance of personal life-cycle rituals (including ascending and abdicating the throne) and his study of the classics with imperial tutors to such mundane affairs as eating meals and sleeping with a wife or concubine.[58] Members of the imperial household were also, of course, bound by similarly detailed codes of conduct according to rank. Within the regular bureaucracy, civil and military officials from grand councilors down to county magistrates adhered scrupulously to an elaborate set of ritual rules and regulations covering every conceivable situation. These situations included formal interviews, inspections, banquets and visits, the receipt of imperial edicts, the administration of the civil and military examinations, local ceremonies such as the community drinking ritual (*xiang yinjiu*), and, of course, the numerous sacrifices undertaken at various levels of government on behalf of the state.[59] Not all such rules and regulations were enforced with equal vigor, but available evidence indicates that even relatively minor infractions of ritual law were often heavily punished.[60] Like officials, non-bureaucratic civil and military degree-holders, because of their privileged ritual status, were subject to minute stipulations regarding dress and ceremonial behavior.[61]

Institutional Buddhism and Daoism had their own separate structures and sets of rituals, but they enjoyed no real economic power or high-level administrative autonomy. Indeed, the Qing government imposed legal limits on the size of the clergy, the number of officially sanctioned monasteries and temples, and the scope of their regular religious activities. Abbots, priests, and nuns were always subject to indirect state supervision and remained at the beck and call of the emperor and his agents.[62]

Yet state control of religion in China was never complete. At the community level in particular, members of the Buddhist and Daoist clergy, as well as ritual specialists of all types, played a major role in local festivals and related rituals virtually independent of the state. In Chinese villages, towns, and cities, unauthorized popular sacrifices (*minsi*) often competed successfully with official sacrifices (*guansi*), much to the dismay of bureaucrats and the throne.[63] But despite the manifest differences between lively popular sacrifices and solemn state ceremonies, the two might share certain features. For example, Stephan Feuchtwang's illuminating comparison of the official Confucian sacrifice of recompense, known as *shidian*, and the popular community rite of renewal, known as *jiao*, indicates that both employed ritual experts, evinced a concern with status differentiation, used certain basic symbols and gestures of respect, shared the same basic cosmology, and had the same generalized social purposes.[64]

Religious temples (*miao*, *si*, *ci*, etc.) in urban and rural areas were often linked with corporate common-interest organizations, including neighborhoods, guilds, and societies (*hui* or *tang*) based on ties of kinship, common surname, home area, profession, scholarly interest, religious outlook, or simple mutual aid.[65] Like lineage and clan organizations, many of these ritually cohesive units performed valuable social services, including education and welfare; but from the Qing government's standpoint (and often in fact), a thin line separated "legitimate" associations and lodges from the subversive "secret societies" that went by the same generic names.[66] Even clan and lineage ritual activities were cause for concern. Although the state recognized the positive role of kin groups in promoting orthodox values, providing social services, and maintaining local control, it feared all well-organized, but non-official, corporate entities, especially those with particularistic loyalties and substantial economic resources and manpower.[67]

Domestic ritual institutions gave the throne no anxiety as long as they reinforced family values and posed no political threat. In fact, the state actively encouraged certain features of domestic life-cycle ritual, ancestor worship in particular. According to the Qing legal code, all members of Chinese society were required to observe funeral and sacrificial rites for their parents, and destruction of one's ancestral tablets might entail the death penalty.[68] Under normal circumstances, officials within the civil and military bureaucracy were required to withdraw from service (*dingyou*) following the death of a parent, in accordance with the general stipulations laid down in the *Liji* regarding the five degrees of mourning (*wufu*). These highly specific mourning categories not only dictated ritual roles and responsibilities within a wide network of kinship relations, but also affected the legal decisions of the state.[69]

Finally, we should include as a category of ritual the vast number of non-official social ceremonies and formalities connected with Chinese life day to day. These formalities, astonishing in their scope and detail, are enumerated not only in classical works such as the *Liji* and the *Yili*, but also in Qing official sources and ritual handbooks of all kinds.

Although most of the Qing period handbooks I have consulted dwell primarily on kinship ritual (births, marriages, family celebrations, mourning ceremonies, prayers, and sacrifices, specialized family nomenclature and terms of address [*chenghu*], household etiquette, etc.), a number deal more broadly and at length with extra-domestic terms of address (both honorific and self-deprecatory, as in the case of kinship), greetings and salutations (*shangli*, *zhongli*, and *xiali*), which specify the kinds of saluting, bowing, kneeling, and kowtowing appropriate to certain situations and relative rank), proper dress, speech, deportment (sitting, standing, walking, eating), a wide variety of written forms (*tieshi*, including invitations and responses, condolences, and congratulations, regular and special correspondence, felicitous phrases, prayers, etc.), ritual terminology for clothing, foodstuffs, ceremonial vessels, and so forth.[70]

Most formal social intercourse in traditional China, like most kinds of family and official ritual, involved food and/or gift-giving (*zengyi*). Accordingly, we find that ritual handbooks often give particular attention to proper arrangements for meals—including advice on seating and, sometimes, stipulations regarding the number and type of dishes to be served to dinner guests on different occasions—as well as specific guidelines on the type and value of gifts to be exchanged in various situations. Many handbooks also provide illustrations, such as depictions of gestures of respect, pictures of articles of clothing and ritual objects, models of ancestral shrines and tablets, diagrams of seating arrangements, charts of family relationships (including the five degrees of mourning), and the layout of formal ceremonies, from weddings to official sacrifices.[71]

The important point is that although not all Chinese were bound by exactly the same ritual requirements, all shared a strong sense of the ritual obligations appropriate to their status and circumstances. J. H. Gray was not alone in observing during the mid-nineteenth century:

> [This] nation which has laid down with such minuteness the distinctions of rank, has elaborated, with an exactness peculiar to itself, a system of etiquette which prescribes the [proper] forms to be observed in official and social intercourse. A Chinese is seldom at a loss to know what polite observances must regulate his behavior.[72]

Similarly, Arthur Smith remarked later in the period that among all social classes in China, there were "very few Chinese who do not know the proper thing to be done at a given time."[73] Chinese ritual may often have been restrictive, but to most Qing subjects it was probably also reassuring.

"Official religion" and life-cycle ritual

Of the several levels of orthodox ritual outlined above I have chosen to focus briefly on two: state sacrifices and domestic life-cycle ritual. Although these two areas of ritual lie, in a sense, at opposite ends of the sociopolitical spectrum, they share a number of important features and, at the same time, intersect with other levels of Chinese ritual in a variety of significant ways. Above all, they reflect two of the most powerful symbols or metaphors in the Chinese sociopolitical vocabulary, the bureaucracy and the family.[74] Before discussing the basic elements of state sacrifice, a few words should be said about the institutional structure of imperial ritual as a whole.

Predictably, the apparatus of state ceremony was extraordinarily complex. Apart from the emperor's "private" ritual activities (including Manchu shamanism), which were administered by the Office of Ceremonial of the Imperial Household Department, most of the important secular and sacred ceremonies of the state were undertaken by the Board of Ritual. Of its four major departments, the Department of Ceremonies (*Yizhi si*) handled regular court ritual, the examination system, regulations regarding official dress and other marks of status, forms of etiquette between various ranks, and forms of written communication. The Department of Sacrifices (*Ciji si*) oversaw state sacrifices, funerals, the dispensing of posthumous rewards, and the presentation of the state calendar (*shixian li* or *shixian shu*) to the emperor. The Reception Department (*Zhuke si*) managed the highly ritualized tributary system of China's foreign relations as well as the giving of gifts to officials and other miscellaneous activities; and the Banqueting Department (*Jingshan si*), as its name implies, was responsible for the preparation of food for certain types of feasts and sacrifices.[75]

Closely associated with the Board of Ritual, and at times during the Qing period directly attached to it, were three "courts": the Court of Sacrificial Worship (*Taichang si*), in charge of all state sacrifices performed directly by the emperor or his appointed deputies; the Court of Banqueting (*Guanglu si*), responsible for preparing six basic grades of Manchu banquets (*Manxi*) and five of Chinese (*Hanxi*); and the Court of State Ceremonial (*Honglu si*), charged with instructing guests in proper ceremonial observances. A Board of Music (*Yuebu*) also fell under the general direction of the Board of Ritual, and various other subordinate offices assumed responsibility for such activities as engraving official seals and translating documents from tributary states.[76]

One often-overlooked metropolitan institution of great ritual significance was the Imperial Bureau of Astronomy (*Qintian jian*). This agency had responsibility for all astronomical and meteorological observations as well as most state matters related to divination. Guided by a number of sophisticated cosmological tracts, officials of the bureau determined "auspicious days" (*jiri*) for state sacrifices and other important ritual occasions and prepared the all-important state calendar, which specified appropriate (*yi*) and inappropriate (*buyi*) times for undertaking various activities at all levels of Chinese society.[77] The *Da Qing huidian* enumerates sixty-seven such activities as matters of imperial concern, from mundane domestic activities and life-cycle rituals to various forms of state sacrifice and prayer, the promulgation of edicts, the bestowal of favors and awards, various diplomatic and military affairs, and events such as banquets. Calendars for other members of the imperial family, as well as for civil and military officials at all levels, reflected similar concerns.[78]

Although production and distribution of the state calendar in its various forms were supposed to be the exclusive prerogative of the Qing government, privately published non-official almanacs (*tongshu*, *lishu*, etc.)—most of them based on the imperial prototype, but designed primarily for commoners—circulated freely. Indeed, they were probably the most widely distributed books of any kind in late imperial China.[79] As long as such works reflected orthodox views and reinforced imperial legitimacy, they could be tolerated; and some were even endorsed by Qing officials. Like state calendars, almanacs designated auspicious and inauspicious days for a wide range of activities, most of which fell under the categories of "people's use" (*minyong*) or "almanac selection" (*tongshu xuanze*) in the collected statutes of the dynasty. Significantly, throughout the Qing period, as in the Ming (and presumably earlier), the activities most frequently mentioned in both official calendars and unofficial almanacs—aside from mundane matters such as travel, domestic chores, and personal health and hygiene—always had to do with ritual: state sacrifices, domestic sacrifices, and ceremonies connected with birth, adulthood, marriage, and death.[80]

Of the five major realms of orthodox ritual prescribed by official statute, by far the most numerous and politically significant were those designated "auspicious rites" (*jili*). Many of these ceremonies had to do with the emperor's personal life and religious concerns as well as with various imperial visitations, banquets, private audiences, and other affairs of the court. The rest were divided into three main categories: (1) great sacrifices (*dasi*); (2) middle sacrifices (*zhongsi*); and (3) lower sacrifices (*junsi* or *xiaosi*). These solemn state ceremonies had two primary and closely connected purposes: one was to invoke spiritual assistance in order to secure blessings and avoid potential disasters; the other was to honor virtue (*chongde*), reward merit (*baogong*), and protect the common people (*youmin*). At each level of worship, official sources such as the *Da Qing huidian* and the *Da Qing tongli* specified with astounding precision the deities to be honored, the size, shape, and location of the sacrificial altars, the arrangement of the spirit tablets, the type and amount of the ceremonial vessels and decorations, the number, rank, duties, and attire of the official participants, the nature and preparation of the sacrificial offerings (including food, wine, incense, and precious objects), the kinds of music and posturing to be performed, the prayers to be offered, the symbolism of the colors, the numbers to be employed, and a myriad of other details.[81]

According to statute, the deities in official religion operated in a systematic hierarchy that paralleled the administrative structure of the empire. At the top, the emperor personally worshiped spirits such as those of Heaven, Earth, and Soil and Grain; and at lower echelons, his designated representatives undertook sacrifices at capital cities down to the county (*xian*) level for a great number of protective deities as well as for the spirits of the "unworshipped dead" (*li*). Tablets of "official" gods were often inscribed not only with their names, but also with bureaucratic designations appropriate to their respective administrative levels. Some received imperially bestowed titles of nobility or other marks of distinction as well. Naturally, the state, with its pervasive powers of patronage and appointment, could honor exemplars of Confucian virtue by building shrines to them and even incorporating them into the formal structure of official religion. These gods might then be promoted or demoted, a process that occurred periodically throughout the Qing period.[82]

Of all the deities in the official pantheon, the city god (*chenghuang*) occupied a position of particular importance at the county level. As the otherworldly equivalent of the county magistrate, the city god was expected to cooperate with his bureaucratic counterpart in bringing peace and prosperity to the area under his "jurisdiction." Although represented by a tablet at the open altars of state sacrifice, the city god was depicted as a regular official in his own temple, which was modeled closely along the lines of a magistrate's *yamen*.[83]

The city god represented a kind of symbolic meeting point between official religion and popular religion. Official worship of the deity (aside from the magistrate's private prayers) involved solemn, dignified ceremonies in which only officials and degree-holders could participate. Popular worship, by contrast, was neither subdued nor exclusive. Individuals prayed to the city god for any and all kinds of blessings (especially good health), and the ceremonies for the deity on his "birthday" and during his thrice-yearly tours of the city were among the largest, most impressive, and most widely observed celebrations in traditional Chinese community life. Naturally, the state sought to keep separate these spheres of ritual activity, but there is evidence to suggest a mingling of official and popular religious worship at this level, undertaken, paradoxically, on the initiative of local Qing bureaucrats.[84]

Although the state naturally sought to organize and manipulate popular religion for its own purposes, it was also subject to a certain amount of popular religious pressure. Officials, for example, were occasionally coerced by the common people into worshiping deities not included in the dynasty's list of official sacrifices. This practice occurred despite the heavy penalty for unauthorized ceremonies of this sort. One important feature of both official life and religious life in Qing China was that even though bureaucrats and gods were considered to be especially efficacious, they were also assumed to be in some sense accountable to their constituents. Thus we find that it was not uncommon for Chinese to curse certain gods for failing to fulfill their responsibilities, such as bringing rain or warding off calamity.[85]

At the other end of the ritual spectrum, Chinese households mirrored the hierarchical, authoritarian structure of the state. From an orthodox Confucian standpoint, after all, the family served as a model for governing "all under Heaven." In the words of the *Daxue* (Great Learning), "The government of the state depends on the regulation of the family." Thus, the virtues of subordination to authority and Confucian self-cultivation were considered equally applicable to both the home and the imperial court.[86] Special terms of address (*chenghu*) and specific ritual privileges and responsibilities within the family left little doubt regarding status distinctions and lines of authority, although certain terms and roles might vary somewhat from place to place and class to class. The precise vocabulary of hierarchical relations expressed in official regulations and bureaucratic guides had its domestic counterpart in handbooks on "family ritual" (*jiali*)—works noteworthy not only for their preoccupation with honorific and self-deprecatory usage, but also for their fine distinctions between male and female ritual spheres and the elaborateness of their kinship terminology for both living and deceased relatives.[87] The most famous and influential work of this sort in late imperial times was undoubtedly Zhu Xi's *Jiali* (Family Rituals), which Patricia Ebrey has translated and annotated. She has also provided a thorough historical analysis of the work, showing how versions of it evolved over time.[88]

Chinese households worshiped a wide range of deities for many of the same reasons that various spirits were worshiped by the state. Among the most popular domestic gods in Qing China were the lord of the earth (*tudi gong*) and the god of the hearth (*zaoshen*), whose presence in effect defined the household. Other deities might also be represented in the home, including the Buddhist goddess of mercy, Guanyin, the god of war, Guangong, and the god of wealth (*caishen*). A number of households had only a single tablet at the family altar known as *wuzi pai*, with five characters representing Heaven (*Tian*), Earth (*di*), the sovereign (*jun*), parents (*qin*), and teachers (*shi*). Despite its broader symbolic representation, ritual devotions to this tablet approximated those of ancestor worship, with regular offerings of incense and food and standard gestures of respect and subservience.[89]

The focus of regular ritual activity in most Chinese homes was the ancestral altar, located by definition in the main room of the household. By late imperial times, family and/or lineage ancestor worship had become nearly a cultural universal in China.[90] As a general rule, immediate agnatic ancestors were to be worshiped daily, and even deceased eunuchs received periodic state-financed sacrifices in the absence of heirs. Enriched by Confucian, Buddhist, and Daoist ideas, ancestor worship buttressed the Chinese kinship system not only by cementing social relationships and underscoring status distinctions and obligations, but also by fostering a profoundly conservative precedent-mindedness at all levels of society. Important decisions within the family, whether made by common peasants or by the emperor himself, usually required the "consent" of the ancestors. Auspicious events were announced to the ancestors, just as major festivals were shared with them. Many, if not most, Chinese believed that ancestral spirits had the capacity to intervene directly in human affairs, generally, but not invariably, in a positive way.[91]

Much of Chinese marriage ritual revolved around the ancestral altar. Not only was the impending event invariably announced to the ancestors, but throughout the "six rites" of marriage (*liuli*) there seems to have been a concerted effort to keep the ancestors apprised of events; and although we know that marriage practices differed widely throughout China, two of the most common rituals—at least in major marriages—were the joint worship by man and wife of the husband's ancestors after the ceremonial transfer of the bride (*qinying*) and the wife's worship of these same ancestors a short time after the transfer (*sanyue miaojian*). These acts, together with the transfer feast, confirmed the symbolic entrance of the woman into her husband's family, her ritual rebirth.[92]

Ancestor worship was also usually associated with events such as births, deaths, adoptions, special birthdays and important festivals. As with marriage, these occasions had at least two primary social purposes: one was to enhance ritual solidarity within the family at major turning points by reaffirming kinship ties; the other was to assert socioeconomic status within the wider community. Although Qing ritual regulations specified in detail the ceremonial observances appropriate to bureaucratic rank and social class in matters such as marriage, mourning, and ancestral sacrifices, most families strained their resources to the fullest on all ceremonial occasions in order to display the proper measure of community status as well as their own personal respect, joy, or sorrow.[93]

Regular lineage ancestral rites were generally more formal and exclusive than domestic sacrifices, although they tended to take place on similar occasions and to serve similar purposes. Often, women, children, and a large proportion of male lineage members were forbidden to participate directly. In a sense, such ancestral sacrifices were like the formal rituals of the "official" religion in their emphasis on the symbolic representation of the larger group by a select (male) elite. Like state worship, lineage sacrifices often entailed elaborate patterns of kowtowing and offerings (guided by designated "officials"), music, prayers, and expensive ritual trappings. Similar elaborate ritual performances might also be held by the lineage at the founder's grave during the Qingming festival—in contrast to the less formal, indeed sometimes casual, approach adopted by individual families at this and other "spirit festivals" (*guijie*) during the year.[94] But whether ancestral sacrifices were formal or casual, large or small, exclusive or inclusive, focused on the grave or the tablet (or both), they all helped to "perpetuate the memory of the traditions and historical sentiments of the group, sustain its moral beliefs, and revivify group consciousness."[95]

Divination was central to much of Qing life-cycle ritual. Most family activities—notably births, adoptions, marriages, burials, and many festivals—required heavy reliance on divining techniques, ranging from personal consultation of the *Yijing* or the state calendar (or the less exalted, but ubiquitous, almanac) to the employment of professional fortune-tellers and recourse to temple oracles. As I have argued elsewhere (including Chapter 5 in this volume), the methods of, and occasions for, divination in traditional China were nearly limitless.[96]

One of the most significant and pervasive forms of divination connected with life-cycle ritual was geomancy (*fengshui*). As is well known, *fengshui* centered on the belief that by making certain complex calculations based on a variety of topological and cosmological factors, a geomancer could either find the most propitious place for a grave or building or improve the fortunes of an existing structure. *Fengshui* specialists operated at every level of society, from the imperial government to local villages. Although they enjoyed comparatively high status in Qing society, they received persistent criticism from certain members of the orthodox Chinese elite. These criticisms were not, however, directed primarily against the general theory of *fengshui*, for all levels of society accepted its basic cosmological assumptions. Rather, the Confucian literati objected to the *practice* of geomancy, which generated social tensions, often led to delayed burials (a serious breach of mourning ritual), and involved the manipulation of the Chinese masses by agents who were not generally part of the orthodox elite.[97] This latter concern seems to have motivated much elite criticism of popular religious practice in traditional China, for ritual specialization could all too easily be turned to heterodox purposes.

Heterodox ritual and "heresy"

As the previous sections of this chapter have indicated (at least implicitly), the core of Chinese orthodoxy resided in the famous "Three Bonds" (*sangang*) of Confucianism: ruler–subject, father–son, and husband–wife. Their central significance in traditional Chinese society can be seen clearly in the Qing dynasty's approach to religious and secular ritual, state worship and domestic ceremonies in particular. Extremely jealous of its ritual prerogatives, the state attempted to monopolize all forms of religious authority and expertise in order to enhance its power, while at the same time seeking through various means, including rewards and punishments, to reinforce the integrity and stability of the family as a key unit of social control. But not all members of Chinese society were amenable to such forms of ritual manipulation. A number were either unwilling or unable to participate fully in the orthodox ceremonies that strengthened the Three Bonds, and their alienation was expressed in a number of ways, from various forms of intellectual dissent to open rebellion. When dissent became disloyalty, "heterodoxy" became "heresy" (*xie*).[98]

From the Qing government's standpoint, heresy was less a matter of ideology *per se* than of practice. Although some officially sanctioned sources defined "heretical teachings" (*xiejiao*) as anything "outside the Five Classics and Four Books" (that is, outside of Confucianism),[99] on the whole, the state displayed remarkably wide latitude in its toleration of other religious and secular idea systems. In 1812, for example, the Jiaqing emperor could state that: "Buddhism and Daoism, although not esteemed by Confucians, … encourage what is good and reprove what is evil. Hence, [Daoist] gods and Buddhas and their temples are allotted a place in the Sacrificial Canon [*sidian*]."[100] What the government seems to have feared from heterodox groups or movements was their usurpation of elite prerogatives through the manipulation of orthodox ceremonial forms and symbols. As the concrete expression of ideology, ritual was an important index of political attitudes and ambitions. Because the state claimed to represent the cosmic order and all realms of authority through its ritual activity, it could not possibly tolerate a challenge to either that order or that authority.[101]

Hence we find that Qing law prescribed severe penalties for activities deemed subversive (or potentially subversive) of the existing ritual system, including destroying sacred altars or shrines, publishing unauthorized calendars or "heretical almanacs," privately worshiping Heaven or the North Star (the exclusive prerogatives of the emperor), keeping at home astronomical instruments and charts, and possessing portraits of ancient kings and emperors or seals of gold or jade. *Yinyang* fortune-tellers (*yinyang shu shi*) were prohibited from entering the houses of civil and military officers and "falsely" (*wangyan*) prophesying fortune or misfortune. Magicians, shamans, and other ritual specialists were forbidden to summon "heretical spirits" (*xieshen*), write charms, carry idols, pray to saints, offer incense, hold night meetings, perform "heretical arts" (*xieshu*), write books on sorcery (*yaoshu*), or promote "heretical formulas" (*xieyan*). Although these and other heretical crimes were seldom rigorously defined and the laws not always enforced, their statutory existence, together with evidence from other sources, suggests the state's preoccupation with matters of ritual legitimacy.[102]

By far the most threatening aspect of heterodox religious traditions was their political potential, their capacity to attract and mobilize disgruntled and alienated members of Chinese society. Ritual both contributed to and reflected the political challenge of heretical organizations. At the most rudimentary level, initiation rites such as the letting of blood and the burning of written oaths gave special solidarity to incipient religious groups. Small wonder, then, that the Qing code stipulated the death penalty for members of sworn brotherhoods formed in this fashion. Significantly, it differentiated between brotherhoods of varying sizes and between those involving blood ritual and those without it. All else being equal, ritual solidarity invited harsher punishment.[103]

The preexisting ritual core of popular religious sects made them a natural focus for heretical movements, and hence a special threat. From the Qing government's point of view, the danger was that ritual activity and a messianic vision would provide structure and appeal to the easily misled Chinese masses. And although some sects, like the White Lotus Society, were invariably identified as "the source of misfortune," any society might be suspect. The Longhua sectarians, for example, although espousing a creed dominated by Confucian values, did not escape persecution by the Qing government. At least one reason was their elaborated ritual and sophisticated organization. On the other hand, the *absence* of certain orthodox rituals could be equal cause for alarm. Certainly this was true of groups such as the Wuwei sectarians, who refused to perform prescribed funeral ceremonies and ancestral sacrifices.[104]

In its relentless effort to root out heterodoxy, the Qing government paid attention not only to religious leaders, organization, and ritual, but also to fields of ritual action. The documentary record abounds with examples of the destruction of "indecent" temples (*yinci* or *yeci*). Such temples usually were destroyed on grounds that they served as meeting points for "ignorant people" and "bad elements" in Chinese society. Clearly, the state feared the existence of any structures that might serve as the ritual focus for politically threatening activity. For this reason, it made a special effort to reestablish official temples and orthodox ritual practices in areas ravaged by rebellion and to transform centers of "wild and perverse teachings" (*kuangbei zhi jiao*) into more conventional places of worship.[105]

Beyond individual temples, associations, and lodges were higher levels of heterodox or heretical organization, culminating in what Philip Kuhn calls "communities in arms." The most striking, widespread, and well-documented example of this phenomenon in Qing times was the Taiping Rebellion. In many ways, as is well known, the Taipings posed a revolutionary challenge to the existing sociopolitical order.[106] Yet they, like many other heterodox contenders for power before them, made numerous concessions to Chinese tradition in the realms of both theory and practice.[107]

One such concession was manifest in their approach toward ritual. To be sure, the Taipings criticized a number of the "impish ceremonies" (*yaoli*) of the Manchus and refused to designate any days as auspicious or inauspicious.[108] Yet it is clear that overall the rebels were heavily imbued with the traditional spirit of *li*. In the *Tianxing daoli shu* (Book on the principles of heavenly nature), for example, the Taipings quote the Confucian *Analects* verbatim in emphasizing ritual as a means of achieving social harmony. In the *Youxue shi* (Ode for youth), it is said: "The principle that governs the two hands is none other than respect, and [that the hands should not do anything] contrary to *li*." The *Taiping qiushi ge* (Taiping songs for world salvation) admonishes parents to teach their children with *li* (ritual) and *yi* (duty). And the only major Taiping tract on ritual, titled *Taiping lizhi* (Taiping regulations on ritual), indicates an abiding concern with status distinctions and matters of ceremonial form. Significantly, much of the Taiping organizational and ritual outlook was derived from the ancient Confucian classic, the *Zhouli*.[109]

Taiping noble and official rank generally followed traditional models, as did much of court ritual and dress (based on Ming precedents), sumptuary laws, taboo characters, and symbolism (including the dragon, phoenix, *qilin*, lion, tiger, the sun and the moon, etc.). Taiping memorials and edicts reflected traditional styles. Certain characters were elevated (*tige*) to show honor and respect, and a rudimentary examination system was established.[110] These imperial trappings, together with the creation of seals, coins, and a state calendar, were deemed essential to the legitimacy of rebel groups, regardless of their sociopolitical message.[111] In a sense, the Taipings were a formidable threat to the Qing government precisely because in ritual form they were so much like it.

It is worth noting, I think, that the Qing documents I have seen on heterodox and heretical groups—including the Taipings—avoid entirely the use of the term *li* to describe the ceremonies and institutions of dissidents. Undoubtedly one reason is that *li* implies orthodoxy and legitimacy, and the use of the term with reference to heterodox or heretical groups would be oxymoronic. What we do find is an official preoccupation with the rectification of "people's minds" (*renxin*) and "popular customs" (*fengsu*), indicating that the source of the problem from the Qing point of view was the delusion of the "ignorant masses."[112] The assumption of the Qing dynasty was that the gentry class would always lay its wager with the throne, an assumption that held true, by and large, until the dynasty proved unwilling or unable to defend either elite interests or official orthodoxy.

Some concluding remarks

In a penetrating but controversial article on the place of the Tianhou cult in official religion. James Watson observed that the "genius" of the Qing government's approach to cultural integration was that "the state imposed a structure but not the content" on Chinese religious life. It promoted, in other words, certain symbols and ritual practices, but did not try to legislate popular religious beliefs. As a result, official religion proved flexible enough to allow people at all levels of society to construct their own (often contradictory) representations of state-approved deities, yet was firm enough to provide "a national culture that appeared—on the surface, at least—to be remarkably integrated, especially when compared to the cultural systems of other premodern societies."[113]

In the past few years, a number of scholars have offered spirited challenges to Watson's so-called "orthopraxy" thesis, questioning in particular the effectiveness of official efforts to "standardize" ritual. They have also called into question Watson's emphasis on external actions rather than inner beliefs as the most important aspect of ritual in the eyes of the imperial authorities.[114] On balance, it seems clear that the Qing state's attempts to standardize ritual practices at the local level were not as successful as Watson has argued: The empire was simply too vast and too variegated; local customs were too strong; and local elites, primarily responsible for encouraging proper behavior at the local level, often had ambivalent feelings about rituals that departed from orthodox models.[115] Moreover, as Evelyn Rawski and others have maintained, beliefs mattered, not only to the state, which always feared that unorthodox thoughts might lead to dissension or even sedition, but also to scholars and others, whose writings, presumably based on deep convictions, shaped Chinese ritual practice in fundamental ways throughout the late imperial era.[116]

Somewhat ironically, the widespread scholarly belief that "ritual arises from moral principles" (*li qi yu yi*) made it possible for officials and other elites in the Qing period to countenance unorthodox ritual practices on the grounds that they proceeded from good motives.[117] And by justifying non-standard rituals in this way, they may have promoted cultural integration by broadening the scope of what was considered acceptable ritual behavior.[118]

In any case, Watson's efforts to explain the extraordinary integration of traditional Chinese society continue to have considerable value. In the first place, we should remember that the state was not completely unsuccessful in encouraging the standardization of community rituals, and that its relentless attempts to "rectify customs" served as a constant reminder of what the state and its local agents considered to be proper behavior. Donald Sutton refers to the various forms of imperially sponsored propaganda as a kind of "unifying rhetoric."[119] Second, even if we grant a certain amount of deviance from canonical models, there remained a remarkable degree of continuity and uniformity in certain types of Chinese ceremonial behavior—domestic ritual in particular. As Patricia Ebrey notes in her analysis of Zhu Xi's *Family Rituals*:

> [We] know that at the end of the imperial period common practice across the country bore many general resemblances to the steps described in Confucian texts. Ancestral rites included periodic offerings of food and drink at domestic altars … and on death-day anniversaries and major festivals. Continuities in wedding ceremonies included negotiations by matchmakers, exchanges of gifts, the highly ceremonial transfer of the bride to her husband's home where the new couple consumed wine and food together and the bride was introduced to her husband's parents, relatives, and ancestors. For funerals and burials, major continuities included ritualized wailing, mourning garments that visibly indicated the proximity of kinship to the deceased, setting food and drink near the coffin until burial, postponing burial days, weeks or months to prolong mourning, restrictions on social activities during deepest mourning, use of heavy coffins, ceremonial funeral processions to the graveyard, and post burial sacrifices to the deceased.[120]

According to Ebrey, the narrowing of class distinctions in these sorts of ritual performances was "a significant factor in creating cultural cohesion within the enormous expanse of China." It allowed people who "spoke mutually unintelligible dialects and knew very little of each other's daily work lives to see each other as Han Chinese." Throughout the imperial era, then, the performance of such family rituals was a significant marker of ethnic boundaries. As Ebrey puts the matter in her concluding remarks, "Every time a family hired a sedan chair for a wedding or paraded to the grave in mourning garments, it was acting out of allegiance to both Han Chinese identity and the Confucian moral order."[121]

To be sure, a tension always existed in China between the realms represented by *li* and *fengsu*, but it was a creative tension and did not usually involve polar opposition.[122] It is now beyond dispute, for example, that the traditional distinction between the Neo-Confucian "rationalism" of the scholarly elite and the "superstition" (*mixin*) of the popular Chinese masses has been much too sharply drawn. Although many aspects of popular religion differed significantly from the religious rituals of the orthodox elite, many did not, or differed only in degree and particulars, not in kind. For all their alleged agnosticism, many elite families worshiped a variety of gods and employed a variety of religious agents.[123] Virtually all members of Chinese society accepted at some level the idea of a spiritual world inhabited by *shen* and *gui* and the need to honor or appease these spirits through sacrifice. Emperors, officials, degree-holders and commoners alike offered prayers on many occasions, particularly during times of crisis. Confucian writings usually downplayed the supernatural aspect of ritual activity, but the evidence strongly suggests that even in the elite mind *shen* and *gui* were not always identified simply as the abstract forces of *yang* and *yin*. And while it is true that Chinese elites periodically tried to influence the outward expression of popular religious beliefs, it is also true that popular beliefs and practices helped to shape the consciousness and behavior of these same elites.[124]

A similar dialectic operated in the realm of divination, which touched Chinese ritual at nearly every turn, at all levels of society. State calendars and popular almanacs, for example, promoted essentially the same worldview and encouraged the same basic ritual activities at the same basic times, despite their different target audiences. Official diviners in the Bureau of Astronomy had their unofficial counterparts in Qing popular culture as a whole, and geomancers (*fengshui xiansheng*) served the state as well as individual households. The Qing government, for its part, tolerated this overlapping of divinatory responsibility as long as its political hegemony remained intact.

Just as most areas of Chinese ritual were related to divination, most drew on a common repository of cosmological and other symbols, many derived from the *Yijing* and virtually all expressed in calendars, almanacs, and the basic vocabulary of geomancy. Although these symbols were seldom interpreted in precisely the same way by all levels of society, they contributed to a shared sense of culture that transcended class and even the divide between orthodoxy and heterodoxy. Among the most common symbols of this sort were *yin* and *yang* (and their various numerical and other conceptual equivalents), the "Three Powers" (*sancai*: that is, Heaven, Earth, and Man), the five elements or activities and their conceptual equivalents, the eight trigrams (*bagua*), and so forth."[125] Also widely shared were the metaphors of bureaucracy, kinship, and the student–teacher relationship. Most Chinese ritual activities took place on similar occasions (notably the lunar phases marked by the first and fifteenth days of the month) and involved the use of similar devotional techniques, including the offering of food, the burning of incense, and the use of written prayers or charms. Most even shared a similar ritual field in that temples and palaces were based on the same basic module as the typical Chinese home.

Other aspects of ritual also contributed to cultural cohesiveness in China. On the one hand, the state used religion and ritual to enhance its power and influence over its subjects. Arthur Wolf writes, for example, that

> [The Chinese government] appears to have been one of the most potent governments ever known, for it created a religion in its own image. Its firm grip on the popular imagination may be one reason the imperial government survived so long despite its failings.[126]

On the other hand, as Emily Ahern has indicated, the Chinese religious and ritual system served a vital teaching function for commoners—including peasants—illustrating vividly "in a variety of contexts how one got power, how one got access to those with power, and how one limited those with power."[127]

Certainly participation in ritual served as a social reward system involving mixed criteria such as literacy, wealth, age and personal morality. Although government statutes distinguished between the ceremonial privileges appropriate to officials (*pinguan*), scholar-gentry (*shushi*), and commoners (*shuren*), in fact, the first two categories overlapped to a significant degree, and there were also opportunities for "longevous" or "philanthropic" commoners to achieve a measure of exalted ritual status outside the regular examination channels and the system of purchased degrees.[128]

At various levels and in various ways, then, ritual served to perpetuate important social and political myths. Among these myths were the value of orthodox education, the possibility of upward mobility, the idea of Chinese cultural unity and the universality of the Confucian way of life, community solidarity and family harmony, and the protective power of the bureaucratic state—not to mention such expansive conceptions as China's cultural superiority, manifest in the tributary system. Everyday ritual activities at all levels conveyed essential social messages regarding power and status, and prayers and religious sacrifices in time of crisis, whether carried out by officials, professional ritual agents, or laymen, provided hope and encouragement to the unfortunate. Traditional festivals, occurring regularly throughout the year, performed the function of periodically reaffirming orthodox values and community ties. And although the particulars of the festival calendar (*suishi ji*) varied somewhat from place to place, a number of festivals were universally observed, both officially and unofficially, contributing to a shared sense of culture at all levels of Chinese society.[129]

Of course, ritual did not always conduce to social order and cultural integration. From the standpoint of the state, the use of ritual by heterodox organizations was invariably divisive, and even such common practices as *fengshui* divination created serious social problems, both within the family and within the wider community.[130] Disputes often broke out in traditional Chinese society over matters related to local temples and ritual land, and questions of proper ritual form periodically precipitated or exacerbated factional strife within the Qing government.[131] Yet on balance ritual undoubtedly played a positive role in helping to maintain a high degree of cultural unity and continuity in China.

The place of ritual in the breakdown of the old order in the late-nineteenth and early-twentieth centuries is beyond the scope of this chapter, although it may be noted that the process had much to do with two spheres of official ritual activity that have been largely neglected in this chapter: the tributary system and the examination system. In any case, even after the destruction of the Chinese world order by Western and Japanese imperialism and the introduction of new ideas into China, many traditional ritual assumptions and practices continued to have remarkable staying power. Throughout the Republican era, despite the iconoclastic onslaughts of the New Culture Movement, popular religious ritual at the community and the domestic level proved extremely tenacious, as did certain assumptions about the propriety of social relations and the ritual role of the state.[132] And even today, after several decades of radical "socialist reconstruction" on the Mainland, the Chinese authorities have still not succeeded in eradicating many so-called "feudal" social beliefs and ritual practices. In fact, most of them seem to be experiencing a revival.

4

THE TEACHINGS OF RITUAL AND THE RECTIFICATION OF CUSTOMS

Echos of tradition in the political culture of modern China

One of the most striking features of traditional Chinese thought is what Thomas Metzger refers to as its "epistemological optimism." In contrast to the generally recognized pessimism of the Western philosophical mainstream—from Plato and Aristotle through Hobbes, Locke, Mill, and others into twentieth-century liberalism—Confucian thinkers in China believed that government was fundamentally a matter of morality, and that there was already "enough absolute moral knowledge available to evaluate everyone perfectly and totally reform the political order." The practicability of this vision was obvious in Confucian eyes. Why? Because the great sages of the ancient past

> had obtained absolute moral-political wisdom which had been transmitted down to the present, even though in a partly unclear way; because the sages' perfect rule in the distant past demonstrated its practicability; because the cosmos existed as an inspiring moral presence, whether in the form of outer order or inner feeling; because the individual lacked an original sin and so could fully grasp and act on that cosmic presence; and because the very nature of the polity was such that any moral person heading it would almost immediately have an enormous transformative effect on the whole society, dissolving immorality throughout China if not the whole world.[1]

Significantly, Metzger identifies "optimistic this-worldliness" as one of the significant continuities between "traditional" and "modern" (i.e. twentieth-century) China—whether manifest in Sun Yat-sen's "Three Principles of the People," in Tang Junyi's Confucian humanism, in Mao Zedong's "sinicized" brand of Communism, or in Yang Guoshu's Chinese-style liberalism.[2]

Taking Professor Metzger's provocative remarks as a point of departure, and drawing inspiration from the writings and teachings of my mentor, Professor

Kwang-Ching Liu,[3] I would like to explore one of the major ways in which Chinese leaders during the Qing dynasty (1644–1912), the Republican era (1912–1949), and the People's Republic of China (1949–present) have attempted to achieve the common "optimistically" conceived political end of moral transformation through the use of similar cultural means. That means has been, and remains, "ritual" (*li*), sometimes translated as "ritual-propriety."[4] For centuries, the Chinese have referred to their country as "the land of ritual and righteousness" (*liyi zhi bang*), equating these two related values with civilization itself.[5]

Ritual and social control in the late imperial era

From earliest times in China, *li* had a cosmological foundation.[6] In the words of the eighteenth-century scholar, Dai Zhen:

> Rituals are the rules and laws of Heaven and Earth. They are perfect rules, and only those who understand Heaven will know them. Etiquette, ceremonies, measures, and regulations are the rules that the sages perceived in Heaven and Earth, and they instituted them as the standard for the myriad generations to come.[7]

The aim of these rules was to preserve order (*zhi*), for the ancient sages knew, as the nineteenth-century statesman, Zeng Guofan, put the matter, that "the mind of Heaven hates disorder."[8] Ritual was thus an ordering mechanism and a transformative process, which worked in two directions: downward from the emperor and his officials and upward from everyone else.[9] The throne inspired the process, but for it to work effectively, individual self-cultivation had to be the starting point. Self-cultivation was never the end for Confucians, however. To be fully human, it was necessary not only to develop one's own nature fully, but also to extend this accomplishment to influence others. The quality that enabled humans to do so was "sincerity" (*cheng*)—a dedication to goodness equated by Confucians with "the Way of Heaven" itself. The *Doctrine of the Mean* (*Zhongyong*) states: "The person possessed of the utmost sincerity is like a spirit (*shen*)."[10]

Proverbially, all Chinese in imperial times knew that "If there is sincerity, then there is spiritual power (*cheng zhi ze ling*)." This "spiritual" capacity of human beings—expressed by terms such as *shen*, *ling* or *jingshen* (essential spirit; sometimes translated as "mental energy")—invested them with enormous mental strength, derived from Heaven itself.[11] The famous Song neo-Confucian scholar Cheng Yi (1033–1107) once wrote that "With the most highly developed sincerity, [the mind] can penetrate metal and stone, and overcome water and fire, so what dangers and difficulties can possibly keep it from prevailing?"[12] Zhu Xi (1130–1200), another renowned neo-Confucian philosopher, put the matter this way: "When [a person] has spirit (*jingshen*), then there is nothing that cannot be accomplished."[13]

The essential point to keep in mind about these conceptions of the human "spirit" is that they were invariably moral in nature, defined primarily in terms of orthodox Confucian virtues. As normative concepts, the terms "spirit," "culture (*wenhua*)" and "civilization (*wenming*)" were inextricably related, and believed to have universal application. To put the matter somewhat differently, (Chinese) culture came to be viewed as the means by which the spirit of (all) human beings could be realized. The writings of modern scholars such as Qian Mu, Tang Junyi, and Wei Zhengtong abundantly document the "spiritual" nature of traditional Chinese civilization and its close connection with ritual.[14]

If one could achieve spiritual develoment through assiduous study and ritually correct behavior, that person would, by definition, contribute to the transformation of society.[15] The famous chain syllogism in the *Great Learning* (*Daxue*) offers a blueprint for the process: When thoughts are made sincere, the mind can be rectified; when the mind is rectified, the self can be cultivated; when the self is cultivated, the family can be regulated; when the family is regulated, the state can be ordered; and when the state is ordered, all under Heaven can be at peace.[16]

Xunzi's essay on "Self-Cultivation" identifies the place of ritualized deportment in the process:

> If all matters pertaining to temperament, will, and understanding proceed according to ritual (*li*), they will be ordered and successful; if not, they will be perverse and violent or slovenly and rude. If matters pertaining to food and drink, dress, domicile, and living habits proceed according to ritual, they will be harmonious and well regulated; if not, they will end in missteps, excesses, and sickness. If matters pertaining to deportment, attitude, manner of movement, and walk proceed according to ritual, they will be refined; if not, they will be arrogant and uncouth, common and countrified. Therefore a man without ritual cannot come to completion; a state without ritual cannot attain peace.[17]

In Qing times, one of the primary devices for encouraging adherence to proper ritual and etiquette empire-wide was the Sacred Edict (*Shengyu*) of the Kangxi emperor, promulgated in 1670. This work grew out of a long tradition of imperial maxims dating from the early Ming period and intended expressly to improve popular customs and to "rectify the hearts and minds of the people."[18] Designed to be read aloud and then explained to all sectors of Chinese society, the Sacred Edict was comprised of sixteen pithy maxims, each only seven characters in length. It reduced imperial orthodoxy to the following basic prescriptions: (1) esteem filial piety and fraternal submission; (2) respect kinfolk; (3) cultivate harmony in your neighborhood; (4) emphasize animal husbandry and agriculture; (5) prize moderation and economy; (6) exalt institutions of learning; (7) eliminate strange doctrines; (8) explain the laws; (9) elucidate the principles of propriety and courtesy; (10) attend to the basic occupations; (11) give moral instruction to sons and younger brothers; (12) stop false accusations; (13) warn against sheltering those who hide from the state; (14) fully remit taxes; (15) unite in collective security units (*baojia*); and (16) dispense with anger and hostility.[19] All of these prescriptions were designed to preserve social harmony and/or to promote the specific political and economic interests of the Qing state.

In 1724, the Yongzheng emperor issued a 10,000-word amplification of the Sacred Edict in order to explicate his father's terse formulations. But these "Amplified Instructions" (*Guangxun*) were written in a classical language incomprehensible to most Qing subjects.[20] As a result, many members of the Chinese elite took it upon themselves to produce more "popular" versions of the Yongzheng emperor's remarks.[21]

The best known and most influential of these works was an eighteenth-century vernacular exposition of the "Amplified Instructions" by a scholar named Wang Youpu. Wang's paraphrase, written in an extremely colloquial style, provides a convenient inventory of the primary social and political concerns of the Qing elite, as well as a sense of the kinds of arguments they employed in order to persuade a popular audience.[22] The tone of the document is openly condescending. It refers repeatedly to the ignorance of the common people, and indicates that many of them are too easily deceived. It admonishes all Qing subjects to be honest, to work hard, and to repay both the filial debt they owe to their parents and the debt of loyalty they owe to their sovereign.

Although Wang's paraphrase is full of Confucian prescriptions for establishing and maintaining a harmonious society, it places heavy emphasis on the harsh legal consequences of improper behavior, and refers occasionally to supernatural retribution for misdeeds as well. It also projects a decidedly negative image of women, blaming them (in the fashion of a number of well-known ritual handbooks and other widely circulated Confucian works)[23] for domestic disharmony, citing common sayings such as "A woman's mouth is like a measure without limits; truth and lies issue from it [without discrimination]."[24]

Many of the Kangxi emperor's sixteen maxims deal with ritualized relations and behavior (filial piety, devotion to ancestors, etc.), but the ninth addresses *li* explicitly. It reads: "Manifest the principles of propriety and courtesy [lit. yielding] in order to enhance popular customs (*ming lirang yi hou fengsu*)." Wang Youpu's paraphrase of this maxim in over 1,700 characters brings together most of the major themes to be found in one or more of the remaining fifteen sections of the "Amplified Instructions." These constantly reiterated themes include: the power of positive example; the importance of moral education; the need for discipline (self-restraint); respect for superiors; kindness to inferiors; the avoidance of lawsuits; mutual assistance among friends and neighbors; frugality; moderation in ceremonial displays, and so forth.

Wang begins his essay by remarking that popular customs determine whether the empire is at peace or not. These customs vary, however. Although all human beings intuitively understand the Five Constant Virtues, their actual feelings and dispositions are not the same. People live in different places, are accustomed to different climates, eat different foods, and speak in different dialects. Some individuals are strong, others are weak; some act hastily while others are slow to move. Customs can be pure in one place and corrupt in another. Because of these differences, the great sages of the past set out rituals to teach and transform the people. As a means of promoting cultural unity across space and time, ritual is "the root of popular customs."[25]

Wang goes on to say that commoners in rural areas may not have studied the rites, and so cannot be expected to be familiar with refined usage. Yet they can certainly figure out how to respect their parents and elders, maintain amicable relations with their neighbors, serve their friends, and be in harmony with their spouses. He urges them to do so by concrete means: showing deference to one another, using proper forms of address; doing away with fighting and swearing; having compassion for the poor; not envying the rich; avoiding the oppression of the weak; not fearing the strong; and above all, showing modesty, courtesy, and deference in all realms of life.[26]

Despite Wang's lip service to the Mencian assumption of a shared good nature for all human beings, his tract reads more like Xunzi. In fact, he offers detailed critiques of all classes of Chinese society, assuming, it appears, that each in its own way will be contentious and self-interested—if not also dishonest and deceitful. Scholars, he says, boast of their literary talents and refuse to yield to one another. Peasants squabble over boundary markers and crops. Artisans try to outdo each other, and merchants vie for customers and profits. Soldiers, beneath contempt, stand ready to kill people on the slightest pretext. None of this would happen, Wang maintains, if all people simply followed the proper ritual prescriptions.[27]

Wang's elaboration of the ninth maxim, like his gloss on other portions of the "Amplified Instructions," includes a number of popular proverbs and quotations from the Confucian Classics, as well as several colorful anecdotes. One of these concerns a peasant by the name of Guan Youan, who exemplified the virtue of yielding (*rang*). Guan, it seems, would never get angry if the animals of other people browsed in his crops. Indeed, he fed the animals and in so doing caused his neighbors to look upon him with favor and admiration. In times of disorder and popular uprisings, rebels would pass his door without searching his house, and people fleeing from danger naturally came to him for assistance. "Thus," Wang tells us, "a single man, by being able to yield, can move many others." The moral of this story is evident: When a person acts in the spirit of propriety and courtesy, people's hearts will be transformed and customs will be improved. "If everyone in the world could do this," he writes, "Heaven would naturally be moved, ... the state and the people would rest in harmony, and peace would reign everywhere."[28]

How did such uplifting stories find their way to the Chinese masses? In promulgating his original Sixteen Maxims, the Kangxi emperor decreed that as part of the dynasty's official propaganda system known as *xiangyue* ("community compact"), lectures on the Sacred Edict were to be given on the first and fifteenth day of each month at designated locations empire-wide.[29] A lecturer would be appointed for this purpose in every region, to be assisted by three or four "honest and prudent persons."

The lecturer had the task not only of explaining the Sacred Edict to the local population, but also of recording the good and bad conduct of various people in the area—which then became a matter of public knowledge through displays posted in special "exposition pavilions" (*shenming ting*). All elders, scholars, and commoners in a given locality were expected to attend the lectures, and students had to memorize the Sixteen Maxims as part of their early education. After the Yongzheng emperor issued his "Amplified Instructions" in 1724, this document, too, became part of the *xiangyue* "curriculum." All aspirants for the *shengyuan* degree were required to write sections of the document from memory as a preliminary test.[30]

Over time, and in response to different regional circumstances, the *xiangyue* system underwent a number of changes. For one, the Qing government made a concerted effort to expand the lecture system to include not only civilians but also military forces (notably the Army of the Green Standard) and various ethnic minorities. Bannermen were spared the formality of bi-monthly meetings, but they reportedly had to memorize both the Sacred Edict and the "Amplified Instructions." (We may doubt that many committed the latter fully to memory.) Meanwhile, enterprising Qing officials attempted to link the *xiangyue* system with preexisting organs of local control, including villages, clans, militia units, and the mutual responsibility system known as *baojia*.[31]

During the late seventeenth century and much of the eighteenth, the *xiangyue* lecture system seems to have functioned relatively effectively. But by the mid-1800s, if not before, it had degenerated into a perfunctory and virtually meaningless political exercise.[32] A succession of emperors, beginning with Qianlong and proceeding through Tongzhi, lamented that the *xiangyue* orations had become a matter of mere routine in many places, and that the system had fallen into complete disuse in some. Nineteenth-century accounts, both Chinese and foreign, suggest that even in urban centers audiences often did not understand the material, and that no effort was made by lecturers to explain it to them.[33] On the other hand, as Victor Mair points out, popular entertainers brought the moralistic themes of the Sacred Edict and the "Amplified Instructions" to the masses in a highly entertaining way—one that seems to have been far more effective in disseminating Confucian ideals than the "pompous, ritualistic, and often lifeless ceremonies presided over by local officials."[34]

Even so, we may wonder how persuasive the message of the Kangxi and Yongzheng emperors was in late Qing China. Perhaps the Qianlong era—a period of great prosperity and comparative domestic tranquility—indeed conduced to "propriety and moderation" as Guan Zhong might have predicted.[35] But during the nineteenth century, China's situation changed dramatically for the worse. Unprecedented population growth created severe social and economic problems, precipitating, among other things, widespread rebellion. Foreign aggression only exacerbated the situation. Yet the Qing rulers characteristically viewed the "deterioration of the people's hearts and customs" as a function of inadequate ritual indoctrination—which they ascribed in large measure to a decline in the *xiangyue* system.

The fundamental problem with the Qing government's approach to moral transformation was not, however, in either its medium or its message. In truth, most members of Chinese society were simply too poor to respond in the required way. In the words of Hsiao Kung-ch'üan: "The dire want and glaring inequities that were the lot of many villagers silently but indisputably controverted almost every one of the injunctions of imperial Confucianism." To tell half-hungry people that they should "prize moderation and economy, in order to prevent the lavish waste of your means" (the fifth maxim) was "worse than a joke; it was sheer mockery."[36]

Meanwhile, foreign imperialism helped to undermine the Confucian social and political order by destroying the emperor's ritual claim to rule "all under Heaven." This led to the dissolution of the concept of "cosmological kingship"—the idea that the Chinese emperor was "a sort of cosmological linchpin functioning at the center of the world and radiating a universal authority on earth."[37] At the same time, foreign aggression generated modern Chinese nationalism. The Sino-Japanese War of 1894–1895 was a turning point in this respect. Increasingly, as China's domestic and international situation deteriorated in the face of foreign firepower, Western ideas and Western values came to dominate the consciousness of China's newly politicized intellectual elite. Sun Yat-sen's "Three Principles of the People" (*sanmin zhuyi*)—democracy, nationalism, and the people's livelihood—became the rallying cry for growing numbers of Chinese who clamored for anti-Manchu, pro-democratic revolution. In 1911–1912, a motley alliance of republican revolutionaries succeeded in toppling the Qing dynasty.

The reconsideration of Chinese tradition: a new culture or simply a new life?

As is well known, parliamentary democracy failed miserably in China following the overthrow of the Qing dynasty. But democratic sentiment and hostility to Confucianism continued to grow during the New Culture Movement (*c.* 1915–1925). In the midst of relentless foreign imperialism, civil strife, and destructive warlord rule, Chinese intellectuals such as Chen Duxiu urged their compatriots to be independent, not servile; progressive, not conservative; aggressive, not retiring; cosmopolitan, not isolationist; utilitarian, not formalistic; and scientific, not mystical. In 1919, Chen wrote: "To uphold democracy, we have to fight against Confucianism, the old tradition of virtue and rituals, the old ethics, and the old politics. To uphold science we have to oppose old arts and old religion." In this hostile climate the "teachings of ritual" (*lijiao*) came under particularly harsh fire. Wu Yu, for instance, denounced *li* for encouraging blind submission to hierarchical authority, equating its destructive effect with "cannibalism" (*chiren*).[38]

Although the thrust of the New Culture Movement often seemed to be toward radical iconoclasm, in fact, the era witnessed a wide variety of approaches to the reconsideration of Chinese tradition. Fierce debates arose over

> fundamental problems such as the relationship between the material and spiritual phases of life, the significance of the scientific method, the merits of materialism and nationalism, the choice between Westernization or preservation of a "national essence" [*guocui*], and the doctrines of socialism, anarchism, and communism.[39]

New styles of discourse, including the widespread use of the written vernacular to replace classical forms of poetry and prose, and the introduction of a great many Japanese-coined neologisms, altered the very nature of China's intellectual terrain. Meanwhile, twentieth-century cosmopolitanism gave expanded meaning to concepts such as "civilization" and even "politeness," which no longer came to be defined exclusively in traditional Chinese terms.[40]

The early 1920s witnessed the birth of two new political forces in China, both ostensibly hostile to traditional values: the Chinese Communist Party (CCP), established in 1921, and the Nationalist Party (Guomindang or GMD), reorganized in 1923–1924. In 1928, after more than a decade of political chaos and unremitting criticism of China's traditional culture by many Chinese intellectuals, Chiang Kai-shek's GMD came to power after turning on its ally, the CCP, in 1927. Espousing Sun Yat-sen's "Three Principles of the People," the Guomindang initiated several campaigns to suppress various kinds of "superstition" (including old cosmological beliefs expressed in traditional-style almanacs and fortune-telling practices), and ordered the abolition of official Confucian rites in the name of "modernity."[41] The rationale was that

> The principles of Confucius were despotic. For more than twenty centuries they have served to oppress the people and to enslave thought. As to the cult of Confucius, it is superstitious and out of place in the modern world. China is now a Republic. These vestiges of absolutism should be effaced from the memory of citizens.[42]

The vestiges were not effaced for long, however. By 1928, Chiang Kai-shek had already begun urging his officers to read the "Four Books" of Confucianism for inspiration; and by 1934, Confucius had been officially recanonized.[43] During the same year, in a period of national crisis both within and without, Chiang, acting in the paternalistic, quasi-imperial style, inaugurated his famous New Life Movement, which called for a return to the four ancient Confucian virtues of ritual (*li*), righteousness (*yi*), honesty (*lian*), and a sense of shame (*chi*).[44] At that time, Chiang maintained that the single greatest cause of China's weakness and inability to contend with foreign imperialism was not inferior weapons, but "the loss of our traditional spirit and fine morality."[45] The New Life Movement was thus designed as nothing less than a "lifestyle revolution" aimed at overcoming the "moral degeneration" and "spiritlessness" of the Chinese people.[46]

The three major manifestations of this spiritlessness, in Chiang's opinion, were: (1) the incapacity to distinguish what was good from what was bad; (2) the inability to differentiate what was public from what was private; and (3) the failure to understand what was fundamental and what was secondary (simply expedient). The result was that officials tended to be "hypocritical and corrupt," the Chinese people as a whole tended toward "aimlessness and numbness," the young tended to be "decadent and lax," adults tended to be "degenerate and ignorant," the rich tended toward "troublemaking and ostentatiousness," and the poor tended toward "foulness and disorderliness." In all, Chiang found himself in charge of a country he described as plagued by filthiness, hedonism, depravity, laziness, barbarism, and lack of reason.[47]

Chiang believed, however, that these ills could be remedied by the reassertion of traditional Chinese values, and the model he proposed for bringing order to the state and peace to "all under Heaven" was none other than the famous chain syllogism of the *Great Learning*. The starting point, then, was the "rectification of each person's mind (*zheng renxin*)." Chiang espoused the entire range of Confucian virtues—including humaneness, filial piety, loyalty, faithfulness, harmony, and peace—but he considered the "four guiding principles" of *li*, *yi*, *lian*, and *chi* to be absolutely fundamental. These principles came straight from the writings of Guan Zhong, but Chiang did not accept Guan's argument concerning the relationship between material conditions and moral values. To Chiang, although spiritual and material forces were complementary, "spiritual forces control material forces, not the other way around."[48]

Of the four "guiding principles," *li* was clearly paramount—the primary means by which to "prevent disorder." Without *li*, Chiang once asserted, there would be no other virtues. The mission of his New Life Movement, then, was to restore China's reputation as "the land of ritual and righteousness."[49] "If we are determined to reform," Chiang steadfastly claimed, "we must first reform our [personal] habits."[50] *Li* provided not only the means to this end, but also the proper mental attitude—a "spirit," Chiang declared, that animated the writings of Sun Yat-sen himself. Drawing inspiration from standard Confucian canonical sources, and also from the ritual-oriented outlook of Zeng Guofan, his Qing-dynasty role-model, Chiang pointed out that the function of *li* was nothing less than to preserve the family and the state from destruction.[51] Like Zeng, the Generalissimo believed that China's internal disorder had to be overcome before external challenges could be successfully met.[52]

Inspired by the time-honored view that ritual distinguished human beings from animals, and civilization from barbarism, Chiang characterized his mortal enemies, the Chinese Communists, as amoral beasts. In language reminiscent of Zeng Guofan's famous denunciation of the Taipings some seventy years before,[53] he remarked that in the Jiangxi Soviet base area:

> Parents cannot love their children, children have no knowledge of filial piety, brothers do not act like brothers, friends cannot be friends, teachers do not know how to teach students, and students show no respect to teachers. In short, superiors do not act like superiors nor do inferiors act like inferiors. Is there any difference between [the Communists] and animals? The place that was once the home of sages and heros, the land of ritual and righteousness, has became a barbaric and horrible place.[54]

Although the New Life Movement had an undeniable Confucian cast, in many respects, it may be viewed as an extension of Chiang's own military mentality.

Indeed, one of the explicit goals of the Movement was the militarization of Chinese society. Chiang's emphasis was clearly on making every citizen "pay attention to order, obedience, and discipline." In fact, *li* often came to be defined less in terms of ceremonial forms and common courtesies than in terms of a "law-abiding attitude"—that is, a focus on obeying the rules. Chen Lifu, in an effort to nuance the term, claimed that *li* meant "discipline" (*jilu*) when applied to groups and "politeness" (*limao*) when applied to individuals. *Yi* (righteousness), traditionally defined as "that which is appropriate," was considered to be "the visible manifestation of *li*." In the words of the official "Outline of the New Life Movement:" "All conduct that accords with *li*—that is, natural law, social rules, and national discipline—is appropriate."[55]

Chiang's speeches in the 1930s leave no doubt that as a military man he abhorred uncleanliness, lack of discipline, laziness, weakness, bad eating habits, poor housekeeping, and shabby appearance. On the other hand, he greatly esteemed the crisp military virtues of orderliness, cleanliness, simplicity, frugality, promptness, and exactness. Unfortunately, Chiang saw far more wretchedness than rectitude in Nationalist China, much more "barbarism" than beauty. In order, therefore, to encourage proper behavior and to provide concrete guidance to all citizens of the Republic, he produced a pamphlet titled *Necessary Knowledge for New Life* (*Xin Shenghuo xuzhi*). This short but comprehensive work—which appeared in all of China's major newspapers on May 15, 1934, and was also incorporated into official publications such as the *Handbook of the New Life Movement* (*Xin Shenghuo yundong shouce*)—contained ninety-six specific rules, divided into Xunzi's four general categories: food (*shi*), clothing (*yi*), residence (*zhu*), and action (*xing*).[56]

Chiang's admonitions appeared in neat four-character phrases, easy to understand and easy to memorize. In addition to certain general precepts regarding the "four guiding principles" and the need for discipline, public-mindedness, patriotism, and frugality, he urged people to have meals at regular times, to keep food and utensils clean, to avoid making noise while eating, and so forth. In public, citizens were to use correct posture, dress neatly (but avoid jewelry and perfume), wear shoes, and keep their clothes clean. At home, family members should rise early, stay well-groomed, plant trees and flowers to beautify the yard, have simple tastes in furniture, exterminate pests, and maintain strict standards of hygiene. Fathers should instruct their sons, elder brothers should teach their younger brothers, and husbands and wives should encourage one another. In their relations with others, people were to be prompt, polite, thoughtful, refined, and considerate. This meant that citizens should always use courteous language and proper greetings. It also meant that there should be no pushing and shoving in crowds, and no spitting, sneezing, or urinating in public. People were urged not to dance, have extravagant wedding and funeral ceremonies, swear, smoke tobacco or opium, gamble, or visit prostitutes.[57]

All these stipulations reflected Chiang Kai-shek's belief that strict adherence to specific standards of external behavior would help change the "inner man" (and woman)—that outward beauty was "a manifestation and forerunner of inward and spiritual beauty."[58] He also believed that adherence to the values of *li*, *yi*, *lian*, and *chi* would lead away from rudeness and vulgarity and toward the general "aestheticization" (*yishu hua*) of Chinese life.[59] If all the Chinese people could be transformed in this way, the nation would be reinvigorated. But the process had to begin incrementally. According to the official priorities of the New Life Movement, social transformation had to proceed from the self to others and from official personnel to the masses. It also had to begin with simple matters and then extend to more complicated ones.[60] Chiang provided one concrete example of a small, but apparently essential, first step: "If we are to have a new life that accords with ... [Confucian values]," he once asserted, "then we must start by not spitting heedlessly."[61]

Of course the transformation of Chinese society in the 1930s involved more than pithy moral admonitions and an effort to enforce what may be called barracks discipline. Chiang Kai-shek himself understood this and repeatedly tried to link the New Life Movement not only with Sun Yat-sen's modern-spirited "Three Principles of the People," but also with traditional mechanisms of local control—notably the *baojia* mutual security system.[62] Chiang also steadfastly urged the creation of a modern Chinese economy based on the idea of "productivization (*shengchan hua*)," that is, an approach to development emphasizing industry and thrift, the utilization of "national goods," and hard work in the service of one's country.[63]

Chiang placed particular stress on patriotism as a means of achieving his political, social, and economic ends.[64] In fact, a number of the ninety-six rules for proper eating, dressing, domestic life, and public conduct in the *Handbook of the New Life Movement* involved explicitly nationalistic strictures: consume Chinese foodstuffs; use Chinese fabrics, respect the Chinese flag, and so forth.[65] Yet despite his nationalistic emphasis, Chiang did not hesitate to celebrate foreign culture in an effort to bring the Chinese people to an appreciation of the need for modernization. "All the activities of foreigners," he proclaimed, "including the way they dress, eat, live, and act are in complete accord with the demands of modern citizenship."[66] Chiang even invoked Christian values (and relied upon the assistance of the Christian Church) in promoting his eclectic moral vision.[67]

The New Life Movement began with a huge fanfare on February 19, 1934, in Jiangxi province, home of the Chinese Communist "bandits." Soon, however, it became a countrywide campaign. On March 11, the Guomindang sponsored a mass meeting, involving some 100,000 people and 142 organizations, in order to inaugurate the movement on "a national scale." July 1, 1934, witnessed the establishment at Nanjing of a national headquarters for the New Life Movement's "Promotional Association," with Chiang Kai-shek as Chairman. Under its guidance, by the end of 1935, propaganda teams in a total of nineteen provinces and five municipalities had given lectures and talks to the public, drumming up support for the Movement at all levels of Chinese society. Young people received particular attention from Guomindang officials, since the prevailing view was that most urban youths had "nothing in mind but pleasure," and that the children in the countryside suffered from ignorance and "demoralization."[68]

According to Chiang's official instructions, the public had to be taught through personal example, lectures, illustrations, written materials, dramatic performances, and movies.[69] In urban centers, local promotional associations arranged for large-scale "citizen's meetings" to show public enthusiasm for the campaign, and "cleanliness campaigns" of various sorts were organized. Many cities, for example, sponsored a "cleanliness week," complete with "health parades."[70] Lantern processions, banners, and slogans written on "gorgeous dragons" and "glittering pagodas" marked many such public displays.[71] Booklets appeared as part of a general series under titles such as *The New Life Movement and National Revival* (*Xin Shenghuo yundong yu minzu fuxing*), *The New Life Movement and Ritual and Music* (*Xin Shenghuo yundong yu liyue*), and *The New Life Movement and Women's Liberation* (*Xin Shenghuo yu funu jiefang*). The movement had its own official red, white, blue, and yellow "shield," full of elaborate color symbolism, and a song with lyrics that linked the "four guiding principles" to the four basic conditions of life. The song emphasized the movement's major themes: orderliness, cleanliness, simplicity, and frugality.[72]

Promotional associations took the lead in forming other groups, including inspection teams, to "study the unsuitable customs and habits of the people," and to encourage adherence to the official ninety-six rules of the New Life Movement. Among the places targeted were hotels, restaurants, tea houses, gambling establishments, opium dens, bookstores, shops, marketplaces, resorts, bathing houses, and bus and railway stations.[73] In the Spring of 1934, a thousand residents of Nanchang were selected as "honorary detectives" to help inspect restaurants, bathrooms, and private homes.[74] The results of these inspections were then put on a public display-board as a means of apportioning praise and blame—a technique recalling the "exposition pavilions" of late-imperial times. Sometimes arrests were made of "anti-social characters," including those who "spit recklessly on sidewalks or in railway stations."

In many areas, small groups gratuitously gave "polite advice" to pedestrians on how they could improve their behavior and appearance. Larger-scale organizations, such as the "New Life Earnest Action Corps," recruited members who pledged in "public compacts" (*gongyue*) to abide by (and to encourage others to conform to) the major stipulations of the Movement in matters such as cleanliness, deportment, courtesy, and thrift.[75] The New Life authorities also organized a number of provincially sponsored women's associations, in an effort to "arouse the interest of China's womenfolk to the vital importance of the movement." Young women students "formed among themselves small groups to persuade female members of the public to abide by the [provincial government's conservative dress] regulations." Hsiao-pei Yen has insightfully discussed the contradictions that attended the New Life Movement's effort to "nationalize" the female body by establishing the "archetype of the healthy and modern Chinese woman."[76]

In cities, towns, and villages simple textbooks, "suitable to the requirements of the common people," expounded on the themes of loyalty to the country, "citizenship, cleanliness, national products, courtesy, [and] the evils of Communism." They also addressed the problems of banditry, opium addiction, and footbinding. Special efforts to educate China's peasants included programs in literacy, elementary hygiene, agricultural improvement, and ideological training. Sun Yat-sen Schools for the Masses, the explicit counterparts of the CCP's so-called Lenin Elementary Schools, were established to eradicate "ignorance and evil," and to teach the people to "regard [the Communist] bandits as shameful." Separate schools connected with the *baojia* system and controlled by local elites had a similar purpose.[77] Overall, the Nationalist authorities felt that "responsibility for pushing the [New Life] movement forward ... [should fall] squarely on the shoulders of the [urban] intelligentsia," but they were at pains to emphasize that even these "so-called enlightened people" in China's cities and towns were in need of moral refinement—perhaps, some argued, "more urgently than the farmers need it to guide them in their appearance and behavior."[78]

Chiang Kai-shek naturally urged his officials to serve as upright and self-sacrificing role models for the New Life Movement. He also proposed himself as the standard of moral achievement and exemplary behavior, claiming rather immodestly that if people wanted to bring credit to the Chinese nation, they must "learn from me."[79] On several occasions, Chiang advanced the view that if only more people were able to follow his example of cleanliness, orderliness, and ambition, there would be a great many human beings capable of "saving the people and reviving the country."[80] Chiang managed to maintain his reputation for personal rectitude and self-discipline during the 1930s, but most of the GMD officials placed in charge of the New Life Movement failed to provide inspiration for the Chinese masses. Some left a decidedly negative impression, and the vast majority seem to have made no difference at all. Located almost exclusively in urban areas, these functionaries had little, if any, contact with the 80 percent of the Chinese people who lived and worked in the countryside.[81]

For the most part, China's rural masses remained isolated, impoverished, and unresponsive to moral exhortations. The regime's so-called "new villages" were new in name only. Even in Jiangxi province, home of the New Life Movement, most government-sponsored educational programs turned out to be "more of an imposition and inconvenience than an aid to restoring peace and harmony."[82] Moreover, as critics of the movement pointed out repeatedly, a campaign that emphasized spiritual conditions without adequate attention to material conditions was bound to fail.[83] Hu Shi, for one, reminded the promoters of the New Life Movement that the first duty of the government was to enable the people to earn a decent livelihood. Many so-called "bad practices," he observed, were simply "the products of poverty-stricken districts." Only after citizens had gained the ability to maintain themselves financially could they be taught "to live a new life."[84] Still, the Nationalists held firmly to the view that "Spiritual values transcend mere material riches," and therefore, "if everyone from the highest official to the lowest wheelbarrow-man would conscientiously practice ... [the four guiding principles] in everyday life, there would be food for all."[85]

Although the New Life Movement continued to receive official GMD support into the 1940s, it had already lost virtually all of its momentum by the late 1930s—particularly after the devastating Japanese invasion of 1937. Graham Peck offered the following evaluation of the Movement in *Two Kinds of Time*:

> When I travelled in China before the war, I had seen that the movement was then cleaning up the surface of life in the big cities ... In the countryside and the smaller provincial towns, however, its work was irrelevant to the point of comedy and tragedy. Uplifting wall slogans were posted in villages where nobody could read, and peasants who could not afford soap were lectured on the duty to wash.[86]

The New Life Movement has often been criticized for its over-attention to minute rules of decorum and for its philosophical superficiality. Most intellectuals in Nationalist China were openly scornful of Chiang's "Return to Confucius" movement, and even young people took little heed of it.[87] The authorities did, however, achieve some limited success in the realm of public health in conjunction with other campaigns. Furthermore, despite its manifest weaknesses, the Movement helped lay the foundations for a government-sponsored approach to the revival of traditional Chinese culture that continues, more or less, to this day on Taiwan. This approach considers fundamental Confucian values to be fully compatible with science and democracy, and conducive to economic growth as well.[88] At present, in the Republic of China (ROC), traditional rituals and religious practices continue to flourish alongside more "modern" aspects of material culture borrowed from Japan and the West, and particularistic concepts such as *renqing* (lit., "human feelings") and *guanxi* ("connections") remain a significant feature of the social landscape on Taiwan.[89]

Communist revolution: new wine in old bottles or just more old wine?

Like Chiang Kai-shek in the 1930s, Mao Zedong aimed at creating "new citizens" in China: individuals who were "physically rigorous, brave, progressive and creative, independent-minded yet disciplined—all out of social devotion to society." For Mao, as for Chiang, soldiers "totally dedicated to voluntaristic action in the service of organizational goals ... provided the paradigm of the good citizen."[90] But unlike Chiang, Mao steadfastly assailed Confucian beliefs and practices from the very onset of his revolutionary career in the 1920s. In Mao's view, Confucius lacked the proper social vision, held incorrect views regarding physical labor, and mistakenly esteemed "old rituals."[91]

Yet for all his revolutionary iconoclasm, Mao was in a certain sense a captive of the very tradition he sought to destroy.[92] Although he never abandoned his theoretical commitment to the utopian goal of a classless and stateless society, in practice, Mao's China, like that of his imperial and Nationalist predecessors, displayed the familiar characteristics of authoritarian, bureaucratic rule: state supervision of political, social, and economic life; an emphasis on political morality over law; a preoccupation with ideological, artistic, and literary orthodoxy; a formalism in both political discourse and personal relations; and a clear stress on collective responsibility, mutual surveillance, and the subordination of the individual to the group. Many of these tendencies were, of course, encouraged and reinforced by Marxist-Leninist ideology and practice.[93]

One important and enduring feature of social life in the People's Republic of China has been an emphasis on "small group ritual"—the direct legacy of CCP political indoctrination techniques developed in the 1930s and 1940s. Throughout Chinese society after 1949, small groups (*xiaozu*) of eight to fifteen members have been organized in various larger units of affiliation (*danwei*), which exist in economic enterprises, schools, neighborhoods, the military and even forced labor camps. Each group has a head, and each must convene on a regular basis to study and discuss political documents, as well as to engage in criticism and self-criticism.[94] The *xiaozu* system represents a refinement of the longstanding Chinese tradition of organizing units of collective responsibility for the purpose of preserving social order through both ritualized moral instruction and mutual surveillance.[95]

Another striking characteristic of social and political life in the PRC, also a product of Chinese tradition, is a sustained emphasis on ethics, politeness, and "spiritual civilization (*jingshen wenming*)."[96] Although the thrust of virtually all major social and political campaigns in the PRC up to 1980 was more on morality than on courtesy, and on "socialist morality" in particular (as opposed to "feudal" or "bourgeois" morality), the profound "crisis of faith" that followed in the wake of the Cultural Revolution caused the CCP authorities to reconsider the role of ritual and politeness as a means of restoring social order, redirecting cynical and alienated young people, improving the general "social atmosphere," and recapturing China's "spiritual civilization."[97] The result was a nationwide campaign, first proposed in February of 1981, known as the "Civilization and Propriety Educational Movement" (*Wenming limao jiaoyu yundong*)—or, more succinctly, the "Five Emphases and Four Beautifications (*Wujiang simei*)."

The Five Emphases referred to a stress on civilization (*wenming*), courtesy (*limao*), hygiene (*weisheng*), orderliness (*zhixu*), and morality (*daode*); the Four Beautifications denoted the beautification of the mind (*xinling*), language (*yuyan*), behavior (*xingwei*), and the environment (*huanjing*). From the standpoint of CCP officials, the first five were the means by which to achieve the last four.[98]

Beautification of the mind aimed at the cultivation of correct thought and moral integrity; support of the Chinese Communist Party and the socialist system; patriotism, uprightness, and honesty; the preservation of personal and national dignity; and the avoidance of either fraud or personal profit at the expense of others. Beautification of language meant courtesy in speech, with kind, gentle, and modest words rather than rude ones, and no forced arguments or slander. Beautification of behavior entailed beneficial service to the people and society; diligence, friendliness, and self-discipline; the avoidance of damage to either collective interests or public property; and an effort not to disturb the social order. Beautification of the environment required personal and public hygiene (with specific admonitions not to spit or litter); cleanliness and orderliness; and an attempt to plant trees and other greenery everywhere.[99]

According to an official circular issued on February 28, 1981, by five key ministries of the State Council (Propaganda, Education, Culture, Public Health, and Public Security), the *Wujiang simei* campaign was designed to achieve three primary goals: (1) to "improve relationships among the Chinese people;" (2) to "uphold stability and unity;" and (3) to develop a positive "social atmosphere" nationwide.[100] In part, the campaign sought to rectify problems directly attributable to the chaos of the Cultural Revolution. But of growing concern to the authorities was the fact that China's newly declared "Open Policy" (*Duiwai kaifang zhengce*), inaugurated by Deng Xiaoping in 1978 to facilitate the so-called Four Modernizations (Agriculture, Industry, Science and Technology, and National Defense), together with Deng's recent economic reforms, had invited "decadent bourgeois ideology" with all of its "bad influences."[101] Even the countryside had become seriously threatened by "the destruction of community implicit in capitalist development."[102]

Exactly one year after the initial February 28, 1981, circular had been issued, Premier Zhao Ziyang declared on national television that the month of March would be designated "Civilization and Propriety Month." His speech, like the official circular a year before, made specific mention of China's ancient and "celebrated" status as "the land of ritual and righteousness." It also indicated explicitly that "spiritual civilization" was important in the development of "material civilization," and thus essential to the achievement of the "Four Modernizations."[103] As Zhao and other prominent Chinese leaders saw it, the aim of this Civilization and Propriety Movement was to combine China's modern "revolutionary tradition" with the best elements of the nation's "fine [old cultural] tradition" in order to create a dynamic new synthesis—one that would add luster to China's national spirit.[104] As the Deputy Director of Shanghai's Municipal Education Bureau put the matter hopefully in an article on moral education in Chinese schools: "China's 5,000-year-old civilization has an international reputation for politeness and propriety."[105]

On the same day as Premier Zhao's speech, some two million Chinese took to the streets in a display of public support, accompanied by more than two-hundred high-level cadres—among them, the CCP General Secretary, Hu Yaobang. Together they swept the streets, took away trash, and participated in other high profile activities. The assumption was that since the living environment reflected the general level of a society's cultural development, it served also as an index of "spiritual civilization."[106] The *People's Daily* (*Renmin ribao*), which soon began publishing a special column titled "Let Everyone be Concerned with Cleanliness, Orderliness, and Courteous Behavior," reported the dramatic event in detail, noting that red flags, colorful banners, and gigantic slogans were noticeable "everywhere in the streets of Beijing." The organizational scale and number of participants struck the *People's Daily* reporter as "very rarely seen."[107]

From this point onward, the Five Emphases and Four Beautifications campaign gathered momentum, although a national committee for the movement was not officially established until March of 1983. By this time, the authorities had pronounced "Civilization and Propriety Month" a resounding success. Long articles in the *People's Daily* and other publications noted that it had achieved a positive effect on Chinese society and had "inspired the party's revolutionary spirit." Although the movement had "not [yet] reached the grassroots level in some places," it had begun to improve "the mental outlook of the Chinese people and social trends in general." In the realms of sanitation, public order, and public service, great gains were "already apparent." In early 1983, Premier Zhao reported that during the past year, "an emphasis on courtesy, hygiene, discipline, and morality" had become a part of the people's "conscious behavior."[108]

Despite paying lip service to China's ancient cultural tradition, the authorities placed heavy emphasis on the CCP's revolutionary tradition—three of its "spiritual" aspects in particular. One was the pre-1949 "Yan'an spirit." Another was the "pioneering spirit" of the early years after 1949—presumably the period up to the Great Leap Forward (1958–1960). And the third was the "spirit of Lei Feng," a deceased young soldier who had been a model of revolutionary self-sacrifice for the Party as early as 1963.[109]

The initial premise of the movement was that China's urban centers would take the lead in promoting cultural change.[110] But the Party recognized that an important goal must be to transform the "old customs and bad practices" of the countryside.[111] Notwithstanding the rural origins of Mao's revolution, prejudices against the agrarian sector in China ran deep. In the fall of 1983, therefore, the Central Committee of the Chinese Communist Party held a widely publicized "Forum on the Building of Civilized Townships All Over the Country," in an effort to bring the Five Emphases movement more fully to the rural landscape.

A major feature of this effort was the promotion of "village compacts (*xianggui minyue*)," locally drafted regulations designed to restore discipline and a sense of "ethical community" in areas threatened by "the divisive forces unleashed by the economic reforms [after 1978]." Within individual villages and townships, small groups assumed responsibility for enforcing these compacts, which dealt with a wide range of social problems, from crime to dispute resolution. The names of these groups are revealing: "Civilization Committees," "Social Ethics Appraisal Committees," "Councils on [Limiting the Cost of] Weddings and Funerals," and so forth. Although initiated as ideological instruments by the CCP, village compacts were intended as organs of local self-government, reminiscent of *xiangyue* and other such rural institutions in both imperial and Guomindang China.[112]

In early 1983, the Party launched a vigorous campaign in support of "socialist spiritual civilization" and against various forms of "spiritual pollution" such as pornography, "money worship," and decadent fashions—the unfortunate byproducts, in the eyes of the CCP, of the Open Policy and new economic developments.[113] By this time, the central government authorities had already added the "Three Ardent Loves (*san reai*)"—that is, love of the Motherland, Socialism, and the Chinese Communist Party—to the Five Emphases and Four Beautifications—presumably in order to tap the wellsprings of Chinese nationalism and to bolster a faltering faith in the Party.[114] Signs and posters appeared throughout the country, proclaiming: "Without the Communist Party, there would be no New China."

The campaign against spiritual pollution lasted only about a year, but it had important political implications. Like Chiang Kai-shek's effort to enhance Guomindang control in the cities and to attack his enemies in the midst of the New Life Movement, Deng Xiaoping used the pretext of "cultural pollution" to curb urban disorder and to undermine a number of his opponents during the Civilization and Propriety Movement. In less than twelve months, Deng managed to crack down on rising crime with a draconian sweep (there were an estimated 6,000 executions in the second half of 1983 alone) and to put in place, at least for the next several years, "trustworthy, pragmatic successors."[115] In the process, the CCP launched a systematic attack on any writings or speeches that ran counter to the country's social system, on "decadent bourgeois ideology" (especially in the form of liberalism or individualism), on misguided popular beliefs, and on all artifacts and fashions that were deemed vulgar, obscene, barbarous or reactionary.[116]

The early 1980s witnessed a spate of books and pamphlets, most designed for popular consumption, with titles such as *Talks on Communist Morality* (*Tantan gongchan zhuyi daode*), *Morality and Moral Education* (*Daode yu daode jiaoyu*), *Morality and Social Life* (*Daode yu shehui shenghuo*), and *The Moral Excellence of Youth—Civilization and Propriety* (*Qingnian di meide—wenming limao*). But it was not until 1983 that the Propaganda Department of the Communist Youth League produced an official handbook for the Five Emphases and Four Beautifications campaign.

This 336-page volume, known as the *Wujiang simei shouce*, laid out in exquisite detail the kinds of behavior the authorities deemed appropriate to all levels of Chinese society. Divided into ten sections, and distributed without cost to all work units, the *Handbook* delineated the general principles of the movement, its "form and content," its ideological underpinnings, and, of course, its prescriptions for proper behavior. These fell under the general categories of "socialist public morality," "occupational morality and standards of behavior," "the morality of marriage and family life," "paying attention to civilized speech," and "courteous treatment of others." The *Handbook* also included large sections devoted to "individual and environmental hygiene" and "greening the motherland and beautifying the environment." Comparatively little was said about the battle against "spiritual pollution."[117]

In several respects, the general admonitions in the *Handbook* differed little from those of previous PRC campaigns. Citizens were encouraged to work hard, obey the law, and protect public property, as well as to oppose gambling, "feudal superstition," and other harmful ideas and activities. The *Handbook* also gave considerable attention to ideological indoctrination through "small group" rituals, stressing not only Communist theory (and practice), but also "the spirit of patriotism" and especially "the spirit of Lei Feng." Citizens were emphatically advised to observe proper hygiene and to present a positive public image. They were to get haircuts and bathe often, and were told not to pick their teeth, sneeze, or dig into their nose with their finger—at least not in front of other people. Naturally, they were not to spit, throw trash, or relieve themselves anywhere they wished. They were also advised against the dangers of smoking and drinking, warned about poisons and disease-bearing animals and insects, and even advised on how to recognize spoiled food and other hazardous products.[118]

Like Chinese ritual manuals for centuries, but with a far wider scope, the *Handbook* provided detailed information on proper usage for all kinds of social situations, from everyday encounters in streets, playgrounds, parks, and schools, to behavior in hospitals, opera houses, stores, restaurants, and hostels. It also emphasized how various people should be treated, from parents, teachers, and political leaders to comrades, women, children, ethnic minorities, and subordinates. It indicated various terms of address and a wide range of polite phrases and gestures to use (as well as a large number of impolite terms to avoid). Finally, it offered a great deal of concrete advice on matters such as marriage and funeral ceremonies, visiting friends, entertaining guests, and so forth. Significantly, the *Handbook* included a few special sections designed to introduce Chinese readers to "overseas courtesies (*guowai limao*)."[119]

To assist in disseminating these and other ideas, the Civilization and Propriety Movement drew upon a vast preexisting propaganda apparatus that included the National Federation of Trade Unions, the Central Committee of the Communist Youth League, the All-China Women's Federation, the Chinese Literary Alliance, the Central Committee of the Patriotic Hygiene Movement, the All-China Student Association, the National Society of Ethics, the National Society for Languages, and the Chinese Society for Aesthetics. In addition, the Party attempted to use a wide variety of popular media—including not only television, radio, and various kinds of written literature, but also drama, music, film and the fine arts—to popularize the movement.[120] Throughout the People's Republic, red banners, billboards, and signs proclaimed the new cultural emphasis.

Meanwhile, the term "civilized" came to be applied (quite literally, by means of prominent placards) to all sorts of organizations—from villages and work units to small shops and school classes.[121] As I recall from several visits to China in the early 1980s, a number of urban streets also bore this designation. Inspections focused on cleanliness and hygiene took place regularly in collective organizations, and failure to be designated *wenming* brought criticism and embarrassment to the group leader—at least in the early stages of the campaign. By 1984, however, the conferring of the title *wenming* had become highly formalized, prompting a number of criticisms from officials that appeared in the Chinese press.[122] Corruption began to invade the system, as large banquets were thrown for inspection teams to buy favor. When new regulations stipulated that meals for guests on inspection days could not exceed "four dishes and one soup," the servings simply became larger.[123]

In early 1985, the *People's Daily* announced that the idea of a "Civilization and Propriety Month" should give way to a year-round awareness of the need to promote civic duty, personal and public hygiene, courtesy, and moral cultivation.[124] This pathetic announcement signalled that the end of the movement was close at hand. Nonetheless, the state still had formidable organizational powers at its command. For instance, a last-ditch campaign against spitting in Beijing during the spring of 1985 enlisted 40,000 people from various local work units to act as full-time superintendents at nearly every street corner in the city. Charged with reminding citizens not to spit, they also had the power to impose fines on those who failed to heed the reminder. Although the fine was not large (50 *fen*, about 20 cents U.S. at the time), the embarrassment and loss of face were sometimes quite considerable.[125]

In its focus on ethics, its emphasis on external behavior as both a measure and a means of self-cultivation, and its assumption that "spiritual civilization" was at least a concomitant of "material civilization," the Civilization and Propriety Movement naturally invited comparisons with the New Life Movement. Not surprisingly, the CCP has denied any such affinities. Indeed, on March 11, 1983, the *Renmin ribao* published an article on the history of the New Life Movement intended precisely to distinguish the ongoing *wujiang simei* campaign from its Nationalist counterpart in terms of inspiration, class character, and ultimate aims.[126] But many Chinese commentators outside the Mainland noted that if only a few minor editorial changes were made—such as replacing the name of Chiang Kai-shek with that of Deng Xiaoping, and substituting the phrase "the privileged class of Chinese Communist bureaucrats" for "landlords and the bourgeoisie"—the *People's Daily* article on the New Life Movement described the *wujiang simei* movement almost perfectly.

Both movements were orchestrated by the government, urban-based, elite-dominated, highly ideological, and linked to specific, nationalistically inspired, modernizing goals. High-minded and idealistic in rhetoric and avowed social purposes, they nonetheless tended in practice to focus on limited, seemingly trivial patterns of behavior, such as spitting and swearing. Although both the CCP and the GMD targeted youths for special attention in their respective campaigns, neither had any real success in attracting their support. Both movements placed a major emphasis on cleanliness and hygiene, and made some strides in urban areas; but on the whole, they achieved little else, and in any case lacked popular appeal, especially in the countryside. Even in the cities, during the heyday of the Five Emphases movement in 1983, I recall that among the many Chinese I talked to, few could explain its specific features.[127] And although Chinese peasants in coastal areas were certainly far better off in the early 1980s than they were in the 1930s, the CCP seems to have found it no easier than the GMD to mold their behavior.

This is not to say that the two campaigns were identical. The specific political and social values of the Chinese Communists in the 1980s differed substantially from those of the Guomindang in the 1930s—as did their general domestic and international situation. After all, the CCP did not have to contend with foreign aggression during the Five Emphases Movement, nor did it face the armies of a formidable political foe from within. Understandably, therefore, militaristic themes did not loom large in its propaganda. Furthermore, the Chinese Communists had a far more effective system of communications in the 1980s than the Nationalists had in the 1930s. From the standpoint of values, CCP officials saw no need to champion "feudalistic" Confucian ethics in the fashion of the GMD, much less Christian ones. In fact, the Three Bonds and Five Constant Virtues came under explicit attack during the Civilization and Propriety Movement.[128] On the other hand, the Chinese Communist authorities felt no compunction when it came to either selectively celebrating the modernizing accomplishments of both Japan and the West (as opposed to their decadent features, of course), or indicating certain affinities between "Communist morality" and traditional (Confucian) virtues.[129]

And occasionally, even explicitly Confucian concepts such as filial piety received favorable notice in the official rhetoric of the *wujiang simei* movement.[130]

Conclusion

In her pioneering and provocative study, *The Last Stand of Chinese Conservatism*, Mary Wright compared the Confucianism of Chiang Kai-shek's New Life Movement with what she believed to be a more genuine and effective cultural-administrative revival during the Tongzhi period (1862–1874) of the Qing dynasty. Although some scholars have taken her to task for idealizing the Tongzhi "Restoration," and others have criticized her for being overly unkind to the New Life Movement, there can be no doubt that a critical element in both reform efforts was an attempt to promote the rectification of customs through what Wright terms proper "social usage" (*li*). What I have tried to do in this essay is to extend Wright's line of analysis to include the Five Emphases and Four Beautifications Movement of the 1980s, in order to highlight certain significant continuities, as well as a number of important changes, in the theory and the practice of state-sponsored ritual in twentieth-century China.

Clearly, the New Life Movement and the Five Emphases campaign were deeply influenced by a long-standing Chinese preoccupation with ethics, spirituality, and the transformative power of ritual. In both cases, China's leaders considered self-cultivation to be the first step in the moral transformation of society, in keeping with the basic injunction of the *Great Learning*.[131] At the same time, they accepted the time-honored premise that the rectification of "popular customs" was the key to achieving the shared goals of peace, order, and unity in Chinese society at large. Both movements aimed at improving the "spiritual appearance" (*jingshen mianmao*) of the Chinese people. The idealistic assumption remained in each that outer appearance and behavior at once reflected and inculcated inner morality.

In a highly paternalistic, authoritarian, and openly condescending way, the Chinese state initiated and directed these "morality movements," relying on elites to spread and reinforce its official orthodoxy (and orthopraxy), and to combat "heterodox" ideas, including various forms of "superstition." Although both campaigns enjoyed an initial measure of broad-based support and a certain level of genuine enthusiasm, they eventually degenerated into a kind of empty formalism. The main difficulty was a lack of fit between the high-minded pronouncements of the state and the grim realities of peasant life in the Chinese countryside. A huge physical and cultural gap separated urban-based bureaucrats from China's rural masses. Moreover, most of the officials involved in the two campaigns failed to provide the kind of leadership that Chinese moral idealism required; many proved to be corrupt. Although they talked a great deal about ethics, propriety, social harmony, and the power of positive example, they were concerned primarily with maintaining order through rigid discipline.

Another reason for the empty formalism of both the New Life Movement and the Five Emphases campaign was the fact that in each case the rituals promoted by the government lacked any sort of cosmological foundation or sanction. Immediately after the establishment of the Republic of China in 1912, the Chinese state condemned the inherited cosmology as a "superstitious" remnant of a bygone era.[132] With this irrevocable break in the official link between ritual, morality and cosmology, the Chinese government found it increasingly difficult to sustain its universalistic claim to be the protector of public morals and propriety—especially in the face of widespread and persistent corruption. In both the New Life Movement and the Five Emphases campaign, references to the "fine tradition of the Chinese people" rang hollow.

The often-employed term "spirit" (*jingshen*) still implied a distinctive Chinese-style morality, but the cosmological ground had become unstable. This was particularly true to the extent that the prospect of social and political transformation through modern science and technology encouraged twentieth-century Chinese intellectuals to relax their time-honored search for transformative spiritual power within the mind.[133] The proper moral spirit had been difficult enough to achieve in China under the best of circumstances, but it was virtually impossible to implement without metaphysical support.

In the absence of any sort of cosmically derived transformative power, Chinese intellectuals became mere mortals. At the same time, however, the deeply rooted assumptions of China's "epistemological optimism" continued to impose a heavy burden upon them. This burden could not easily be relieved by recourse to the alien assumption of "rule by law." Good administration remained more a matter of individual responsibility and "human talent" (*rencai*) than of either legal statutes or institutional structures. This attitude can still be seen in China today, despite the fact that "law and the legal system [now] ... form part of a growth industry."[134]

In short, traditional techniques for "rectifying customs" have had obvious limits in twentieth-century China. Yet the Chinese authorities on the Mainland continue to evince a deeply rooted concern with issues of "civilization" and "propriety," as well as a more modern, Meiji-style sensitivity to the perceptions of outsiders. For example, during China's abortive bid in 1993 to host the Olympics in the year 2000, at least one large sign in Beijing, focusing at least nominally on the Seventh National Games, admonished local residents to "establish a civilized city and behave like civilized people."[135] At about the same time, the Chinese government issued a number of tips for PRC travelers abroad which recalled the admonitions of both the New Life Movement and the Five Emphases campaign.[136] Similar signs and state-sponsored admonitions can still be found in China today (2011).

Meanwhile, at a more mundane level, and beyond the state's control, private Chinese citizens continue to grapple with issues of proper social usage. Among the many works currently available for guidance is a huge compendium on "taboos" (*jinji*) that apply to every conceivable realm of life. Significantly, this book, which deserves further study in its own right, provides advice based not only on Chinese standards of propriety (both "traditional" and "modern") but also on ritual practice abroad. Nonetheless, although it acknowledges the inescapable impingement of foreign values on the People's Republic, it persists in identifying China alone as *liyi zhi pang*.[137]

5

DIVINATION IN THE QING[1]

In a widely read essay titled "An Inquiry into Fate" (*Yuanming*), the Qing scholar Quan Zuwang (1705–1755) expressed the ambivalence of both his time and his social class toward the problem of destiny and divination. On the one hand, he had to acknowledge that "the ancients spoke a great deal about fate." On the other, he was quick to point out that talk of good and bad fortune by astrologers and other soothsayers was "unworthy of belief." Citing Huang Zongxi (1610–1695) and other authorities, he went on to say that although divination systems based on simple calculations of time of birth could not possibly yield a satisfactory discussion of destiny, there were simply too many examples of wise men who believed in fortune-telling techniques to dismiss divination out of hand.[2]

Quan's fundamental position, derived from orthodox neo-Confucianism, was that although Heaven endowed man with his nature, it had no actual control over his actions. Surely, he argued, the will of Heaven was that every man be a sage and that virtuous government by sage-rulers would bring peace to the world. Heaven also undoubtedly wanted all people on earth to be healthy, strong, and properly provided for. Yet not everyone was a sage; villains often ruled despotically. The virtuous might well suffer from cold, hunger, and premature death; villains might live long and comfortable lives. "These things," he claimed, "Heaven can do nothing about."

Quan's purpose was, of course, to show that what appear to be blessings (*fu*) and calamities (*huo*) are nothing more than ephemeral phenomena, "like morning dew and melting icebergs." History ultimately reveals good and evil, and humans are therefore bound to strive relentlessly for sagehood. Superior men, he asserted, refer to fate only in certain unavoidable situations—such as Confucius did in the face of Boniu's fatal illness (*Analects* 6:8) or in the case of Gongbo Liao's slander of Zilu (*Analects* 14:38). Such remarks were not at all, he emphasized, like the predictions of fortune-tellers.

Of course Quan knew perfectly well that Confucius said much more about fate, as did Mencius and a great number of other prominent thinkers.[3] But to include these references, or even to allude to the range of opinion that they represented, would have made his essay too intellectually untidy; for in truth, Confucian scholars, both during the Qing period and at earlier times, were profoundly uncertain about exactly how to know fate (*zhiming*), as Confucius advised (see *Lunyu* 20:3)—not to mention what to do with the knowledge acquired.[4]

This much was indisputable: techniques of divination had abundant sanction in the classical literature. Indeed, all canonical works in the Confucian tradition mention the phenomenon. The hallowed *Yijing* (Classic of Changes) actually began as a fortune-telling text, and it continued to represent the orthodox tradition of divination throughout the imperial era (see below). Works such as the *Zuozhuan* (Commentary of Zuo) contain numerous references to the use of the turtle shell and milfoil stalk, as well as other mantic techniques, including dream interpretation, to determine good and bad fortune (*jixiong*); and the *Liji* (Record of ritual) tells us simply: "By means of divination the former sage-kings caused the people to have faith in the seasons and the times, to respect spiritual beings, to fear laws and orders, and to elucidate and settle doubts and suspicions."[5]

To be sure, the rational tradition of neo-Confucianism downplayed the supernatural world of ghosts and spirits (*guishen*)—despite their prominent appearance in the classical literature, and at least a few Qing scholars, such as Wang Fuzhi (1619–1692), took pains to point out that prevailing cosmologies derived from the Classic of History (*Shujing*) were in fact distortions concocted by later generations of fortune-tellers and numerologists. Moreover, we know that the editors of the massive Qing literary project known as the *Complete Collection of the Four Treasuries* (*Siku quanshu*) made a special effort to disassociate mantic works based on the *Yijing* from the classic itself, in order to distinguish the orthodox Confucian tradition from the Daoist tradition of magical practices.[6]

In everyday life, however, the distinction between the two often became hopelessly blurred, if not obliterated. This was particularly true for those at the end of the Chinese social spectrum whose interests and outlook were seldom represented in the *Siku quanshu*—commoners, especially peasants. For these unnamed actors in China's ongoing drama, the line between Confucian orthodoxy and Daoist magic was not worth drawing—at least not in most day-to-day affairs. As a result, in the Qing period—and indeed throughout China's imperial history—peasants and other commoners persistently sought to know as much about the future as possible, and to do as much as they could about it, by whatever means. For the traditionally disadvantaged throughout the realm, including large numbers of women, any advantage worth possessing was worth investigating.[7]

This essay examines the phenomenon of divination in the Qing dynasty from a popular as well as an elite perspective, and seeks to identify a relationship between the two. It attempts to show not only the prevalence of divination in late imperial times, but also the variety of forms that it took and the social and political significance that it had. Much of the data for this study comes from Yuan Shushan's

(1881–1952) monumental *Zhongguo lidai buren zhuan* (Biographies of diviners in China by period), first published in 1948, which consists largely of excerpts from local Chinese gazetteers, official histories, and other such sources. Of the more than 3,000 biographical entries in Yuan's book, about a third are from the Qing period. Descended from a prominent family of doctor-diviners, and considered to be one of the "Three Great Masters of Fate Prediction" (*Mingxue san da mingjia*) in twentieth-century China, Yuan was a rare specimen: a Chinese scholar who gave serious attention to divination as a social phenomenon, and who chronicled out of his own personal interest both the techniques and the lives of its major exponents.[8]

Qing attitudes toward divination

Until the 1980s and 1990s, comparatively little modern scholarship was done on either the theory or the practice of Chinese divination in late imperial times, aside from the work of Yuan Shushan and a few others.[9] One reason is that most forms of fortune-telling were regarded by twentieth-century Chinese intellectuals as nothing more than popular superstition (*mixin*), unworthy of serious study. This attitude is reflected in the following quotation from a highly regarded anthology of translated Chinese writings produced in the mid-1960s:

> While educated Chinese have paid homage only to Heaven and their ancestors, and sometimes to Confucius, Buddha, Lao Tzu, and a few other historical personages, the common people have believed in the existence of thirty-three Buddhist Heavens, eighty-one Daoist Heavens, and eighteen Buddhist hells, and put faith in astrology, almanacs, dream interpretation, geomancy, witchcraft, phrenology, palmistry, the recalling of the soul, fortune telling in all forms, charms, magic, and many other varieties of superstition.[10]

In fact, however, divination was not simply a preoccupation of the common people. Virtually everyone in traditional China believed in divination. The problem was not whether to believe in it, but whom to believe. In the words of a popular proverb: "Do not say that King Wen's hexagrams are ineffective [*buling*]; fear only that the fortune-teller's reading is untrue [*buzhen*]."[11] It was specialization and professionalism that stigmatized divination, not the theory itself. From an elite standpoint, fortune-telling—like medicine, husbandry, gardening, and a number of other specialized, non-scholarly pursuits—was a minor employment (*xiaodao*),[12] to be learned by Confucian gentlemen for the benefit of their own families, perhaps, but not left to petty people (*xiaoren*), who were concerned only with profit.[13] This elitist view of divination was also reflected in a well-known adage in Qing China attributed to the Sung scholar Zhang Zai: "The *Yijing* is for consultation by the exemplary person [*junzi*], not the petty person."[14]

But the complexities of calculating fate (*suanming*, *shuming*, etc.) kept professional divination specialists in great demand—not only in religious temples and private homes but also in businesses, and even the offices of Qing officials. These complexities had to do with an enormous array of cosmic variables, including the phases of change associated with *yin* and *yang*, the sun and the moon, the four seasons, the five agents (*wuxing*) and their numerous correlations, including the five visible planets (*wuxing*; lit "five stars"), the six cardinal points (*liuhe*), the eight trigrams (*bagua*), the nine palaces (*jiugong*), the ten heavenly stems (*shigan*) and twelve earthly branches (*shierzhi*), the twenty-four solar periods (*ershisi qi*), the twenty-eight lunar lodges (*ershiba xiu*), the sixty-four hexagrams (*liushisi gua*), and over two hundred auspicious and inauspicious stars (*xing*).[15]

The way in which these and other cosmic forces interacted with one another varied according to different divination systems, and, not surprisingly, many of these systems overlapped or intersected.[16] But the most sophisticated exponents of fortune-telling in Qing times generally took all such variables into account. Complicating an already complex picture of the future were Buddhist concepts of karma, and sectarian Daoist ideas of merits and demerits (*gongguo*). These notions, although disparaged by many Qing intellectuals, were widely held and, predicated on the assumption that the past, present, and future, were linked in a morally grounded chain of causation. In a sense, they accorded with the traditional idea of fate following behavior (*suiming*)—one of several early Chinese concepts of fate in addition to regular fate (*zhengming*), also known as fate received (from Heaven; *shouming*). Meanwhile, the Maitreya Buddhist idea of a Heavenly destiny (*Tianyun*), like Shao Yong's notion of rhythms of fate (*yunhui*), implicated not only individuals but also the entire world in predictable patterns of cosmic change.[17]

Although considerable debate existed among both scholars and professional fortune-tellers over the question of exactly how to know fate, and over which cosmic variables were most important, there was widespread agreement in Qing society on most of the variables themselves, and a shared faith that fate could in fact be known. In a certain sense, Chinese divination may be compared to modern stock-market speculation based on commonly held economic theories or principles but reflecting different particular emphases. As with traditional Chinese doctors (or modern Western stockbrokers, for that matter), a number of competing mantic specialists might be consulted in any given situation.

Assuming that one's fate could be known, the question was then: What could be done about it? Did one simply accept destiny passively? Philosophical Daoism argued yes: "Rest with fate (*anming*)," argued Zhuangzi. But the answer for orthodox Confucians, as Quan Zuwang's essay suggests, was quite the opposite. In the Confucian tradition, it was not enough simply to know fate, or to rest satisfied with fate; one also had to establish fate (*liming*)—that is, devise a moral strategy for contending with predestined situations. In the words of the great Qing scholar Tang Jian (1778–1861): "He who knows fate will cultivate the Way [Dao]; he who [merely] relies on fate will do harm to the Way."[18]

It is true that both classical sources and a large number of popular proverbs from the Qing period suggest the inevitability of fate in realms such as birth and death, sickness and health, wealth and blessings, marriage and children, official rank and scholarly achievement.[19] But the evidence suggests that on the whole this type of fatalism did not cripple self-reliance. Individuals continually strived to achieve status and material benefits, as well as health and longevity, through both self-exertion and the use of religious agents, including Daoist priests, Buddhist monks, magicians, sorcerers, and shamans.[20]

Of course, as a matter of principle Chinese elites distrusted all such religious specialists and regularly denounced their "deceitful" and "self-interested" behavior. The same tended to be true of diviners. As if this were not enough of a problem for exponents of the mantic arts, the Qing period witnessed a number of particularly vigorous and powerful attacks by Chinese scholars on inherited cosmological assumptions. Some authorities have argued, for example, that "technical and empirical studies" (*kaozheng xue*) by savants of the seventeenth and eighteenth centuries in areas such as philology, astronomy, and geology had a profoundly subversive effect, contributing not only to a marked decline of the traditional cosmology but also to "more or less popular criticism" of at least a few of the more prevalent forms of Chinese divination.[21]

Although there can be no doubt that *kaozheng* scholarship had important implications in many realms of intellectual life, I have found no evidence to suggest a decline of Qing interest in fortune-telling compared to previous eras. On the contrary, it flourished throughout the period, and may even have experienced a surge of popularity in the nineteenth century parallel to the revival of Buddhism in the midst of precipitous dynastic decline.[22]

Qing interest in divination can be measured in a variety of ways. One is the attention devoted to the subject in the section on arts and occupations (*yishu dian*) in the massive, imperially commissioned encyclopedia titled *Gujin tushu jicheng* (Compendium of writings and illustrations, past and present), first published in 1725. This work contains several large subsections (*bu*)—over 280 chapters (*juan*) in all—devoted to various theories, techniques, and famous exponents of divination.[23] Other sources of the period, from local gazetteers and the *Da Qing huidian* (Collected statutes of the Great Qing dynasty) to legal cases and anecdotal accounts such as the *Qingbai leichao* (Classified Anecdotes from the Qing), also attest to a widespread interest in various fortune-telling techniques on the part of the Chinese at all levels of society.[24] Short stories, plays, and novels of Qing times also reflect the social importance of divining techniques—even if fortunetellers themselves are often made out to be the objects of elite derision.[25]

The observations of foreigners provide another index of the prevalence of divination in late imperial times. The remarks of S. W. Williams, a longtime resident of nineteenth-century China, are typical in tone and substance:

> No people are more enslaved by fear of the unknown than the Chinese, and none resort more frequently to sortilege to ascertain whether an enterprise will be successful or a proposed remedy avail to a cure. This desire actuates all classes, and thousands and myriads of persons take advantage of it to their own profit.

William Milne asserted in 1820 that "astrology, divination, geomancy and necromancy everywhere prevail [in China]." John Nevius, another experienced observer of China, stated in 1869 that a belief in fortune-telling was "of universal prevalence" among the Chinese. Similarly, A. P. Parker claimed in 1888 that fortune-telling in China was "universally believed in," and at the end of the century, Arthur Smith remarked that "the number of Chinese who make a living out of ... [divination] is past all estimation."[26]

To be sure, most Westerners in Qing China had a deep-seated hostility to Chinese mantic practices, which they viewed as an impediment to "progress" of one sort or another. How could they possibly evaluate Chinese divination objectively? Missionaries, as attuned to "superstition" as to sin, might well have exaggerated the situation in China to justify their own "civilizing" enterprise. Yet where evidence exists from the Chinese side, whether in the form of official documents, letters, diaries, anecdotes, proverbs, popular fiction, or scholarly indictments of fortune-telling, it invariably confirms the accuracy of Western accounts. Taken together, Chinese and Western sources indicate that divination was an extraordinarily significant social phenomenon in Qing times.

The most influential work on divination in China for well over two thousand years was, of course, the *Yijing*. Nearly everyone in premodern times had occasion to use the document in some form, at some time—from the emperor and his advisers at court, to the lowliest peasant consulting a street-stall diviner. A number of Qing intellectuals chose to emphasize the moral value of the *Yijing* over its utility as a book of divination (a long-standing debate), and several scholars—notably Mao Qiling (1623–1716) and Hu Wei (1633–1714)—questioned the authenticity of certain post-Han accretions to the work.[27] But the *Yijing* retained enormous scriptural authority as an explanation of time, change, and causality. The highly influential neo-Confucian compilation known as the *Jinsi lu* (Reflections on things at hand), informs us, for example, that the *Yijing* is

> comprehensive, great, and perfect. It is intended to bring about accord with the principle of ... [human] nature and destiny, to penetrate the causes of the hidden and the manifest, to reveal completely the nature of things and affairs, and to show the way to open resources and to accomplish great undertakings.[28]

The fundamental assumption of the *Yijing* was that the universe was intelligible—that fate could in fact be known. One of its "Ten Wings," the "Explaining the Trigrams Commentary" (*Shuogua zhuan*) states:

> In the distant past, the way the sages made the *Changes* was as follows: It was to be used as a means to stay in accord with the principles of [human] nature [*xing*] and of fate [*ming*]. It was for this reason that they determined what the Dao of Heaven was, which they defined in terms of *yin* and *yang*, what the Dao of Earth was, which they defined in terms of hard and soft,

and what the Dao of Man was, which they defined in terms of humaneness and right behavior.

According to the *Shuogua zhuan*, by following the Way (Dao), correctly ordering their behavior, investigating principle to the utmost (*qiongli*), and exhausting their nature (*jinxing*), the sages arrived at a complete understanding of destiny.[29] This understanding was available to others as well, if their virtue and sincerity could be properly cultivated.

The sixty-four hexagrams of the *Yijing* and their constituent trigrams and individual lines, together with written judgments (*tuan*) and line statements (*yaoci*), reflected and explained various predestined situations evolving out of the natural patterns and processes of eternal cosmic change. According to the "Great Commentary" (*Dazhuan* or *Xici zhuan*) on the *Changes*, the *Yijing* served as a kind of medium that allowed the person consulting it to establish a spiritual link with Heaven and to be able divine the future by means of the milfoil stalks, through the interpretation of omens, by observing the planets and stars, and by means of numerical devices such as the "Yellow River Chart" (*Hetu*) and the "Luo River Writing" (*Luoshu*).[30] In short, for most Chinese of the Qing period, the *Yijing* was the starting point for an understanding of fate, however it might be conceived.

Qing divination practices

Scholars and commoners alike believed that most Chinese mantic techniques were related in some way to the *Yijing*.[31] The pervasive use of trigrams and hexagrams as explanatory symbols in a variety of fortune-telling schemes naturally reinforced this belief. So did the many correlative systems that linked the symbols of the *Yijing* with the forces of *yin* and *yang*, the five agents or phases of change, and the system of stems and branches of the ancient sexagenary cycle.[32] Among the most popular of these numerologically oriented systems were Six Direction divination (*liuren*), the method of calculation known simply as the Great One (*Taiyi*), and hidden stems expertise (*qimen dunjia*).[33]

Other popular divining schemes—all at least a thousand years old (and most likely more than two thousand years old) by Qing times—included fate extrapolation (*tuiming*), the interpretation of auspicious or inauspicious omens (*jixiong zhi zhao*), dream divination (*mengzhan*), physiognomy (*xiangren, xiangmian, kanxiang*, etc.), word analysis (*chaizi* or *cezi*), meteorological divination (*zhanhou*), geomancy (*kanyu, fengshui*, etc.), spirit writing (*fuji, fuluan*), the use of bamboo or wooden blocks (*zhijiao, beijiao*), and the interpretation of messages connected with bamboo sticks (*qianbu, linqian*, etc.). Individuals also consulted official state calendars (*shixian shu*) and popular almanacs (*huangli, tongshu*, etc.) for daily guidance, and occasionally asked deceased ancestors and other spirits (*shen*) for advice concerning the future.[34]

The eight natal characters (*bazi*)—one stem and branch combination each for the year, month, day, and hour of birth—were considered especially revealing. Although several different techniques of fate extrapolation based on birth dates existed during Qing times, the most commonly used was the system of Xu Ziping, developed during the Song dynasty, in which eight characters were grouped into four units of two each (known as the four pillars, *sizhu*). These characters reflected both cosmological and social relationships. Each stem and branch, for example, was designated *yang* or *yin*, and correlated with one of the five agents, as well as a particular season and a twelve-stage developmental cycle extending from prenatal life in the womb (*tai*) to extinction (*jue*). The four pillars were generally correlated with one's ancestors, parents, spouse (if married), and son(s). Many fortune-tellers placed particular emphasis on the idea of an operational force (*yongshen*) that determined whether the correlations would be creative or destructive, harmonious or antagonistic—not only for the individual under consideration but also for other individuals with whom the principal party might be involved.[35]

Regardless of the method employed, and whether or not all eight characters were taken fully into account, few Chinese of any social class were inclined to dismiss birth dates as irrelevant to future affairs—particularly in matters such as marriage. Evidence from local gazetteers, as well as from handbooks on etiquette and the testimony of contemporary observers, both Chinese and foreign, indicates that practices such as fate extrapolation, the selection of lucky days (*xuanze*) and the exchange of the eight characters in marriage ritual were extremely prevalent throughout the Qing Empire, from Chuxiong county in remote Yunnan province, to Liaoyang in Manchuria (Liaoning).[36] It is true that Wolfram Eberhard's study of over 3,700 married couples from the Jung clan of Kwangtung province in the period from 1600 and 1899 concludes that the members of this clan, at least, "did not let themselves be influenced by the auspicious and inauspicious combinations of signs in their selection of marriage partners." But his investigation, which also contains data from modern-day Taiwan, considers only the year of birth (associated with one of twelve cyclical animals) as an index of marital compatibility, thus oversimplifying a very complex process by which families contemplating marriage negotiated the future of their children.[37]

A sensitivity to omens and portents was pervasive in traditional China—especially at the top, where emperors and dynasties depended on Heaven's mandate (*Tianming*) for legitimacy. Auspicious signs (*furui*, *xiangrui*, etc.), such as the appearance of colored vapors, bright lights, rare animals, unusual plants, strange stones, sweet dew, and so forth, reflected heaven's pleasure and portended joyful events; calamities (*zaiyi*), such as unanticipated eclipses and similar celestial anomalies, earthquakes, floods, and other natural disasters indicated Heaven's displeasure, and perhaps even loss of the mandate. The Qing dynasty, like all of its imperial predecessors, dutifully recorded all unusual signs from Heaven, and the Manchu emperors took them very seriously.[38] About a third of the imperial amnesties granted from the early seventeenth century to the late nineteenth century were justified as a means of restoring harmony to the natural order after portents indicated that it had been disrupted by human excesses.[39]

To be sure, the Kangxi emperor—arguably the most scientific and skeptical of all the Manchu rulers—might assert that, although the dynastic histories were full of accounts of omens, they were of no help in governing the country, that "the best omens were good harvests and contented people."[40] But the "Respecting Heaven" (*jing Tian*) sections of the *Da Qing shichao shengxun* (Sacred instructions for ten reigns of the Great Qing) provide numerous instances in which he, like both his predecessors and successors, registers great concern over portents, is moved to self-examination because of them, and admonishes his officials to rectify their administration as a result. Scholars might disagree over whether the sovereign's self-cultivation could actually prevent anomalous events from occurring, as the great Song philosopher Zhu Xi (1130–1200) seems to have believed, but few doubted that a relationship existed between portents and the need for an immediate imperial response to the messages they conveyed.[41]

At less exalted levels, heavenly portents were also taken seriously, along with less dramatic signs, such as birdcalls (*niaojiao*), the residue or shape of lamp wicks (*denghua*), and various body sensations (itching, tingling, sneezing, etc.).[42] The *Qingbai leichao* devotes considerable attention to the interpretation of various kinds of natural and supernatural omens, as do other Qing sources, both Chinese and Western. Many anecdotes involve scholars and at least a few famous officials.[43]

Dreams had special significance in Chinese society and were classified in treatises on the subject according to various broad categories, such as those pertaining to heaven, the sun and the moon, thunder and rain (and related phenomena), mountains and streams (a category including roads, earth, and rocks), shapes and sounds, food and clothing, utensils, wealth and property, writing implements and scrolls, the civil service examinations, supernatural things and occurrences, longevity, various plants and animals, giving and receiving, and certain vague indications (*fanyu*).[44] Undoubtedly the most popular work on dream divination in the Qing period was the *Zhougong jiemeng* (Dream interpretations of the Duke of Zhou), published in a great many versions and widely disseminated in almanacs empire-wide, usually in simplified form.[45]

Qing accounts of dream divination abound. Often people prayed for dreams (*qimeng*) and spent the night in local temples in hopes of receiving clues to the future. Justus Doolittle writes of Fujian in the mid-nineteenth century:

> Many people, in case they find great difficulty in deciding what course to take in regard to an important subject under consideration, visit some popular temple and, having burned incense and candles, beg the divinity worshipped to favor them with a dream shedding light on the subject of their perplexity, which they briefly state.[46]

Of the literally dozens of stories on dream divination in the *Qingbai leichao*, no less than ten focus on "praying for dreams." Significantly, most of the individuals involved in seeking them are highly reputable scholars—among them, such Qing notables as Li Guangdi (1642–1718), Jiang Yi (1631–1687), Zhang Ying (1638–1708), and Xu Ben (1683–1747).[47]

Most Chinese believed that the physical appearance of individuals could yield important clues regarding both character and destiny.[48] Elites as well as commoners held this view. John Gray writes, for instance, "It is surprising to find what a number of respectable and influential men throughout the empire resort to professors of the art [of physiognomy]."[49] The *Qingbai leichao*, local gazetteers, and other sources indicate that a great many Qing civil and military officials relied on physiognomers for advice, and a number of Chinese scholars actually practiced the art—including such high-ranking individuals as Ji Huang (1711–1794), Zeng Guofan (1811–1872), Shen Baozhen (1820–1879), Liu Mingchuan (1836–1896), Li Wentian (1834–1895), and Zhao Shuqiao (d. 1900).[50]

The social importance of impressive features in Qing China can scarcely be underestimated. In the first place, physiognomic stereotypes unquestionably influenced the way people at all levels responded to one another. These stereotypes were reinforced not only in popular proverbs but also in folktales and novels. At least some almanacs, like all physiognomy manuals, depicted different types of faces, as did public displays that advertised the skills of professional physiognomers "with painted representations of the human countenance."[51] Facial stereotypes could even affect a person's bureaucratic career. A middle degree holder (*juren*), for example, could reportedly enter Qing official life at a comparatively high level after failing the *jinshi* examination three times if he possessed what was considered to be the right kind of face.[52] Professional physiognomers generally focused primary attention on the hands and basic facial characteristics of their clients (ears, forehead, eyebrows, eyes, nose, cheeks, mouth, chin, and complexion), but they might also pay attention to other parts of the body, as well as movements, vocal sounds, and even urine and excrement.[53]

Closely related to physiognomy in its emphasis on visual interpretation (*xiang*) was the art of fathoming or dissecting written characters, a technique employed not only as a separate category of divination but also in conjunction with dream analysis.[54] The great Qing scholar Ji Yun (1724–1805) once described the omens of word analysis as developing from the same sort of spiritual communication as the selection of milfoil stalks in *Yijing* divination or the use of oracle bones.[55] For the convenience of illiterate or semiliterate commoners, professional fortune-tellers provided a limited number of characters to be chosen by clients and interpreted, but elites generally wrote their own word choice down. The logic of word analysis, like that of dream interpretation, physiognomy, and most other forms of Chinese divination, was inveterately associational, allowing, perhaps even compelling, diviners to link all relevant concepts, cosmological as well of personal, to the client. Stories surrounding the dissection of characters circulated widely in Qing China, and many involved famous scholars and officials such as Zhang Ying, adviser to the Kangxi emperor.[56]

The heightened attunement of the Chinese to shapes influenced other Chinese divination techniques, from the analysis of "chops" or seals (*yinxiang*) to meteorological divination, known generically as *zhanhou*. *Zhanhou* techniques, like those of seal interpretation, were not, of course, as obviously psychological in orientation as word analysis, which involved a degree of conscious or unconscious choice, and they relied to a far greater extent on explicit stem-branch correlations and astrological configurations in the process of making predictions. Furthermore, the category *zhanhou* included a number of nonvisual divination methods, such as the analysis of wind sounds (*fengjiao*, lit. "wind corners") as part of its repertoire. Nonetheless, techniques such as observing the rain (*xiangyu*), watching the vapors (*wangqi* or *wanghou*), and divining from clouds (*zhanyun*) required the same careful attention to visual detail required of physiognomy, word analysis, and geomancy.[57]

Geomancy or, more properly, siting—the Chinese practice of selecting auspicious locations for tombs, houses, and other structures—took many forms and went by a variety of names in the Qing period.[58] There were numerous schools and subschools of the art, but all had a common goal: to harmonize dwellings for both the living and the dead with the immediate physical environment as well as the larger cosmic scheme. Geomancy thus shared with traditional Chinese medicine a concern with maintaining a harmonious *yin–yang* equilibrium in the midst of constant change. A given location, like the human body, represented a microcosm of the universe, which naturally required a balance of cosmic energy (*qi*) for proper functioning. As specialists in siting, geomancers, often known as *fengshui* masters (*fengshui xiansheng*) were, in effect, "doctors of the earth," and a great many were conversant in traditional Chinese medicine.[59]

As with physiognomy, *fengshui* divination both reflected and encouraged a certain aesthetic logic, but the calculations of geomancers usually involved much more than simply a visual analysis of topography. Cosmological calculations required a compass, or *luopan*—a complex device used by most *fengshui* specialists that included virtually every major symbol used by the traditional Chinese in dealing with space and time.[60]

Although geomancy had to do with all sorts of structures related to the living and the dead, its primary focus was on the deceased. Specialists in "tomb divination" (*xiangmu* or *zhanmu*) were therefore ubiquitous in Qing times. Classical sanction for a concern with the proper place and time for burials may be found in such hallowed sources such as the *Chunqiu* (Spring and Autumn annals), the *Liji* (Record of ritual), the *Zhouli* (Zhou rituals) and the *Xiaojing* (Classic of filial piety). The *Xiaojing* states, for example: "[The filial son] determines the burial place [of his parents] by divination and puts them to rest" (*bu qi zhaizhao er ancuo zhi*).[61]

But, as Qing critics of geomancy repeatedly pointed out, for a thousand years or more gravesites in China were chosen primarily to bring benefits to survivors rather than comfort to the departed. In the straightforward words of the *Huangdi zhaijing* (Yellow Emperor's siting classic), "If a site is proper (*an*; lit., peaceful), a family will have generations of good fortune; if not, then the family will decline."[62] Yet despite relentless condemnation by scholars and discouragement by the state (especially when it created social tensions, as it often did),[63] geomancy flourished throughout the empire—particularly in the south-eastern provinces of Jiangsu, Zhejiang, Jiangxi, Fujian, and Guangdong.[64]

Practitioners of geomancy in Qing times ranged from famous scholars and high officials such as Zhao Zhixin (1662–1744), Li Wentian (1834–1895), and Zhang Zhidong (1837–1909) to commoners whose names are forever lost to history.

According to the *Qingbai leichao*, some emperors were considered conversant in the art, and, despite regulations in some monasteries prohibiting the practice of geomancy and certain other forms of divination, there were a number of Buddhist and sectarian Daoist *fengshui* practitioners, including the eminent Chan abbot Xuyun (1840–1959).[65]

The clients of *fengshui* specialists likewise spanned the entire Chinese social spectrum. Emperors, empresses and members of the imperial household employed them, as did civil and military officials at all levels of administration. Throughout the empire, individual families, clans, villages, and even whole cities engaged specialists in the art of siting, since filial sons wanted to do all they possibly could for their deceased parents (and themselves), and larger entities could hardly afford to neglect geomantic factors that might affect their collective scholarly fortunes. Among the many famous officials who utilized *fengshui* specialists in their personal lives were Chen Hongmou (1696–1771), Wei Yuan (1794–1856), Tao Zhu (1779–1839), Zeng Guofan, and Ding Richang (1823–1882).[66]

Geomancy tended to be a rather subdued and private form of divination, but many other forms were more public, and at least a few entailed colorful and highly visible dramatic action. Spirit writing, in particular, involved elaborate ritualized preparations—including the offering of incense, ceremonial bowing, and prayers of petition—prior to the actual descent of a spirit into the diviner's stick. The ritual of spirit writing, which, like all the other forms of divination I have been discussing, can still be seen in Taiwan and Hong Kong today, consisted of furious bursts of stick writing by the medium (often a Daoist priest) in response to written questions solicited orally from petitioners. The spirit's esoteric written message was then read aloud, transcribed, and given to the petitioner.[67]

During the Qing period, spirit writing enjoyed widespread popularity. According to Xu Dishan, in Ming and Qing China there was a spirit-writing altar (*jitan*) in virtually every prefectural and county-level city. And as interest in *fuji* divination grew, specialists began to operate in private homes and ancestral halls as well as temples. There were even spirit-writing clubs for various strata of Chinese society. Xu tells us that particularly in provinces such as Jiangsu and Zhejiang, where a "scholarly air" (*wenfeng*) prevailed, the attitude was simply that, "if one did not believe in *fuji* spirits, one could not be successful in the examinations." He goes on to say that the questions asked of spirits regarding the examinations might affect one's entire career and that even after scholars had attained official positions, they would still think of *fuji* divination whenever faced with a difficult decision.[68]

In part, Qing scholars esteemed *fuji* divination for its self-consciously moral emphasis—although most other Chinese mantic messages also had explicitly ethical content. They also enjoyed the elaborate wordplay of spirit writing. In many instances, interpretation of a *fuji* revelation, like that of a dream, required skills closely akin to, if not identical with, conventional word analysis. Sometimes the conversation between a *fuji* spirit and his interlocutor took the form of an attempt to match couplets (*duidui*); and scholarly patrons were occasionally moved to impose specific stylistic requirements on the poetry of a spirit. But the primary reason *fuji* divination attracted the attention of Qing scholars was its utility in answering questions of ultimate importance to them. These included matters of life and death, sickness and health, problems of civil and military administration, and career issues. Not surprisingly, a number of scholars asked *fuji* spirits for assistance in the civil service examination system—either advice on how to write better essays or information concerning the questions themselves.[69]

The *Qingbai leichao*, Xu Dishan's collection of anecdotes on spirit writing, and other sources provide a wealth of information on the use of *fuji* divination in the Qing period. Like professional fortune-tellers, *fuji* spirit mediums predicted disasters such as droughts, floods, epidemics, and rebellions. They also advised individuals on how to conduct their lives for maximum personal advantage, assisted officials in their administrative affairs (including legal decisions), and helped them defeat pirates and bandits. Among the many high-ranking Qing officials who relied on *fuji* oracles were Peng Dingqiu (1645–1719), Ye Mingchen (1807–1859), and Zeng Guofan.[70]

Women did not normally engage in spirit writing. They did, however, perform most other shamanistic functions—primarily for members of their own sex. These services included healing, fortune-telling, and the performance of exorcisms and magic. A number of women specialized in communicating directly with the dead—perhaps because so many unplacated spirits (*gui*) in traditional China were female ghosts who had an axe to grind with the oppressive patriarchal society that had caused them so much misery in life.[71]

Qing elites tended to be particularly critical of women shamans and their allegedly "foolish" woman followers. But all spirit mediums proved threatening to the scholarly class—not only because of their willingness to flout Chinese social norms openly but also because of their social power. They could, for example, criticize local power holders or ask officials for the redress of grievances (*shenyuan*) in the name of the gods. And even when shamanistic oracles worked to the advantage of the elite—such as when a local spirit writer advised thousands of angry and recalcitrant peasants that they should honor the wishes of the local magistrate, who had canceled a procession honoring the Qingpu (Jiangsu) city god on his "birthday"—the scholarly class could hardly fail to recognize the political implications of such utterances.[72]

Of course a number of Chinese fortune-telling techniques, such as the use of bamboo blocks (*jiao*) in homes and temples, and divination by means of bamboo sticks (*qian*) at religious shrines, required no supernatural intervention and little, if any, formal education.[73] Not surprisingly, such techniques tended to be undertaken primarily by unlettered women, for whom trips to local temples represented one of the few forms of nondomestic recreation available to them. But women were not the only people to employ such humble methods of divination. The great Qing scholar Yan Yuan (1635–1704), for example, resorted to *qian* sticks in an effort to locate his father, from whom he had been separated since the age of three, during the Manchu conquest. Other prominent Qing elites who are known to have practiced *qian* divination at one time or another include Wang Shizhen (1634–1711), Han Tan (1637–1704), Bi Yuan (1730–1797), and Enming (1846–1907).[74]

Divination and official orthodoxy

If individual Chinese were preoccupied with fate, the Qing government was positively obsessive about it. The reasons are not difficult to find. In the first place, imperial legitimacy was viewed, in large measure, as a matter of predicting the future correctly—whether in terms of anticipating events such as eclipses, interpreting portents, or selecting auspicious days for conducting all-important state rituals.[75] Because failure to perform these functions satisfactorily called into question the regime's right to rule by virtue of the heavenly mandate, activities such as regulating the calendar (*zhili*) and fixing the time (*shoushi* or *shiling*) assumed special importance to every dynasty. Not surprisingly, throughout the imperial era, divination often played a significant role in dynastic transitions—not least in the changeover from Ming to Qing.[76]

The state calendar (*shixian li* or *shixian shu*), compiled by the Imperial Bureau of Astronomy in conjunction with the Board of Ritual, and issued annually with great fanfare after review by the emperor himself, was the single most visible index of imperial legitimacy.[77] Although based in part on mathematical calculations, the Qing calendar was primarily a cosmological document, reflecting, like an elaborate horoscope or a geomancer's compass, the elaborate interplay between variables such as *yin* and *yang*, the five agents, the seven astral administrators (*qizheng*, i.e., the sun, moon, and five planets), the eight trigrams, the nine palaces (*jiugong*), the ten heavenly stems and twelve earthly branches, the twenty-eight asterisms, and well over two hundred "stars" (*xing*)—usually personified as benevolent or malevolent "spirits" (*shen* and *sha*, respectively). A rectangular, color-coded "diagram of the position of the spirits for the year" (*nianshen fangwei tu*) was the most common graphic device used to communicate the influence of these and other cosmic variables to readers of the calendar.

The Manchus, like all of their imperial predecessors, jealously guarded the emperor's astronomical and calendrical prerogatives. For example, the administrative regulations of the Six Boards prohibited the circulation of any astrological or calendrical books that might be used "to predict order and disorder" in the empire. The *Collected Statutes* of the dynasty outlawed all unauthorized versions of the official calendar, and certain editions of the work carried an explicit warning on the cover to the effect that those who forged copies of the calendar were subject to decapitation, and that those who informed on such persons would receive an imperial reward of fifty taels. Calendars that did not carry an official seal were considered to be private and, therefore, strictly speaking, illegal.[78]

Privately published almanacs, many based closely on the imperial calendrical model and all reflecting the same orthodox cosmology, circulated freely throughout the Qing period.[79] They generally had auspicious red covers rather than imperial yellow ones, and they possessed a number of pages of written text in addition to the standard columns, charts, and diagrams of their official counterparts. Many contained popular wood-block prints not to be found in calendars, such as depictions of the two dozen or so difficult or dangerous passes (*guan*) that might be encountered by Chinese children, or annotated pictures of the twenty-eight asterisms and their general influence on life situations. And every almanac I have seen (not to mention at least a few works erroneously designated calendars [i.e., *shixian shu*]) included illustrations of the spring ox (*chunniu*) and its herdsman (*shentong* or *mangshen*), whose rich color symbolism—supposedly indicating agricultural prospects for the new year—was evocative, instructive, and nearly universal in traditional China.[80]

Other common elements in almanacs—particularly in the late Qing period—were folk stories, morality tales (such as the "Twenty-Four Examples of Filial Piety," *Ershisi xiao*), and popular religious materials. Almanacs might also include practical advice on matters relating to agriculture, business, health, proper etiquette, and, of course, omens and techniques of divination. By the early twentieth century, a number of these works had become conduits for new information on recent educational and political changes, modern science and technology, and news regarding foreign nations. Photographs of Chinese political leaders, such as the notorious Prince Qing and other Manchu nobles, began to adorn their pages.[81]

At the core of all state calendars and annual almanacs was a month-by-month, day-by-day breakdown of the entire year, designed to coordinate and control certain aspects of Chinese political, social, ritual, and economic life. Generally, each column, whether for the month as a whole or for the individual days within each month, indicates the particular spiritual influences, agents, asterisms, and other cosmic variables affecting that particular time. The positive spiritual influences (*shen*) normally appear at the top of each column, and the negative spirits (*sha*) at the bottom. For each day—with only a few exceptions in any given year—there are a certain number of activities designated appropriate (*yi*, i.e., auspicious) or inappropriate (*buyi*, inauspicious). Designated activities of this sort were categorized and included in the Qing collected statutes. Sixty-seven matters fell under the rubric "imperial use" (*yuyong*), thirty-seven under "people's use" (*minyong*), and sixty under "almanac selection" (*tongshu xuanze*).[82]

Of the sixty-seven categories of imperial concern identified in the Qing collected statutes, a great number dealt with general administrative and ritual matters, such as various forms of sacrifice and prayer, the submission of documents, the promulgation of edicts, the bestowal of favors and awards, personnel matters, diplomatic and military affairs, and events such as banquets. On a more personal level, the calendar provided guidance in choosing days for educational activities, domestic rituals, health, business decisions, and the solicitation and adoption of suggestions and advice. Among mundane activities regulated by the calendar we find bathing and grooming, cutting out clothes, household cleaning and decoration, establishing a new bed, discarding things, moving things (*banyi*), traveling, breaking ground, well digging, construction and repair, hunting, fishing, trapping, planting, cutting wood, and herding animals.

The thirty-seven activities designated for the people encompass virtually all of the major personal and mundane matters indicated above. This category also includes a few additional mundane items, such as visiting and receiving relatives and friends, as well as important ritual and administrative acts such as carrying out sacrifices, submitting a memorial to the throne, assuming an official post, and arranging for burials (the last not expressly specified as an imperial concern). The almanac category, although listed last, seems to stand midway between the two previously mentioned sets of concerns. It includes most of the major categories of activity noted above but omits those that are obviously the exclusive prerogatives of the emperor. At the same time, it is especially specific about certain personal and mundane affairs. It indicates, for example, particular stages of marriage ritual, the treatment of particular afflictions, and the use of particular medical remedies. It also mentions a few activities not included in either of the preceding categories, such as constructing, opening, and repairing storehouses and granaries, fermenting liquor, smelting metals, and crossing the water by boat or other means.[83]

Relatively few activities were designated "inappropriate" in the hundred or so Qing calendars and almanacs I have studied. By far the most common categories of concern were traveling (*chuxing*); moving one's residence (*yixi*); beginning construction (*dongtu*; lit. "stirring the earth"); planting or sowing (*zaizhong*); and sewing (*zhenci*). Particularly striking in this regard are the number of days that calendar-makers considered inauspicious for travel—as many as eighty-two in one year (1886), out of a total of 354 days. In this same calendar, forty-two days were designated inappropriate for planting; thirty-seven for beginning construction; thirty-five for sewing; and thirty-three for moving. As a rule, when the evil star Lunar Repression dominated a day, both traveling and moving one's residence became inadvisable—and often planting as well.[84]

How seriously, one might ask, did Qing elites take the divinatory stipulations of calendars and almanacs? Positive evidence is difficult to come by in Chinese sources, because most people seem to have taken the predictions of these publications for granted—rather like weather forecasts in the modern West, which are seldom worthy of a recorded remark. We know, however, from diaries, correspondence, dedicatory inscriptions, and other sources that Qing scholars and officials were acutely aware of auspicious and inauspicious days. We also know that almanacs remained among the most widely circulated books in the entire Chinese empire in late imperial times. At least one late Qing ritual handbook asserts categorically that "no household is complete without an almanac."[85]

Furthermore, Western observers continually remarked on the prevalence and pervasive influence of calendars and almanacs. A report by the Society for the Diffusion of Useful Knowledge in China (November 21, 1838), which referred to "the almost universal demand" for these books among all classes of Chinese, echoed the views of a great many other well-informed foreigners—from Matteo Ricci in the early seventeenth century to Robert Hart in the late nineteenth. A. P. Parker, an astute student of Chinese calendrical science in the 1880s tells us;

> [T]he astrological part [of the official calendar] is universally believed in, though there seems to be considerable difference in the practice of the details by different persons—some considering it necessary to be careful

> about the times and places of carrying out the most important affairs of life, such as marriage, burial, house-building, &c., while others believe it necessary to be careful as to the time and place for the most commonplace details of everyday life, such as opening a shop, entering school, going on a journey, giving an entertainment, sweeping the floor, shaving the head, taking a bath, &c.

Similarly, W. A. P. Martin, who had extensive contact with Chinese bureaucrats as head of the Interpreter's College (*Tongwen guan*) in Beijing, once remarked: "No man [in China] thinks of beginning a journey, laying a corner-stone, planting a tree, marrying a wife, burying a parent, or any of a thousand functions in public or private life, without consulting this convenient oracle." Further, he noted:

> The late archimandrite Palladius told me that he found ... [the Qing] calendar useful, as it enabled him to select an unlucky day for his visits to the Russian legation, four miles distant, when he was sure to find the streets unobstructed by marriages or funerals.[86]

As long as almanacs and unofficial versions of the calendar reflected orthodox views and reinforced imperial legitimacy, they could be tolerated; but if not, the state suppressed them with a vengeance. This was naturally true of calendars produced by major contenders for political power, such as the Yongli emperor of the Southern Ming during the latter half of the seventeenth century, or the Taiping rebels of the mid-nineteenth century.[87] But it was also true of certain privately printed heretical (*xie*) almanacs such as the early nineteenth-century handbook titled *San Fo yingjie tongguan tongshu* (Comprehensive Almanac for Responding to the Kalpas of the Three Buddhas [Past, Present, and Future])—used by Lin Qing of the proscribed Eight Trigrams sect to, in the Jiaqing emperor's words, "deceive the people" and "violate the authority of Heaven." After discovery of this document, the emperor ordered all copies destroyed, and engaged in a relentless search for the authors and publishers.[88]

The Qing authorities adopted the same approach toward other millenarian cults, such as the White Lotus Society, which drew on orthodox cosmological symbols—including the eight trigrams, the nine palaces, the ten stems and twelve branches, and the twenty-eight lunar lodges—but placed them in a heretical framework: the advent of the Maitreya Buddha. Prophecy, under such circumstances, was a form of divination that had enormously threatening political potential.[89] From the standpoint of the Qing government, to surrender authority concerning the future to non-orthodox agents of any sort was to compromise the very foundations of imperial rule.

What the Manchus feared most, of course, was the capacity of individuals outside the state's direct sphere of control to manipulate and mobilize the masses. The preface to the dynasty's official divination manual known as the *Qinding xieji bianfang shu* (Imperially approved treatise on harmonizing the times and distinguishing the directions) reflects this persistent anxiety: "Specialists of the occult arts [*fangshi*] talk of good luck and bad luck, calamities and blessings, and frighten the people into having no standards of behavior."[90]

To be sure, divination did not have to be dangerous to invite criticism. As I have already indicated, throughout the late imperial era—and, in fact, well before—a number of influential Chinese intellectuals assailed divining practices such as fate extrapolation, physiognomy, the selection of lucky and unlucky days, and especially geomancy. The famous late Qing compendium known as the *Huangchao jingshi wenbian* (Writings on statecraft from the present dynasty) contains several critical essays on these and related topics by scholars such as Gu Yanwu (1613–1682), Lu Shiyi (1611–1672), Feng Jing (1652–1715), and Quan Zuwang.[91]

These essays reflect the major concerns of the Qing elite with respect to divination. One widely shared fear was that blind acceptance of fate on the part of commoners would diminish their impulse toward moral improvement and self-reliance. Another, shared with the state, was that false prophecy could easily delude people into creating disorder (*huoren zuoluan*)—that is, into instigating rebellion. A third concern was that many mantic practices generated or exacerbated social tensions and led to breaches of etiquette (such as late burials, in the case of geomancy). Finally, in the eyes of many scholars, fortune-tellers were simply unreliable. Gu underscored the divergent and often misleading interpretations of diviners; Lu questioned the validity of astrological predictions and portent interpretations because they were unscientific ("Western studies," he reports, "say absolutely nothing about divination"). And Quan, following Huang Zongxi, argued that "Many people are born on the same day, yet some are fortunate and some are not; some will enjoy longevity and others will die young." And Feng, in an essay titled "Refuting the Theories of *Yinyang* Soothsayers" (*Bo yinyang jia yan*), maintained that skill with esoteric numerical calculations was "not equal to an understanding of the principles of things."[92]

Significantly, although Lu's reference to Western science was rather unusual, none of these criticisms of divination were new. In fact, they followed closely the lines of attack crafted centuries before by individuals such as Xunzi in the Late Zhou, Wang Chong in the Han, and Lü Cai in the Tang—all of whom are excerpted in various sections on divination in the *Tushu jicheng*. Like their ancient predecessors, not to mention their more orthodox neo-Confucian contemporaries, Qing *kaozheng* scholars had no quarrel with the idea of divination itself, for the practice had far too long and illustrious a pedigree in China's classical tradition to ignore. Rather, their hostility was directed toward specific misapplications of the practice—especially by professional fortune-tellers who were not by definition members of the orthodox elite. As in the case of both ritual and religion, the Qing elite made a sharp (though somewhat artificial) distinction between their own enlightened beliefs and the crude customs of the popular masses. The bias against divination was essentially a class prejudice, masked by the rhetoric of Confucian morality.

In any case, scholarly criticisms of divination had little effect. Even the charge that fortune-tellers made inaccurate predictions fell on essentially deaf ears. The very complexity of most Chinese divination systems provided conceptual flexibility, and contributed to their self-confirming character. Wrong prognostications did not invalidate the idea that fate could be known (an article of Confucian faith, after all, encouraged by the *Yijing* itself); they merely indicated that someone had misinterpreted the huge number of cosmic variables involved, or that either the diviner or the client was insufficiently moral or insufficiently "sincere."[93] Furthermore, as I have indicated, many Qing scholars employed the same mantic methods that they decried. A popular book by Wang Weide, published in 1709 and titled *Bushi zhengzong* (Orthodox milfoil divination), castigates members of the Chinese elite for affecting a disdain for divination, only to use it surreptitiously through the agency of friends or servants.[94] The imperial government displayed the same sort of hypocrisy, denouncing the practices it regularly relied on. The official commentary to the Qing legal code, for instance, described geomancy as "basically absurd and false, unworthy of belief," yet we have seen that the throne and members of the imperial household, as well as civil and military officials, used *fengshui* specialists all the time, as they did many other kinds of diviners and divining techniques.

The persistence of divination

Hypocrisy aside, how do we account for the longevity and pervasiveness of divination in China? One answer might be China's comparative disinterest in modern (Western) science. While "scientific" *kaozheng* scholarship in the seventeenth and eighteenth centuries brought a new epistemology to China, and with it, "a fundamental pattern of intellectual change," it certainly did not have the same effect on magic and divination that Keith Thomas ascribes to the roughly contemporary "scientific and philosophical revolution" in Europe.[95] Nor have all premodern societies embraced the mantic arts with equal enthusiasm. In fact, two of the cultures of the world least interested in divination appear to be the traditional Australian Aborigines and the American Plains Indians.[96]

A more likely explanation for the staying power of divination in China is that it both embodied and reflected many of the most fundamental features of traditional Chinese civilization. Although it always had a certain heterodox potential, it was not fundamentally a countercultural phenomenon. On the contrary, it remained an integral part of the most important state and domestic rituals—from official sacrifices to life-cycle ceremonies—throughout the imperial era. Enjoying abundant classical sanction and a long history, Chinese mantic practices followed the main contours of traditional Chinese thought.

Most forms of Qing divination were eclectic, "spiritual," associational, intuitive, tradition-bound, and highly moralistic. These qualities fit comfortably in a syncretic society whose dominant class esteemed ancient Confucian values, relied heavily on correlational logic, believed in a spiritual link between Heaven, Earth, and Man (which made divination possible, after all), and saw knowing the future as an activity in which "the rational operations of the intellect were not sharply disconnected from what we [Westerners] would call intuition, imagination, illumination, ecstasy, aesthetic perception, ethical commitment, or sensuous experience."[97] The *Yijing* epitomized both the philosophical and the mantic traditions of China, establishing a crucial epistemological link between the two. Although it possessed a kind of biblical authority, it pointed to a very different path of cosmological and ontological understanding.

In more mundane terms, divination had a decidedly visceral appeal. Qing fortune-tellers invariably employed a colorful and universally resonant symbolism that remained deeply imbedded in the consciousness of Chinese at all levels of society. Allusions to historical, classical, and mythological figures in prophetic statements evoked powerful responses in clients, as did references to certain symbolic colors, numbers, plants, and animals. Furthermore, fortune-tellers invariably surrounded themselves with culturally familiar paraphernalia. Even the most rudimentary fortune-telling table on the street would generally be adorned with writing materials, books, and calligraphic inscriptions—the marks of scholarly refinement and moral cultivation. More elaborate settings in homes or at divination parlors might boast religious icons or spirit tablets, as well as incense and candles. Visual representations of cosmic power such as the *Taiji tu* (Diagram of the Supreme Ultimate) and the eight trigrams, which often adorned divination handbooks, almanacs, and fortune-telling stalls, were also ubiquitous as decorative elements (and charms) in elite and commoner households alike.[98]

The rituals of divination were likewise satisfying and culturally familiar to all clients. Ceremonies such as the burning of incense, which invested divinatory procedures with a mystical if not a magical aura, had a truly universal appeal. The use of writing by diviners not only enhanced their social prestige but also gave them cosmic leverage, because so many Chinese believed that written words had magical power. Meanwhile, the theatrical performance of fortune-tellers contributed to their public allure. In several respects patronizing a diviner was like watching a play, or even participating in one. Dramatic forms of divination such as spirit writing and other types of spirit possession held audiences literally spellbound; but even the more subdued rituals of physiognomy, fate extrapolation, and word analysis attracted passers-by on the streets, in marketplaces, and at temples throughout China. Although private fortune-telling parlors existed, and a number of people consulted diviners in the privacy of their homes, Chinese divination was fundamentally a public affair. This was yet another reason why most Qing subjects simply took it for granted.[99]

The close link between divination and traditional Chinese medicine probably contributed to the tenacity of both. Despite the great diversity of medical and mantic theory in late imperial times, doctors and diviners shared many of the same cosmological assumptions about systematic correspondence as well as demonology.[100] And, although the Confucian classics and a number of popular proverbs emphasized that health and longevity were predestined, few individuals in Qing China accepted their fate passively. Most people believed that they could modify their fate, either by moral or magical means. In fact, the two sources of power were closely related. For the elite, the same spiritual capacity that made foreknowledge possible, gave those who had developed their sincerity to the utmost the ability to transform their own lives in concert with Heaven and Earth. Theirs was a kind of cosmological mind magic, sanctioned by no less authority than the *Zhongyong* (Doctrine of the Mean). For commoners, charms and the advice of soothsayers might do the trick.[101]

On the other hand, the idea of inescapable destiny occasionally served as a convenient explanation for adversity and disappointment. Thus, we find that even Zeng Guofan, a staunch Confucian and certainly no slouch, once remarked during a difficult phase of the Taiping Rebellion that fate determined 70 percent of a situation, leaving only 30 percent to man's exertions. And Guo Songtao (1818–1891), a hard-working and dedicated scholar who had the misfortune to assume a career in late Qing foreign affairs, went so far as to ask that his obituary end with the following remark: "I don't believe in books; I believe in luck" (*buxin shu xin qiyun*).[102]

Another point to keep in mind is that in Qing China, divination did not stand as starkly opposed to either science or religion as it did in seventeenth- and eighteenth-century Europe. To be sure, Chinese scholars in the late imperial era were well aware of the expanding parameters of knowledge about the natural world, and of an increasing ability to predict accurately. But scientifically minded individuals of the Qing period had neither a religious belief in order of the sort that inspired their European contemporaries, nor did they hold the conviction that in time all phenomena would yield their ultimate secrets. The typical belief was that:

> natural processes wove a pattern of constant relations too subtle and too multivariant to be understood completely by what we would call empirical investigation or mathematical analysis. Scientific explanation merely expressed, for finite and practical human purposes, partial and indirect views of that fabric.[103]

The influential Song dynasty scientist Shen Gua (1031–1095) articulated this notion in the following way:

> Those in the world who speak of the regularities underlying phenomena, it seems, manage to apprehend their crude traces. But these regularities have their very subtle aspect, which those who rely on mathematical astronomy cannot know of. Still, even these are nothing more than traces. As for the spiritual processes described in the *Classic of Changes* ... [which] penetrate every situation in the realm, mere traces have nothing to do with them. This spiritual state by which foreknowledge is attained can hardly be sought through traces, of which in any case only the cruder sort are attainable.[104]

Divination thus was a form of science in Qing times. Although the period witnessed a revival of interest in mathematics, mathematical astronomy, and geography, stimulated in part by the Jesuit educational effort, astrology remained integral to the Chinese scientific tradition. The Jesuits themselves practiced divination in the Imperial Bureau of Astronomy.[105] Chinese fortune-telling techniques, like all such self-confirming methods, could not be falsified, but they at least involved generalizations based on observation as well as a kind of experimentation that required verification (*zheng*). Furthermore, and more importantly, divination categorized and explained experience in culturally significant ways. The elaborate schemes used by fortune-tellers to analyze heavenly phenomena, earthly forms, personality types, and so forth were undoubtedly more generally well known and persuasive than other systems of scientific explanation available in Qing dynasty China. To the extent that science can be viewed as an ordering device for managing data, divination served Chinese scientific purposes nicely. Certainly this was true, for better or worse, of the hallowed *Yijing*.[106]

The spiritual preoccupations of diviners did not necessarily preclude scientific investigation. It is true that experts in wind, rain, and cloud divination—like other types of Chinese fortune-tellers (and most of the rest of the Qing population)—believed in the influence of supernatural forces. But they were also close and insightful students of meteorology. Exponents of *qimen dunjia* and other numerological systems, although concerned primarily with cosmological calculations to determine auspicious times and locations, often studied military science as part of their training, and specialists in military divination (*bingzhan*), for their part, were widely employed.[107] Geomancers used mystical compasses to identify lucky sites and times for building and making repairs, but they also knew a great deal about landforms and hydraulic systems—information of value in public works as well as military affairs. In all, geomancy probably exerted a more profound influence on the physical environment, and the way the Chinese responded to it, than most other natural sciences of the time.[108]

An especially significant reason for the prevalence and persistence of divination in Qing China was the multifaceted social role it played. In the first place, it contributed to social order by regulating the rituals and rhythms of daily life. Few devices were more powerful as mechanisms for structuring society than the stipulations regarding lucky and unlucky days in Chinese calendars and almanacs. Divination also conduced to social stability by depersonalizing difficult decisions and validating status distinctions—most notably in situations related to marriage and mortuary ritual. It is true, of course, that practices such as geomancy could prove socially disruptive, as elite critics had long claimed, but the social advantages of these practices, both for individuals and for corporate entities, far outweighed the liabilities. For this reason, elites themselves availed of divination, including the employment of *fengshui* specialists, under all sorts of circumstances.

Moreover, Chinese fortune-tellers served as the functional equivalent of modern-day psychologists. As therapists and personal counselors, they helped individuals in China to cope with their anxieties, whether inspired by bureaucratic problems, the examination system, or more mundane concerns. Divination clarified the source and nature of difficulties, alleviated doubt, and invested lives with longed-for meaning. It also empowered people with a special kind of cosmic knowledge and perhaps endowed them with greater self-confidence and a more profound sense of personal identity. The optimistic thrust of techniques such as *qian* divination and dream book interpretation provided hope in time of uncertainty and fear, as did geomancy, with its alluring promise of cosmic control.[109] Exponents of fate extrapolation and physiognomy gave individuals a glimpse of their long-term future, as well as concrete advice on how to contend with seemingly inescapable problems. Methods such as word analysis did the same for more immediate issues; personal consultation of the *Yijing* offered ways of "resolving doubts" that emphasized introspection and personal initiative. By various means, not all equally effective, divination in Qing China restored "value and significance to lives in crisis."[110]

But fortune-tellers were more than personal therapists. In a society such as China's, where so many aspects of life and thought hinged on compromise and conciliation, and where intermediaries were essential to all forms of social intercourse, diviners proved to be cultural middlemen *par excellence*, mediating not only between the client and the cosmos, and between Confucian, Buddhist, and Daoist versions of reality, but also between contending elements within their own local communities, from quarreling couples to feuding clans. Some relied primarily on their psychological skills and verbal ability to settle disputes. Others made use of their comparatively intimate knowledge of the personal histories and local connections of many community members.[111] Still others, notably shamans, in calling on supernatural authority to develop and maintain group consensus, brought "order out of confusion."[112]

Diviners also helped bridge the gap between commoners and the elite in Qing dynasty China. By summoning up visions of orthodox heroes and urging clients to embrace conventional values, fortune-tellers reinforced the dominant literati culture. For this reason, among others, they periodically received commendation by local officials for their skill as prognosticators. Peng Tianlun, for one, got an honorific plaque that read, "He is able to know the future" (*keyi qianzhi*)—a phrase instantly recalling the rare and admirable quality of foreknowledge touted by Mencius.[113] But the process of cultural transmission was not one of downward movement only. Popular values also found their way to the upper levels of Chinese society in the course of divination. Although the exact nature of this two-way interaction cannot be documented with precision, we know that at least some fortune-tellers had both elites and commoners as clients and that a number of non-elite diviners enjoyed close relations with the literati by virtue of their unusual and much-coveted skills.[114] Furthermore, it is clear that certain divinatory media, spirit writing in particular, conveyed messages that were suffused with elements from the Chinese folk tradition as well as Confucian high culture.[115]

The special talents of diviners, from weather prediction and siting to the evaluation of personnel, made them valuable not only to Qing officials but also to the leaders of their own communities. Fortune-tellers helped to maintain the mechanisms of local defense and control, undertook famine relief, managed schools, and supervised public works projects. Some used their special talents to help neighbors find lost or stolen property; others provided free advice and medical assistance to their local areas in times of crisis. These altruistic activities helped diviners to overcome the common stereotype of being devious and selfish and brought them more fully into the mainstream of Chinese community life.[116]

Finally, in attempting to account for the remarkable staying power of divination in China, we should bear in mind institutional factors: For one, the Qing religious establishment did not actively attempt to suppress divination in the fashion of the Christian Church in the West. Although Buddhist monastic codes formally prohibited divination by monks, Chinese temples and monasteries supported a wide range of divinatory activity, undertaken by clerics and professional fortune-tellers as well as private individuals. One of the most famous Buddhist monks in late imperial China—Ouyi Zhixu (1599–1655)—engaged in divination throughout his monastic life, using popular mantic texts such as the *Zhancha shane yebao jing* (Sutra for divining the karma of good and evil actions).[117]

Chinese religious establishments did not have the institutional power to challenge established mantic practices, even if they had the will, for the church in China remained ever subordinate to the imperial state. And the Qing government, for its part, reinforced the inherited cosmology and sanctified orthodox mantic practices at all levels. Edicts and other official announcements constantly referred to auspicious and inauspicious dates, times, events, and omens; the state calendar institutionalized divination empire-wide by designating certain days as appropriate and inappropriate for various activities; and bureaucrats availed themselves of divination in all kinds of civil and military situations. Without fully realizing it, the Qing government was as much in the grip of the future as it was of the past. For the emperor to dispense with divination would be to abandon his cosmological claim to kingship—an abdication of his role as mediator between Heaven and Earth.[118]

When the Qing dynasty finally fell in 1911–1912, state-sponsored cosmology suffered a mortal blow. But the overthrow of the Qing and the destruction of the imperial system had far less to do with the decline of correlative cosmology than with the rise of modern Chinese nationalism. Put another way, it was a new, politically inspired assault on the idea of cosmological kingship, rather than the early Qing critique of correlative cosmology, that spelled the end of the official worldview. The production of a new state calendar for the republic in 1912, with its vigorous denunciations of the old Qing calendar's superstitions (*mixin*), does not warrant the conclusion that the inherited cosmology had already died. In fact, the revolutionary government's self-conscious repudiation of the inherited cosmology may be viewed precisely as an effort to eradicate a still-dangerous potential source of imperial authority.[119]

Seen in this light, the assertion that notions such as *yin* and *yang* and *wuxing* were unworthy of attention by "May Fourth period champions of the new culture" because of their "dwindling influence and utility" is misleading.[120] Indeed, one of the most prominent and fierce critics of the "old culture" during the so-called May Fourth Movement (*c.* 1915–1925), Chen Duxiu (1887–1945) wrote in 1918 that only by first casting away heretical practices such as geomancy, fortune-telling, spirit writing, and the use of charms, spells, and alchemy, could the Chinese people begin to put their minds right (*zheng renxin*).[121] And, as late as 1934, the renowned Lu Xun wrote a piece titled "Fate," in which he castigated the Chinese people for their superstitious belief in monks, priests, shamans, astrologers, and geomancers.[122] Chen and Lu Xun were not attacking straw men, for old-style mantic techniques and cosmological assumptions remained in force, not only during the New Culture Movement but throughout the republican era.[123]

Today, in environments such as Hong Kong and Taiwan, fortune-tellers continue to enjoy extraordinary popularity—despite the pervasive influence of modern science and the attractiveness of certain Western intellectual fashions. So do traditional-style almanacs. Jonathan Yeh tells us, for example, that the best-selling book in Taiwan, which "has been at the top of the list for decades if not hundreds of years ... and sells more than six million copies islandwide each year," is the old-style "farmer's almanac" (*nongmin li*).[124] Even on the Chinese Mainland, after four decades of suppression, evidence abounds of traditional forms of divination in cities as well as the countryside.[125] In part, we may assume that the sustained appeal of fortune-tellers is related to the same set of cultural, social, and psychological factors that operated with such force in Qing times. But we may also hypothesize that the old cosmology has remarkable staying power.

Although proportionally there are no doubt fewer Chinese in any one environment today who take the predictions of fortune-tellers as seriously as their forbears in the Qing dynasty, large numbers continue to find attractive the ancient notion of a spiritual resonance between Heaven and Man, a special kind of cosmic connection. And, although some modern Chinese feel the need to choose between what they have come to view as two utterly incompatible cosmologies, many do not.[126] To them, recourse to fortune-tellers and the use of charms are no more inherently unmodern or unscientific activities than the Western habit of praying to God for guidance and assistance.

In other words, the "modern" Western tendency to distinguish between scientific "rationality" and primitive "superstition" may prevent us from appreciating the degree to which the two types of thought might be made to fit under certain sets of circumstances. As Stanley Tambiah and others have pointed out, Western culture has no more of a monopoly on "reason" than so-called "primitive" societies do on mystical intuition. The sharp lines drawn in the early modern (seventeenth century) West between religion and more "primitive" sacramental magic, between prayers and spells, and between sovereign deity and manipulable divine being(s), begin to blur on closer inspection and particularly in cross-cultural perspective.[127] Empiricism has its limits. For example, the discovery by modern astronomers of new stars and planets that have no place in traditional Chinese astrology did not negate the idea of celestial influences on fate, because the vast majority of Chinese star spirits are unembodied, and their ultimate power is determined by an elaborate interplay of cosmic forces that Western science does not recognize. Significantly, many of these forces—notably *yin* and *yang* and the *wuxing*—seem to operate with predictable and powerful effect in traditional Chinese medicine.

Addendum

In the nearly twenty years since I wrote the above essay, a great deal of valuable scholarly work has been done on virtually all facets of Chinese divination, quite apart from the thousands of articles and books published every year on the *Yijing* in various Asian and Western languages.[128] Although comparatively little has been written recently about divination in late imperial times, broad surveys of the various "magical" and mantic arts often give at least some attention to the Yuan, Ming and Qing periods,[129] as do a number of works on specific divinatory techniques, including physiognomy,[130] geomancy,[131] spirit writing,[132] numerology and time selection,[133] astrology,[134] military divination,[135] dream divination,[136] word analysis,[137] and prophetic texts.[138]

For a variety of reasons, the 1990s proved to be a particularly fruitful period for divinatory research on both Taiwan and the Mainland. Although a number of authors in the PRC persist in referring ritualistically to the Chinese mantic arts as "superstition" (*mixin*) or "pseudo-science" (*wei kexue*), they, along with scholars from other parts of Asia and the West, have nonetheless investigated the subject of divination thoroughly and enthusiastically, producing an enormous amount of valuable articles, monographs, reference works, and collectanea.[139] Of particular value to Western scholars of the mantic arts are Chinese-language dictionaries that help to clarify technical terms, esoteric expressions and difficult allusions.[140]

One factor contributing to the burst of interest in Chinese divination from the 1990s onward has been the discovery of new source materials, including a wealth of extraordinarily important archaeological artifacts and written texts dating from the late Zhou and early Qin-Han periods. Another has been a concerted effort on the part of Chinese scholars to gather these materials together in convenient compilations.[141] A particularly valuable work in English that draws heavily upon this recently gathered archaeological evidence is Michael Loewe and Michael Nylan, eds., *China's Early Empires: A Re-appraisal* (2009).[142]

Moreover, during the 1990s and into the first decade of the twentieth century there has been a growing scholarly interest, both in Asia and elsewhere, in Chinese "popular" culture.[143] What this work has done, among other things, is to open up a productive space for investigating the complex interaction between elite and commoner beliefs and practices in China at various times and places. Meanwhile, scholars throughout the world have done a great deal of excellent work in areas of Chinese culture closely related in one way or another to divination, including mathematics, science and technology, medicine and various other "therapeutic arts," including *taiji quan* and *qigong*.[144]

How has the recent explosion of books, articles, compendia and other materials on the Chinese mantic arts shaped the contours of our understanding of late imperial China? Although this question is far too large to answer satisfactorily in brief compass, in the following few pages I would like to make some general observations and also to offer a few preliminary suggestions for further research.

The first observation—somewhat self-serving and personally satisfying, I must admit—is that on the whole, recent scholarship on Chinese divination seems to confirm (or at least not openly dispute) the major points I have tried to make in my earlier writings about the mantic arts in late imperial China:[145] (1) In the Qing, and throughout the imperial era, divination remained a central cultural activity at all levels of Chinese society, operating in a great many realms, from politics and social life to science and medicine; (2) as in the past (for some two thousand years), critiques of correlative cosmology and divinatory practices by Qing dynasty scholars focused rather narrowly on the faults of one or another scheme, rather than on the inherited cosmology as a whole; (3) there were two primary ways of understanding the cosmos and predicting events that operated more or less simultaneously in the minds of Qing scholars—one more "rational" and one more "spiritual," but neither necessarily antithetical to the other; and (4) the tenacity of divination as a social practice in late imperial China was a function not only of its explicitly moral dimension, its generally assumed efficacy, its aesthetic appeal and its psychological role, but also its abundant classical sanction (particularly in the form of the hallowed *Yijing*), its harmonious relationship with institutional religions, and its close association with both state power and "secular" ritual.

But the spate of recent studies on divination in premodern China have clarified many relationships (between people and groups as well as ideas and practices), and have added a great deal of color, complexity, depth and detail to the overall picture—not to mention a much higher level of theoretical and analytical sophistication.[146] This body of work reveals, among other things, that the roots of almost all Chinese divinatory practices and correlative schemes go back much further than many of us thought. Manuscripts and other artifacts discovered at various Zhou, Qin and Han archaeological sites over the last two decades or so—recently published and even more recently analyzed—are fundamentally transforming our understanding of the place of the mantic arts in early imperial China, and their relation to other technical, religious and philosophical traditions.[147]

We have also learned much more about mantic theories and practices in later periods of Chinese history. Consider, for example, efforts to catalogue and study the substantial corpus of Tang and Five Dynasties materials on divination located at Dunhuang.[148] Unlike other large collections of material discovered at this remote site—including Buddhist, Daoist and Confucian texts, encyclopedias, literary compendia and so forth—these resources have never been evaluated in a systematic and comprehensive way. In 2003, however, a team of Chinese and Western scholars, under the general editorial direction of Marc Kalinowski, published an impressive volume titled *Divination et société dans la Chine médiévale: Une étude des manuscrits de Dunhuang de la Bibliothèque nationale de France et de la British Library*. This work catalogues these diverse, fragile and often incomplete Dunhuang divinatory materials (some 250 items in all) in a way consonant with traditional Chinese bibliographical categories, but it also does a great deal more. A model of collaborative scholarship,[149] the book provides a remarkably comprehensive study of the place occupied by divination and the "traditional sciences" in ninth- and tenth-century China.

Each of its ten chapters consists of two parts: (1) an interpretive essay in which the author or authors discuss the background, content and significance of their respective topic or topics, and (2) a detailed content analysis of the texts, together with some useful illustrations of the materials that have been discussed. These materials not only provide us with a wealth of information on little-known mantic practices, such as the "Confucius' Horse Head Method" (*Kongzi matou bufa*), but they also allow us to relate "medieval" divinatory traditions more closely to those of the Qin, Han and Six Dynasties periods. We are thus in a much better position to evaluate critically not only the pioneering work of scholars such as Van Xuyet Ngo (1976) and Kenneth De Woskin (1983), but also later scholarship based on recent discoveries of early manuscripts and other artifacts at Mawangdui, Baoshan, Wangshan, Shuihudi, and elsewhere.[150]

Furthermore, the Dunhuang texts make possible a more sophisticated analysis of the mantic systems that developed in the Song and later periods, a job that remains to be done more thoroughly. By their careful attention to these difficult Dunhuang materials, the authors of *Divination et société dans la Chine médiévale* have already been able to show that certain practices thought to have arisen in the Song actually had their genesis in an earlier period. Moreover, their explanations of these texts have helped to clarify terms and concepts that have perplexed many students of Chinese divination in the past, myself included.

A particularly admirable feature of the Kalinowski project is its effort to elucidate Chinese local history by examining the "social dynamics" of Dunhuang during the late Tang and Five Dynasties era. The various authors show, for example, how diviners operated at the lower levels of Chinese government in this remote area, and how local traditions of divination differed from those operating in the metropolitan bureaucracy and imperial court.[151] These authors also explore the ways divinatory theories and practices were transmitted at the local level and how mantic traditions intersected with institutional Buddhism and Religious Daoism.

Among Western studies of divination and religious life in post-Tang times, Edward Davis's *Society and the Supernatural in Song China* (2001) is noteworthy for its effort to show that "folk" activities, including spirit possession, shamanism, and exorcism, were not confined to the "common" people, and that they were important coping mechanisms for all strata of society.[152] Benjamin Elman's *A Cultural History of Civil Examinations in Late Imperial China* (2000) does something similar for the Ming and Qing periods.[153] Although most of Elman's research, both before and after publication of this book, has focused on the "rational" and "scientific" traditions associated with "evidential research" (*kaozheng xue*), he devotes most of Chapter 6 ("Emotional Anxiety, Dreams of Success, and the Examination Life")—more than seventy pages in all—to the way literati of Ming and Qing times "turned to religion and the mantic arts" in an effort to contend with the pressures of examination competition.

According to Elman, "reading fate" (*kanming*) became "an obsession" for anxiety-ridden Ming-Qing examination candidates, as they "sought some auspicious sign for their prospects in the prohibitive examination market."[154] Among the specific divinatory techniques he discusses are word analysis, dream interpretation, fate extrapolation, day-selection, astrology, numerology, the use of milfoil stalks and spirit writing. The widespread recourse of Chinese scholars to these techniques, together with their indulgence in various Buddhist and Daoist religious practices designed to improve their chances in the examination, indicates to Elman "how far off the historical mark our earlier accounts of 'Confucian agnosticism' or elite areligiosity [*sic*] have been."[155]

Research of this sort reinforces my view that some of the stark dichotomies that have been constructed in the past to make sense out of China's historical and cultural experience may have to be reconsidered. Obviously the line between what "elites" thought (and did) and what "commoners" thought (and did) was seldom sharp and clear. Nor was the divide between what was considered "orthodox" and what was considered "heterodox." Similarly, the distinction between certain "schools" of Chinese thought—for example, Han Learning versus Song Learning—which often appears so neat in Asian and Western works on Chinese intellectual history, blurs substantially on closer examination. So does the contrast between "tradition" and "modernity." In fact, meaningful lines cannot always be drawn clearly and confidently between what is "Chinese" and what is "foreign."[156]

In my opinion, then, we need to devote less attention to pigeon-holing Chinese scholars and more attention to appreciating the full range of their thought, including their inconsistencies.[157] Take, for instance, Fang Yizhi (1611–1671), who is well known as a defender of Western-style science, and a person who "generally accepted European explanations of natural phenomena, such as a spherical earth, Tychonic cosmology, and human physiology." He is also commonly considered to be an opponent of both Cheng-Zhu moralism and Lu-Wang introspection as well as an ardent advocate of "concrete investigations" (*zhice*; lit. "fathoming substance")—an approach to the "investigation of things" (*gewu*) that yielded "a descriptive knowledge of the natural world."[158]

Far less well known, however, is Fang's preoccupation with the highly speculative, cosmologically grounded theories associated with the "images and numbers school" (*xiangshu xue*) of the *Classic of Changes*. Such theories emphasized elaborate correspondences between the various features of the natural world (both physical and metaphysical) and the hexagrams, trigrams, individual lines and diagrams of the *Yijing*. These correspondences included correlations with the cosmic forces of *yin* and *yang*, the five agents, musical notes, heavenly bodies, divisions of time and space, and so forth. According to *xiangshu* theory, numbers—and the images they represented—revealed the principles, relationships and processes that yielded, in turn, not only an understanding of nature but also, ultimately, control over it.[159]

As Zhu Bokun has discussed at length, Yizhi and his father, Fang Kongzhao (fl. *c*. 1630), took it upon themselves to sum up and synthesize the development of "images and numbers" thinking in China from the Northern Song period onward. Their aim was to "comprehend the principles of things" and to "explain natural phenomena"—all "in accordance with the times" and in order "to meet the practical needs of the people."[160] Drawing upon orthodox Cheng-Zhu neo-Confucianism, as well as Western science and the *qi*-based speculations of certain late Ming thinkers, they developed an interpretation of the natural world that relied heavily upon numerologically oriented documents of highly disputed provenance—notably the *Hetu* (Yellow River Chart) and the *Luoshu* (Luo River Writing).[161] In the words of the elder Fang, "everything [in the realms of Heaven and Earth is expressed in] the *Hetu* and *Luoshu*." In the same vein, his son wrote that "the images and numbers of the *Hetu* and *Luoshu* provide verification for all [things]."[162] These views are difficult to reconcile with the idea of "concrete investigations" and "evidential research."

Another fascinating and revealing example of intellectual complexity and flexibility in late imperial China is Jiang Yong (1681–1762)—a wholehearted supporter of Jesuit astronomy and mathematics, a loyal follower of Zhu Xi (although a critic of his mathematics and science), a famous evidential scholar, and an ardent exponent of divination.[163] Unlike many *kaozheng* advocates, including the well-known and influential astronomers, Wang Xichan (1628–1682) and Mei Wending (1633–1721), Jiang did not see the need to privilege Chinese tradition over Western science, or to argue for the "Chinese origins of Western learning" (*Xixue Zhongyuan*). A man of wide-ranging and eclectic interests, such as ritual, phonology, mathematical astronomy and harmonics, he is best known as the author of several influential publications, including the *Lishu gangmu* (An outline of books on ritual) and the *Shuxue* (Study of mathematics). Jiang came from a family associated with popular Buddhism and his chronological biography was actually published by a Buddhist institute in the early twentieth century, but we do not know whether he actually embraced Buddhist beliefs.[164] We can be certain, however, that he practiced divination and that he wrote at least two books on *Yijing*-based numerology.

Jiang's *He Luo jingyun* (Quintessence of the Yellow River Chart and the Luo River Writing; 1756), probably the only one of these two books still extant,[165] reveals his deep commitment to various traditions of "Chinese" learning, several of which bear striking resemblances to the eclectic, *Yijing*-oriented thought of the Jesuit missionary and well-known "Figurist," Joachim Bouvet (1656–1730).[166] As the title suggests, Jiang's *He Luo jingyun* focuses on the two major charts associated with the *Changes*, inspired by Zhou Dunyi's (1017–1073) statement that "the Sage shows the essence (*jing*) of his ideas by drawing the trigrams and hexagrams, and explicates the substance or significance (*yun*) by interpreting the trigrams and hexagrams."[167]

Jiang's work ranges broadly, from *kaozheng*-influenced textual exegesis—he includes, for example, a long and illuminating analysis of *Yijing* use in the *Zuozhuan*—to medicine, mathematics, astrology, geography, and various divination systems, including *qimen dunjia* and *fengshui*. Not surprisingly, given Jiang's broad outlook and interests, among the many illustrations to be found in the *He Luo jingyun* is a comprehensive three-page chart purporting to show that the principles for all things had their source in the *Yellow River Chart*.[168]

Jiang believed that Western science was generally more accurate than Chinese science. This did not mean, however, that he thought Chinese civilization was inferior to that of the West. Quite the contrary. The achievements of Cheng-Zhu Confucianism were enough to establish China's cultural superiority in Jiang's eyes. In his pragmatic view, Western astronomy and mathematics, with all their precision, held the promise of assisting the Chinese in fathoming principles (*li*), thus eventually yielding a fuller understanding of the Confucian Dao. Western astronomy could not, of course, displace Cheng-Zhu orthodoxy in the moral realm, but it could reveal cosmic patterns and regularities more effectively than Chinese astronomy, thus serving a cultural as well as a scientific purpose. To put the matter another way, Jiang's Confucian-based understanding of the broader cultural significance of astronomy led him to defend Western techniques.[169]

John Henderson identifies Jiang as a critic of correlative cosmology, which in some respects he certainly was.[170] But Jiang embraced a number of beliefs, abundantly documented in the *He Luo jingyun* and his other works, that seem utterly incompatible with the Western science he so ardently defended. One was the view, expressed by Mei Wending himself, that China was located on the "face" (*mian*) of the earth, and thus in a superior geographical and cosmological position. Although ideas of this sort, often linked with *fenye* (field allocation) schemes,[171] had been under attack since the early Qing, Jiang claimed that *fenye* divination could be specifically and uniquely applied to China, despite the fact that the constellations on which it was based were shared by all people on Earth. For, he reasoned, just as the *qi* of a human being was manifested in the five facial organs, so must the light of Heaven interact with the *qi* associated with China, the "facial" part of the Earth.[172] In other words, for Jiang, the cultural superiority of China was founded not only on Cheng-Zhu Confucianism but also on the metaphysics of *qi*. This strain of cosmological thinking, also reflected in the "evidential" writings of Jiang's contemporary, Li Fu (1675–1730), proved to be remarkably tenacious, even in late Qing times, as Wang Erhmin has convincingly demonstrated.[173]

For Jiang Yong, as indicated briefly above, the myriad things had their origins in the *Hetu* and *Luoshu*, which he persisted in believing, against mounds of solid *kaozheng* evidence, originated with the early *Classic of Changes*. Jiang notes in his preface to the *He Luo jingyun*:

> The *Hetu*, *Luoshu*, trigrams and hexagrams, and individual lines all emanate from the same source, [they reflect] common trends, and are mutually interactive; hence, concepts such as *gougu* [a traditional system of triangulation] and *chengfang* ["multiplication squares"] in mathematics, the five sounds and six notes (*wuyin liulü*) in music, the positions of the seven luminaries (*qiyao*) in astrology (*tianwen*), the *najia* and *nayin* systems of five agents specialists, the resonant and pure consonants in phonetics, the *li* and *qi* of the geomancers' compass, the *doushou* and *qimen* methods of the "day-selection" (*zeri*) experts, and even the foundations and principles of medicine, including the five movements and six "breaths" of Heaven and the veins of the human body, all emanate from the *Hetu*, *Luoshu*, trigrams, hexagrams, and lines.[174]

In Jiang's view, all celestial and terrestrial phenomena fit into certain mathematical patterns and regularities which, when revealed in the *Hetu* and *Luoshu*, guaranteed the comprehensibility and commensurability of the natural order. Thus, for example, the degrees of the celestial sphere, the zodiacal signs, and the twenty-four fortnightly periods were all derived from the *Hetu* and *Luoshu*. So were mathematical harmonics and the pitch-pipe, which Jiang linked, in turn, with standard units of length, capacity, weight and even money.

Other correlations, including the harmonic sounds (*fansheng*) of the *qin* or "lute," focused on the "two poles" (*liangyi*) of *yin* and *yang*, the "three powers" (*sancai*) of Heaven, Earth and Man, and the five agents (*wuxing*). These, in turn, became linked to a crosswise diagram of the Former Heaven sequence of the eight trigrams (*Xiantian bagua hengtu*), a similar arrangement of the sixty-four hexagrams, the fortnightly periods, and the *nayin* system. Even phonology found its way into Jiang's vast correlative system. Since language was unique to man, it corresponded naturally to the five agents, as manifested in the five major human organs. Not surprisingly, Jiang found a way to connect the thirty-six classificatory characters used in Chinese phonology with the five agents and then integrated this system into the *Hetu*.[175]

In short, from Jiang Yong's standpoint, the myriad things were related to each other as part of a comprehensible Dao that was all-encompassing. Their mutual relationships allowed him to remove the social and intellectual boundaries between Chinese and Western science and to define their range of applications. Thus, as Chu Pingyi has insightfully observed, Jiang did not recommend that the traditional Chinese calendar be abolished, nor did he argue for the universal validity of Western calendrical methods. To Jiang, these different systems of knowledge were all parts of the Dao, each with its own application. If the goal was to make an accurate calendar, nothing was better than Western astronomy. But, if one sought to understand the classics, ancient Chinese techniques and strategies were naturally best.[176]

Of course, a great many far more detailed studies of individuals such as Jiang Yong will have to be undertaken before we can generalize with any confidence about how the scholars of premodern China viewed the cosmos and their place within it. But looking again at Yuan Shushan's monumental *Zhongguo lidai buren zhuan* (Biographies of diviners in China by period) and other such sources, it is quite evident that well after the allegedly devastating critiques of scholars such as Huang Zongxi, Hu Wei and Mao Qiling, *xiangshu* techniques associated with charts such as the *Hetu* and *Luoshu* continued to be used by a wide variety of people, many of them scholars, in a wide variety of ways.[177]

In the meantime, we might explore comparative questions of the sort I raised at the end of *Fortune-tellers and Philosophers*: What would a systematic comparison of the attitudes toward divination on the part of such highly influential thinkers as Socrates and Zhu Xi (both of whom believed in the limited use of oracles) yield? How do Judaic, Christian and Islamic approaches to divination compare historically with those of Confucians, Daoists and Buddhists? To what extent can we compare the social purposes of divination in the urban and rural environments of Asia, Africa, Europe and the Americas, both past and present? What affinities does Chinese dream divination share with the phenomenon of induced dreams in the ancient civilizations of Egypt and the Middle East or in more recent cultures? How closely in both theory and practice does the astrological matching of horoscopes for prospective marriage partners in India correspond to the Chinese system? In what particular ways does Ifa divination in Nigeria, with its spiritual framework, moral concerns, poetic verses, and binary divinatory ritual, resemble use of the *Yijing* and related Chinese mantic techniques?[178] Answers to some of these questions have begun to emerge in recent scholarship,[179] but much more remains to be done.

In the past decade or so a great deal of material has been published on the relationship between Daoism and the mantic arts in various periods of Chinese history,[180] but comparatively little published work has been done on Buddhism's relationship with divination—aside, of course, from some excellent work on Tibetan fortune-telling and the pioneering work done by the Dunhuang team discussed above.[181] There are a few recent studies from the Mainland that touch on the divinatory practices of China's "other" ethnic minorities, for example, the Miao, but they are generally rather superficial and often exhibit a rather patronizing tone, despite protests to the contrary.[182]

What other topics might be explored? Certainly more attention should be given to the relationship between divination, ritual and other kinds of performance, as well as games and gambling, in late imperial times.[183] We also need to look at divination more closely from the standpoint of its social, political, economic, cultural, and psychological role; I am under no illusion that I have come anywhere close to covering the ground, even for the Qing. Overall, we still know more about divinatory texts than we know about the people who used them. We should thus take much more fully into account the education and training of diviners, the background and interests of their clients, the political and social problems fortune-tellers encountered (and created), the criticisms they endured, the debates they engaged in, the discursive strategies they employed, and the personal connections they tried to cultivate and maintain.

I am quite certain, for instance, that a biography of the prolific twentieth-century scholar/diviner, Yuan Shusan, would shed much useful light on such issues, as would biographies of practitioners in earlier periods. And now that Yuan's *Zhongguo lidai buren zhuan* is more readily available in the form of a Taiwan reprint, there is simply no excuse for scholars to continue to ignore it. Having gone through more than twelve hundred of his biographical entries at one time or another, I can give assurances that there is still much to learn from Yuan's life and from the lives of those individuals who appear in his many stimulating works.[184]

6

JESUIT INTERPRETATIONS OF THE *YIJING* IN GLOBAL PERSPECTIVE

Preliminary remarks

This chapter argues that the use of the *Changes* by the Jesuits in eighteenth-century China, like the "accommodation strategy" of the Society of Jesus more generally, should be viewed in broad historical and comparative perspective. What I seek to show is that the Jesuit effort to emphasize affinities between the Bible and the *Yijing* in Qing dynasty China was part of a much larger process by which the *Changes* came to be transmitted to other cultures, and that this process of transmission and transformation, sometimes described as "globalization,"[1] sheds useful light on questions of cross-cultural contact and cross-cultural understanding. Although the Jesuit hermeneutical strategy described as "Figurism"[2] was severely condemned by the other Catholic orders, eventually proscribed by the Church, and maligned even within the Jesuit establishment itself, this interpretive approach was part of a long tradition of *Yijing* exegesis and textual transmission—one that not only pre-dated the Jesuits by several centuries but also proved remarkably tenacious well after the Society of Jesus had been disbanded. Indeed, evidence of its tenacity can still be found today, both East and West.

Introduction

During the past three thousand years or so, the *Yijing* or *Classic of Changes* has gradually become a global property. By stages, in a process that is still rather poorly understood, the *Changes* gradually spread from China to other realms of East Asia—notably Japan, Korea, Annam (Vietnam) and Tibet. The Jesuits brought knowledge of the classic to Europe during the eighteenth century, and from there it traveled to the Americas, finding a particularly receptive audience in the United States from the 1960s onward. A recent annotated bibliography on the *Yijing* lists more than 500 books and dissertations and about the same number of articles devoted to the *Changes* in English alone.[3] Meanwhile, in Asia itself, the work continues to flourish, even on the Chinese Mainland, where an *Yijing*-fever (*Yijing re*) erupted in the 1980s and continues to this day. Every year for the past two decades or so, literally thousands of books, articles, essays and conference papers have been written on the *Changes* in dozens of different languages. How do we account for the remarkable transnational spread and enduring popularity of this document, both East and West?

Clearly the "globalization" of the *Yijing* was in part the product of its alluring "special features" (*tese*, *tezhi*, etc.): its exalted position as "the first of the [Chinese] classics;" its challenging basic text; its elaborate numerology and other forms of symbolic representation; its utility as a tool of divination; its philosophically sophisticated commentaries; its psychological potential (as a means of attaining self-knowledge); and its reputation for a kind of encyclopedic comprehensiveness. The spread of the *Yijing* was also facilitated by the self-concious strategies employed by those who sought to use it in various environments for their own political, social, intellectual or evangelical purposes. In the process the *Changes* invariably became "domesticated," but not always in ways that its advocates might have originally envisioned.

The Jesuit approach to the *Yijing* provides a particularly interesting glimpse into the process of globalization—first, because it involved not only the transmission of the Chinese classic to the West but also the transmission of the Bible to China through the *Changes* (and other canonical writings); and second, because the Jesuit strategy had certain striking unintended consequences. That is, although the Jesuits developed certain positive interpretive strategies that appealed to kindred spirits both in China and Europe, by engendering a powerful negative reaction in each place, they also helped to shape in significant ways alternative approaches to the *Yijing*.

Before exploring the specific contributions of the Jesuits, and the later history of their "biblical" approach to the classic, it may be useful to review briefly a few basic features of *Yijing* hermeneutics.

Interpreting the *Changes* in late imperial times

As indicated in the opening chapter of this book, the extraordinarily ambiguous language of the *Yijing*'s "basic text" gave interpreters of the document wide latitude throughout the imperial era. Chinese scholars argued endlessly over how to understand hexagram names, how to interpret various images and symbols, how to gloss individual words and phrases, and even how to punctuate the text. Virtually no passage of the *Changes* was so clear as to preclude debate. Consider the judgment and line statements of the Gou hexagram (number 44), translated below according to one prominent Song dynasty understanding of the text.

> [Gou, Coming to Meet] shows a woman who is bold and strong. It will not be good to marry (such) a woman. The first line shows how the subject (of the hexagram) should be kept (like a carriage) tied and fastened to a

> metal drag, in which case with firm correctness [*zhen*] there will be good fortune. (But) if he moves in any direction, evil will appear. He will be (like) a lean pig in captivity which is sure to keep jumping about. The second line shows its subject with a container of fish. There will be no error, but it will not be well to let (the subject of the first line) go forward to the guests. The third line shows one from whose buttocks the skin has been stripped so that he walks with difficulty. The position is perilous, but there will be no great error. The fourth line shows its subject with a container, but no fish in it. This will give rise to evil. The fifth line (shows its subject as) a melon gourd covered by a willow. If he keeps his brilliant qualities [lit. jade talisman (*zhang*)] concealed, (something good) will descend from Heaven. The sixth line shows its subject locking horns with others. There will be occasion for regret, but no error.[4]

Virtually every substantive word in this passage is susceptible to multiple readings. This ambiguity unquestionably contributes to the overall richness and subtlety of the text, but it also complicates the search for meaning—not to mention the problem of adequate translation. Take the phrase: "He will be (like) a lean pig in captivity which is sure to keep jumping about" (*leishi fu zhizhu*). Putting aside the animal symbolism, which is itself quite problematic, we are still left with a sense of uncertainty about what is actually going on. *Zhi* ("to walk"), when pronounced *di*, refers to the hoof of a pig, while *zhu* can mean both to limp and to amble, as a horse might. In Qing dynasty usage, *zhizhu* connotes embarrassment, or doubt over what to do. The term *lei* ("lean" or "emaciated") may be a loan word for *lei* ("to tie with a rope"), which would, of course, reinforce the notion of captivity; but, as the late Qing scholar-official Yu Yue (1821–1907) pointed out, *fu* (captive) may well be a loan word for *ru* ("to suckle")—in which case the meaning of the passage (a pig suckling its young) would change considerably, even if *lei* were still understood as a loan word.[5]

Let us take another less complicated but still problematical example: The third line statement of the Meng hexagram (number 4). A pre-imperial understanding of this admonition was probably something like "Don't take the woman as a wife. [She] sees a man of metal; [he/she] has no body. Nothing about this is favorable." The critical expressions here are "man of "metal" (*jinren*) and "has no body" (*buyou gong*). How could a man be "made of metal," and what does it mean to "have no body"? According to a third-century commentary by Wang Bi (226–249), these two phrases refer to: (1) a man who is "strong as metal;" and (2) a woman who "discards her self-possession"—in other words, one who "takes the lead in seeking the man." In Wang's eyes, this phrase strongly suggests that such an aggressive person would turn out to be a disobedient wife. Cheng Yi (1033–1107) provides an even more negative interpretation of the woman in question. In his view, *jinren* refers to a "wealthy man" (*jin* can mean both "gold" and "metal"), whom the woman wants only for his money. And since she gravitates to the convenient *yang* line occupying the second place rather than the more appropriate *yang* line in the sixth place (these are the only two *yang* lines in the hexagram), Cheng considers her to be not only "greedy" but also "opportunistic."[6] How have Wang and Cheng come to these conclusions?

In attempting to resolve textual questions of this sort, Chinese intellectuals in late imperial times mined every available reference work, from dictionaries and encyclopedias to compilations such as Li Dingzuo's (eighth-century) *Zhouyi jijie* (Collected explanations of the *Zhou Changes*), which focused on Han dynasty interpretations of the classic, and Li Guangdi's (1642–1718) massive *Zhouyi zhezhong* (A balanced [edition of the] *Zhou Changes*; 1715), which contains more than 200 pre-Qing commentaries on the *Yijing*—including eighteen from the Han, five from the Six Dynasties, eleven from the Tang, ninety-eight from the Song, twenty-two from the Yuan, and sixty-one from the Ming.[7]

One of the most important resources for understanding early Chinese texts such as the *Changes* was a Han dynasty work, Xu Shen's (*c.* 58–147 CE) *Shuowen jiezi* (Explanation of characters for the discussion of writing). This highly influential etymological dictionary, which makes a special point of tracing the invention of Chinese writing to Fuxi's eight trigrams, analyzes nearly ten thousand characters by dividing them into their constituent parts. Of the several techniques Xu employs to this end, the one known as *huiyi* (combining ideas) is of particular relevance to this chapter.

The hallmark of *huiyi* analysis is the division of a written character into its constituent elements, followed by an attempt to show how the relationship between these elements reveals the meaning of the word. For example, the *Shuowen jiezi* explains that the Chinese character for "to split" 析 (*xi*) is "derived from *mu* 木 (wood) and *jin* 斤 (axe)," and that the word for "faithful" or "sincere" 信 (*xin*) is "derived from *ren* 人 (human being) and *yan* 言 (words)"—i.e. a person standing by his or her word. Not surprisingly, Chinese commentators on the *Yijing* used a similar relational logic in analyzing or explaining trigram and hexagram names. For instance, the trigram/hexagram pronounced Dui was often associated with verbal communication because the character *dui* 兌 forms part of the word *shuo* 說, "to speak" or "to explain." This sort of word-analysis had its counterpart in Chinese popular culture in the widespread divining practice known as "dissecting" (*chai*) or "fathoming" (*ce*) characters, but even elites often used this interpretive approach in their essays as a rhetorical device.[8]

As indicated briefly in the initial chapter of this book, the open-ended nature of so many passages in the *Changes* encouraged Chinese thinkers of nearly every philosophical and religious persuasion to cite it for authority and also to comment upon it. From the Wei-Jin period (265–420 CE) onward, for example, Buddhist apologists often used the trigrams, hexagrams and commentaries of the *Yijing* to explain their ideas. Perhaps the best-known example in late imperial times is Ouyi Zhixu's (1599–1655) *Zhouyi Chanjie* (A Chan [Zen] interpretation of the *Zhou Changes*), which often analyzes hexagrams in terms of the various states of *dhyana* (meditation, absorption, etc.) and *prajna* (wisdom, insight, etc.).[9] In the process, Zhixu added significantly to the symbolic repertoire of each hexagram. For example, in discussing the initial hexagram of the *Yijing* he tells us that:

> Qian means strength. In Heaven it is [i.e. symbolizes] *yang* [or the sun]; on Earth it is firmness [*gang*]. In human beings it is intelligence [*zhi*] and a sense of duty [*yi*]; with respect to the essence of mind [*xing*; i.e. the Buddha nature] it is clear light [*zhao*; i.e. *prabhaasvara*]; in spiritual practice it is observation [*guan*; i.e. visualization]. In the physical world it is what provides cover [*fou* (for the sentient beings who inhabit it?)]. In the physical body it is the head, the higher ruler. In the family it is the patriarch; in the country it is the king; in the empire it is the emperor.[10]

Advocates of Religious Daoism, Judaism and Islam in China made similar efforts to promote their respective religious traditions using the authority of the *Changes*.[11] As one of many examples, Ming dynasty inscriptions on a stele from the synagogue at Kaifeng suggest that the ancient Hebrew script was derived from the signs and diagrams developed by the great Chinese cultural heroes Fuxi and Yu in ancient times.[12] But a shared appreciation for the *Yijing* as a classic text did not necessarily lead to any sort of philosophical agreement. Buddhists and Confucians often viewed the *Changes* in radically different ways. For instance, the former tended to see the Fu hexagram ("Returning;" number 24) as a reference to the notion of achieving *samadhi* (access to *dhyana*) through a process of "returning, extinguishing and resting,"[13] while the latter generally considered the hexagram to be symbol for the eternal process of "producing and reproducing" (*shengsheng*)—a gloss designed explicitly to undermine the Buddhist concept of "emptiness" (*kong*) in favor of neo-Confucian "principle" (*li*).[14]

Despite such differences, the prestige of the *Yijing* grew ever greater from the Song period onward, inspiring thousands of essays, inscriptions, memorials, eulogies, poems and works of rhyme-prose (*fu*), as well as numerous works of art, from sophisticated literati paintings to popular crafts. It also provided a rich analytical vocabulary, based in part on trigram and hexagram symbolism, that proved extraordinarily serviceable in all realms of Chinese political, social, scientific, artistic and literary life.[15] We should not be surprised to find, then, that during the imperial era the *Yijing* exerted a profound cultural influence upon several other countries on the periphery of China.

The *Changes* in East Asia and Tibet

Although the specific circumstances under which the *Changes* found its way to these countries naturally differed, there seem to be certain common patterns in the way that it traveled. In the first place, because elites in Korea, Japan, and Vietnam were familiar with the classical Chinese language, there was no need to "translate" it—except, perhaps, to render it in vernacular prose to make it accessible to commoners. And since the *Yijing* continued to occupy an exalted position in China for some two thousand years, there was never a time when it lacked prestige in these peripheral areas. Initially, scholars in each of the three societies embraced the *Changes*, using the work for their own varied purposes. Then other sectors of society followed suit. In this way, the *Yijing* gradually became "domesticated," undergoing sometimes radical transformations in the process.[16]

Japan provides an excellent illustration of these themes. The *Yijing* found its way to Japan as early as the sixth century CE, but it was not until the seventeenth century that interest in the document blossomed. From the beginning of the Tokugawa shogunate in 1603, to the fall of the regime in 1868, more than a thousand books were written on the *Changes*—a number that is not much less than the total number of books written on the *Yijing* during the more-or-less contemporary Qing period in China—which had a population about fifteen times greater than Japan's. The *Yijing* was employed primarily to bolster and amplify Tokugawa Confucianism, but it was also used to validate or undergird other Japanese cultural traditions—including both "native" Shinto and "borrowed" Buddhism.[17]

As a result, the *Changes* penetrated all levels of Japanese society. As in China, political leaders regularly sought its spiritual and practical guidance and elites studied it assiduously. Manuals for artisans explained crafts in terms of the *Yijing*, and in polite society its symbolism appeared everywhere—from artistic, literary and musical criticism to popular drama, the tea ceremony, flower arranging and even board games. It also played a significant role in Japanese science, medicine, military affairs and martial arts.[18]

Over time, the *Yijing* became increasingly assimilated to the indigenous culture of Japan, at least in some circles. Thus we find Jiun Sonja (1718–1804), a Shinto priest, arguing that:

> The images of the River Chart [*Hetu*, which by some accounts provided the model for the eight trigrams], were manifested through the Okitsu Mirror [a round bronze object kept at the sacred Ise shrine] … [and the authors of the *Changes*] copied our ancient divination of Takam-ga-hara [the plain of the high heaven where Izanagi and Izanami lived] in formulating its text and style. The whole book is completely borrowed from us [the Japanese].[19]

Hirata Atsutane (1776–1843), for his part, went so far as to assert that the ancient Chinese culture hero Fuxi, putative inventor of the trigrams, was actually a Shinto deity.[20]

Like Hirata, and perhaps influenced by him, the nationalistic Korean scholar, Sin Ch'aeho (1880–1936), attempted to "domesticate" the *Yijing* in a similar way, arguing, on the basis of forged texts, that Fuxi was in fact a Korean prince who had learned the *Changes* from Hang Wong, an early Hangguk ruler. And even earlier, in a similar expression of cultural pride, Chong Yagyong (1762–1836), suggested that the brother of the last Shang dynasty king, the Viscount of Ji (Jizi)—who allegedly left China for Korea and was viewed by many Koreans as a kind of ancient "patriarch"—might have written part of the basic text of the *Changes*. Another strategy of domestication in Korea was to invent a book derived from, but different than, the *Yijing*—rather like Yang Xiong's (53 BCE–18 CE) *Taixuan jing* (Classic of great mystery), or the apocryphal Han treatise known as the *Qianzuodu* (A penetration of the laws of Qian) in the Chinese tradition. The most prominent example in Korean history is the nineteenth-century derivative work known as the *Correct Changes* (Chongyok).[21]

As yet, I have found very little information on such strategies of domestication in pre-modern Vietnam, although a contemporary Vietnamese scholar, Nguyen Vu Tuan Anh, has recently suggested the possibility of Lac Viet (Ancient Vietnamese) origins for certain important features of *Yijing* interpretation, including a unique Lac Viet version of the Later Heaven sequence of the eight trigrams.[22] In any case, it is clear that use of the indigenously developed *Nom* script in works such Dang Thai Phuong's *Chu dich quoi am ca* (Songs of the *Zhou Changes* in national pronunciation; 1815) had the effect of domesticating the document, even though such works explained the cryptic text of the *Yijing* explicitly in terms of the Chinese exegetical tradition known as *xungu*. As the preface of Dang's book puts the matter, "Our [Confucian] learning in Vietnam is the same as that of the Chinese, but our pronunciation [of the words in the texts] is different."[23]

In many respects, pre-modern Korean and Vietnamese intellectuals approached the *Yijing* in the same spirit as Japanese scholars. In both of these cultural environments, the work retained its aura as a venerated Chinese classic; and in both it had wide application at all levels of society as an explanatory device, extending into the realms of language, philosophy, religion, art, literature, science, medicine, and social customs.[24] In both Yi dynasty Korea (1392–1910) and Le dynasty Vietnam (1428–1789), scholars seem to have appreciated the Images and Numbers (*xiangshu*) approach to the *Changes* more than the Meanings and Principles (*yili*) approach, but, as I have indicated in the first chapter of this volume, the sharp distinctions that are often drawn between these two exegetical traditions tend to blur significantly when individual thinkers and their texts are examined closely.[25] This is as true for Japanese, Korean and Vietnamese approaches to the *Yijing* as it is for Chinese approaches.

In the case of Tibet, the process of transmission involved substantial modification—in part, no doubt, because unlike Japanese, Korean and Vietnamese elites, most Tibetans knew little or no classical Chinese. The *Yijing* (*Yigying* in Tibetan) first came to Tibet as a respected Chinese "classic" during the early Tang dynasty, in the seventh century CE. Soon thereafter, Tibetan diviners began using the trigrams of the *Changes* in more or less the Chinese fashion. Later they also borrowed some of the numerological diagrams of the *Yijing*—notably the *Luoshu* (Luo River Writing). But they also created new divinatory symbols, including four-lined tetragrams (there was, of course, a Chinese precedent for this sort of permutation with Yang Xiong's well-known *Taixuan jing* (Classic of supreme mystery), which was based on five-lined pentagrams. Like the Japanese and Koreans, the Tibetans seem to have been particularly eager to assimilate the *Yijing* to their indigenous culture. Some Tibetan commentators emphasized affinities between the *Yijing* and Tantric Buddhism, and other scholars in both the Buddhist and Bon traditions transformed Confucius—the putative transmitter of *Yijing* divination (and other forms of fortune-telling) into their own religious figures. Moreover, in at least some cases, the eight trigrams acquired significantly different symbolic identifications in Tibet than their traditional Chinese ones. Zhen, for example, usually associated with Thunder, came to be linked in certain Tibetan divination systems with "meteoric iron."[26]

A distinctive feature of the process by which various East Asian peoples borrowed from Chinese culture was their periodic use of emissaries—individuals and groups who visited China and brought back Chinese texts and traditions to their home countries in a self-conscious and sometimes quite systematic way. Westerners, too, sent missions to China, and they brought back all kinds of information, but these missions proceeded from very different motives, and had a different focus from their East Asian counterparts.

The Jesuit enterprise

The Jesuit missionaries shouldered a heavy burden. Their primary duty was, of course, to bring Christianity to China (and to other parts of the world), but they also had to justify their evangelical methods to their colleagues and superiors in Europe (including, of course, Rome). A kind of "double domestication" thus took place. In China, the Jesuits had to make the Bible appear familiar to the Chinese, while in Europe they had to make Chinese works such as the *Yijing* appear familiar (or at least reasonable) to Europeans.[27] In neither case were they ultimately successful, but not for lack of effort.

As is well known, from the time of Matteo Ricci onward, the strategy of "cultural assimilation" involved assiduous study of the Chinese language, as well as a concerted effort to learn as much as possible about Chinese philosophy, religion, literature and customs.[28] The Jesuits also engaged in some inventive strategies of accommodation, such as Giulio Aleni's attempt to equate the City Gods (*chenghuang*) of Chinese tradition with guardian angels.[29] But recent scholarship on the Jesuits reveals that their encounter with Chinese civilization involved far more complex cultural negotiations than the standard narrative of "adaptation" and "acculturation" suggests—negotiations involving the individual personalities, educational backgrounds and national identities of the missionaries as well as the elaborate interplay between the "corporate culture" of the Society of Jesus and Chinese elite and popular culture.[30]

One important factor in this process, which deserves further study, was the ability of particular individuals to satisfy the specific political, intellectual, and psychological needs of the Chinese throne. This had been the case, for example, with the Shunzhi emperor (r. 1644–1661), whose spiritual "grandfather," Johann Adam Schall von Bell (1592–1666), reportedly brought the young Qing dynasty ruler closer to conversion than any other Chinese emperor.[31] In the case of the great Kangxi emperor (r. 1662–1722), it is quite clear that his deep interest in mathematics, science and cartography, together with his particular personal and intellectual preoccupation with the *Yijing* as a practical divinatory tool and a book of Confucian wisdom, gave individuals such as Father Joachim Bouvet (1656–1730) extraordinary leverage with the Manchu monarch—at least for a while.

In 1688, Bouvet arrived in Beijing, and two years later he entered the service of the Kangxi emperor, becoming one of his mathematics instructors and spending as much as two hours a day tutoring the Qing monarch in algebra and geometry. Along the way he learned Chinese and Manchu, translating technical tracts and participating in various mathematical, scientific and cartographic activities.[32]

Before leaving for Asia, Bouvet had already laid the groundwork for an approach to esoteric knowledge that the Kangxi emperor would find congenial. Bouvet tells us:

> I brought with me [to China] a special knowledge of the Hebrew Mosaic cabbala [kaballa, etc.], and of the Pythagorean and Platonic philosophy, which are the true elements of the whole hieroglyphic wisdom of the Chinese, or rather [that] of the Old Patriarchs.[33]

Thus armed, Bouvet became the initiator and a leading exponent of "Figurism" in China (also known as "Bouvetism," "Symbolism," "Enochism," "Mythologism," and "Kinism" ["Jingism," from the common Jesuit transliteration of the Chinese term for "classic," "*jing*"]).[34] Other prominent Figurists in China during the early Qing period were Jean-François Fouquet (1665–1741), Joseph de Prémare (1666–1736) and Jean-Alexis de Gollet (1664–1741).

As Claudia von Collani and others have noted, the "Figurist" approach to the Bible was based on three interpretive traditions within European theology: (1) typological exegesis, designed to reveal hidden meanings in the Old Testament that unlocked the mysteries of the New Testament (Witsius Cocceius [d. 1669] and his followers are a European example of this sort of orientation); (2) "Ancient Theology" (*prisca theologica*), predicated upon the idea of a "divine revelation" by pagan saints (including Melchizedek, the Queen of Sheba, the Three Wise Men from the East, Pythagoras, Socrates, Plato, Orpheus, Zoroaster and others); and (3) the Jewish mystical tradition of the Kabbala, which, during the Renaissance, had produced, among its various permutations, a Judaeo-Christian version that drew upon Ancient Theology and neo-Platonism.[35]

The writings of Bouvet, Foucquet and Prémare provide numerous examples of their Figurist approach to the Confucian classics, which often involved rather far-fetched etymological speculations. Dissection of the Chinese character for Heaven 天 (the number two 二 combined with the word for Man 人), for instance, indicated, among other things, a prophecy of the second Adam, Jesus Christ; the character for boat, *chuan*, could be broken down conveniently into the semantic indicator for a "vessel that travels on water" 舟 (the character *zhou*) on the left side, and the characters for "eight" 八 and "mouth(s)" 口 on the right—signifying China's early awareness of Noah's Ark, which contained, of course, the eight members of Noah's family.[36] In Figurist discourse, a wide variety of Chinese philosophical terms closely associated with the *Changes* came to be equated with the Christian conception of God, including not only Tian and Shangdi, but also Taiji, Wuji, Taiyi, Dao, Li and even *yin* and *yang*.

Certain trigrams and hexagrams were similarly revealing. The three solid lines of the Qian (Heaven) trigram, like the three "ones" of the character for three

三 (*san*), indicated an early Chinese awareness of the Trinity. The first hexagram, Qian, naturally referred to Creation, and the hexagram Xu (Waiting, number 5), with its stark reference to "clouds rising up to Heaven" (in the Commentary on the Big Images), symbolized the "glorious ascent of the Savior." The hexagrams Bi (number 12, Obstruction) and Tai (number 11, Peace) referred, respectively to "the world corrupted by sin" and "the world restored by the Incarnation," and so on.[37] Efforts to link Chinese cultural heroes to biblical figures produced a number of creative connections: Peng the Ancestor (Pengzu), for example, became Adam, and Fuxi, putative inventor of the eight trigrams, was the mysterious Patriarch Enoch, who reportedly "walked with God," learning about creation, mathematics, astronomy and the end of the world.[38]

The effort by Bouvet and his colleagues to find hidden biblical messages in the *Yijing* was fully consistent not only with the belief of contemporary Chinese *kaozheng* scholars that the *Changes* held the key to an "authentic reconstruction of antiquity," but also with the spirit of the Great Commentary, which described the *Yijing*'s discursive style as one that "twists and turns but hits the mark" (*yan qu er zhong*)." The things and events dealt with, the commentary states, "are obviously set forth, but hidden implications are involved."[39] Bouvet's steadfast goal was to recover the "original meaning" of the *Changes*, and the hexagrams, trigrams and cryptic judgments and line statements of the "basic text" gave him a great deal of interpretive latitude.

Of all the Figurists (about one-third of all French Jesuits in China eventually fell into this category), none enjoyed the emperor's favor as much as Bouvet, and none became as preoccupied with the *Changes* and with prophecy. It is not clear exactly when Bouvet began to study the *Yijing* in earnest (probably in the early 1690s), but we can assume that familiarity with the classic and its commentaries took him a number of years, even taking into account the French Jesuit's sharp intellect and the assistance of dedicated individuals such as the Catholic convert and *juren* scholar, Lu Ruohan, who helped Bouvet render his ideas on the *Changes* into classical Chinese.[40]

In focusing his attention primarily on the imagery, allusions and numerology of the *Yijing*, Bouvet was following a path blazed by Chinese Christian writers such as the late Ming convert, Shao Fuzhong (*jinshi*, 1596), whose book, *Tianxue shuo* [On the Heavenly learning], draws upon the Great Commentary, hexagram analysis and the writings of Shao Yong and others in comparing concepts and images in the *Yijing* with various Catholic doctrines such as the Trinity and the Immaculate Conception.[41] Other Chinese scholars, including a Fujian Christian named Zhang Geng and Bouvet's rough contemporary, Zhou Zhi, wrote similar tracts identifying affinities between Catholic theology and the *Changes*.[42]

Bouvet's effort to link the religious traditions of China and the West through the *Classic of Changes* can be found not only in his correspondence with various European scholars (see next section) but also in his many Chinese-language writings—most of which seem to date from 1711 to 1716. One succinct statement of his major theme appears near the beginning of his *Yigao* (Draft [essay] on the

Changes): "The inner ideas [*neiyi*] of the *Yijing* are very similar to the teachings of Christianity."[43]

In a preface to his *Yiyao* (Key to the *Changes*)—after emphasizing not only the extraordinary comprehensiveness of the classic but also the role it can play in guiding people to the kind of behavior that leads to a "heavenly reward"—Bouvet draws on the "Great Commentary" to make his case for the unity of all mankind:

> The people of China and of the four corners of the world originally came from one ancestor and are [thus] brothers. They are all born of the same great parent [*da fumu*], who gave birth to Heaven, Earth and Man ... Since all people belong to one family and are connected to the one great parent, they should all follow the same Way, the same principle, the same learning and the same teachers ... From this [line of reasoning, we can see that] the Chinese and foreigners are all connected to the Heavenly Lord. If the ancient Chinese had the true tradition of Heavenly learning [*tianxue*] and the learning of the mind [*xinxue*], how could it be that the myriad states [outside of China] in the four corners of the world [have no such teachings and beliefs] at all?[44]

He goes on to argue that the truths of the *Changes* and the Bible are essentially the same, and he devotes the rest of his essay to a more or less systematic effort to demonstrate this point.

Bouvet's *Yikao* (Examination of the *Changes*; also called *Yiyin* [Introduction to the *Changes*]) repeats several of the arguments contained in the *Yiyao*, making the point that although the West did not ever have a document with exactly the same "illustrations and script" (*tuwen*) as the *Yijing*, it did have similar texts and illustrations, which contained the same subtle messages about the beginning and end of the universe. This work also draws upon standard Chinese reference works, such as Ma Duanlin's famous *Wenxian tongkao* (Comprehensive examination of [Chinese] literature; 1224), to demonstrate early linkages between China and ancient Rome.[45]

In his *Yiyin yuangao* (Original draft of an introduction to the *Changes*), Bouvet argues explicitly that Catholicism is not simply the religion of Westerners; it is also an authentic teaching "passed down by the former Sage [Confucius]." People therefore ought to follow it, he says—Chinese and Westerners alike. If they do, "no one will fall" (*xiazhui*); all will presumably be saved. This teaching, Bouvet argues, will be to the great advantage not only of people from afar, but also to the Chinese themselves.[46]

One of Bouvet's greatest and most persistent desires was to demonstrate a relationship between the numbers and diagrams of the *Yijing* (especially as expressed in the *Hetu* and *Luoshu*) (Figures 6.1 and 6.2) and the systems of Pythagoras, the neo-Platonists, and the Kabbalah.[47] This is evident not only in his Chinese-language writings, but also in his broad-ranging Latin manuscripts. For example, in identifying the black and white Ain Soph ["God"] symbol at the top of the Ten

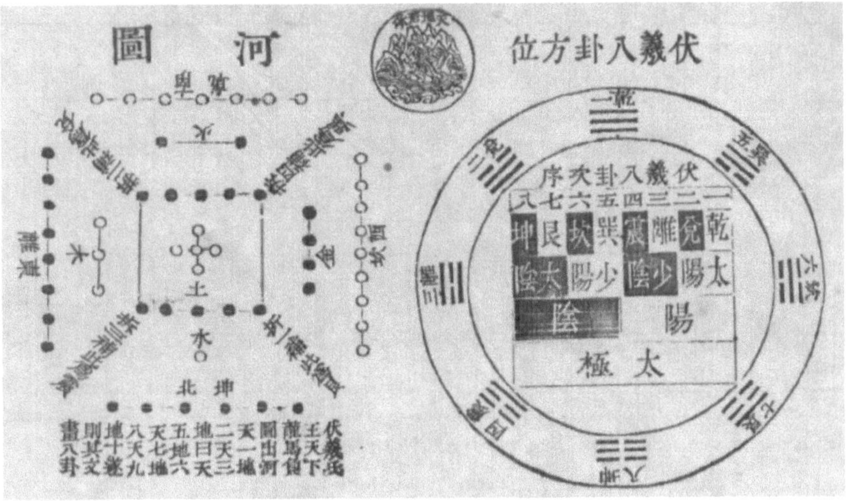

FIGURE 6.1 The numerology of the *Hetu* and the Former Heaven sequence of the eight trigrams from Illustrations of the Unity of the Three Powers

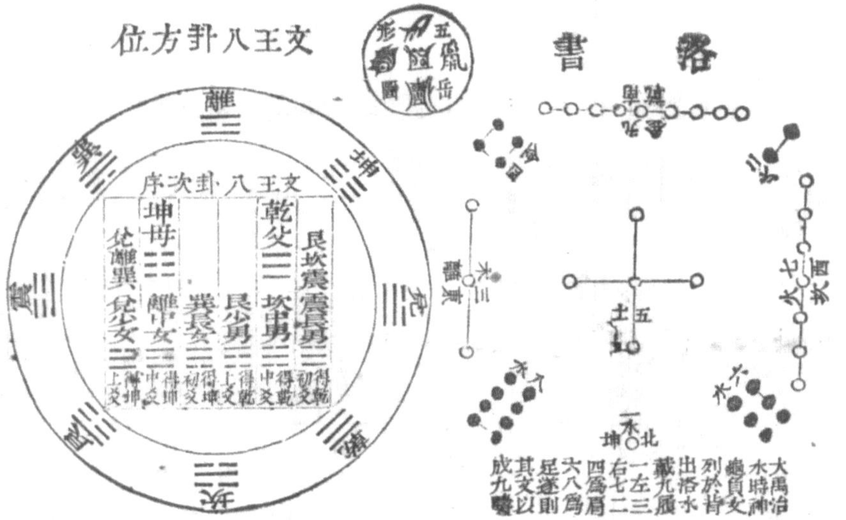

FIGURE 6.2 The numerology of the *Luoshu* and the Later Heaven sequence of the eight trigrams (from the same source)

Sephiroth of Moses (i.e. the so-called Tree of Life) with the "Diagram of the Supreme Ultimate" (*Taiji tu*), Bouvet remarks:

> Once upon a time in the first schools of the ancient Chinese wise men the doctrine existed about the one and triune God, founder of all things, about

> the incarnation of the Son of God and the reformation of the world through him: in a word the doctrine is very similar to evangelical doctrine, and not different from it except in symbol and prophecy and therefore clearly similar to the ancient Kabbala of the Hebrews which, because the founder of all things and the Lord Restorer of the world ordered everything in number, weight and measure, as it enigmatically comprehends the whole arcane mystery of the divine works, under the figurative clothing either of the ten elementary numbers or of the 22 letters of its mystic alphabet, so the hieroglyphic wisdom of the ancient Chinese monuments preserves this same whole hidden mystery of divine wisdom, under the symbols of the 22 characters of its mystic cycle *Kia Y* [*Jiayi*], and the coverings of the same 10 elementary numbers of the mystic figure *Ho tu* [*Hetu*].[48]

Similarly, he writes that the first two hexagrams of the *Yijing*, Qian and Kun,

> are the two principal characters of God creator and redeemer; and of which first Kien [Qian] with the numerical power 216, the triple of the Tetragram number 72, is the symbol of justice and later with the numerical power 144, double the number 72, is the symbol of mercy: and both together taken up with the power of the same tetragram number 72 quintupled, are the symbolic mark of the two principal virtues of the divine Redeemer, outlined in the hieroglyphics of the Chinese just as in the sephirotic system of the Hebrews.

In short,

> [God] made everything in number, weight and measure (Sap. XI, 21), … [perfecting] these in wisdom; from there it follows by necessity, that the numbers are, so to speak, the fundamental base of all true philosophy, or of the sacred wisdom of the old patriarchs, first infused in the very first-formed parent of human beings.[49]

The opening chapter of Bouvet's *Yijing zonglun gao* (Draft of a general discussion of the *Classic of Changes*) articulates his basic hermeneutical strategy, which sounds quite conventional from the standpoint of traditional Chinese exegesis:

> The *Yi* as a classic contains principles (*li*; also rendered "patterns"), numbers (*shu*), images (*xiang*; also rendered "figures"), and charts (*tu*). The charts cannot be separated from the images, the images cannot be separated from the numbers, and the numbers cannot be separated from the principles. The principles generate the numbers, the numbers generate the images, and the images generate the charts. When the charts are prepared, the images become manifest; when the images stand out, the numbers are set forth; when the numbers emerge, the principles show. Therefore, for knowing the principles,

> there is nothing better than understanding the numbers; for understanding the numbers, there is nothing better than observing the images; and for observing the images, there is nothing better than exploring the charts.[50]

Bouvet thus quite naturally held great stock in diagrams of various sorts. Some of these, like Shao Yong's "Chart of the Fuxi [Former Heaven] Arrangement of the Sixty-Four Hexagrams" and the triangles and "magic squares" (*zongheng tu*) from Cheng Dawei's (1533–1606) *Suanfa tongzong* (Systematic treatise on arithmetic, 1592),[51] were, of course, Chinese in origin. Others, Bouvet devised himself. In general, Bouvet's "mystical mathematical vision"—aptly described by David Mungello as a combination of "the Pythagorean-Platonic tendency toward mathematics" and "the Aristotelian-Thomistic tendency toward classification"—focused on three types of numerical progressions, together with the rules of proportion and geometry, and the "laws of statics."[52] These progressions, rules and laws, expressed in the lines, trigrams, hexagrams and charts associated the *Yijing* (notably the *Hetu* and *Luoshu*), became the foundation for most of Bouvet's illustrations, and the means by which to mirror nature, and thus to understand all natural processes.

Perhaps the most important single diagram to Bouvet was his triangular *Tianzun dibei tu* (Chart of heavenly superiority and earthly subordination). As we can see from the versions reproduced in Claudia von Collani's excellent 1985 study of Bouvet's life and times,[53] and also from several closely related diagrams and discussions from earlier periods, contained in documents housed in both the Vatican Library and the Bibliothèque Nationale,[54] the *Tianzun dibei tu* is a version of the so-called *Jia Xian sanjiao*, known in the West as "Pascal's Triangle."[55]

Bouvet's title was inspired by first four characters of the Great Commentary of the *Changes*, and his diagram sought to integrate the numerology of the *Hetu* and *Luoshu* in a single mathematical "grand synthesis," similar in certain respects to Shao Yong's "Chart of the Fuxi [Former Heaven] Arrangement of the Sixty-Four Hexagrams."[56] Like Shao's "Former Heaven" chart, but with less schematic specificity, the "*Tianzun dibei tu*" attempts to convey "the quintessence of heavenly patterns and earthly configurations" (*tianwen dili zhi jingyun*), illustrating not only the evolution of, but also the mutual interaction between, the hexagrams and their constituent trigrams and lines.[57] And like Shao's numerical calculations, Bouvet's diagrams were supposed to yield an understanding of good and bad fortune as well as an appreciation of the larger patterns of cosmic regularity and cosmic change.[58] But unlike Shao's round and square "Former Heaven" chart, Bouvet's diagrams tended to be triangles. This, it seems, for at least two main reasons: First, triangles occupied a prominent place in the mathematics that Bouvet had been teaching to the Kangxi emperor at court, and they were especially well suited to portraying the mathematical progressions of which the French Jesuit was so very fond.[59] Second, triangles lent themselves particularly well to the depiction and analysis of various triadic relationships, not least (but also not solely) the Holy Trinity.[60]

Bouvet's diagrammatic speculations on the Second Coming of Christ suggest certain similarities with Shao Yong's stages of cosmic change as articulated in the

Huangji jingshi (Supreme principles that rule the world). But contrary to Shao's idea of endlessly repetitive cycles of 129,600 years, Bouvet naturally predicted the "end of time"—a termination that would come about much sooner than Shao's cyclical one, with no possibility of, or any need for, further cyclical patterns. According to Bouvet, the 55 numbers of the *Hetu*, when added to the 45 numbers of the *Luoshu*, yielded 100—the number of cycles representing the duration of the world. By his calculations, the product of the first eight natural numbers resulted in one cycle of 4,032,00 days, equivalent to 120 "sabbatical years" or 91 "solar years").[61]

Initially, Bouvet divided the 100 cycles of the duration of the world into a "Former Heaven" (*Xiantian*) period and a "Later Heaven" (*Houtian*) period, in which the "Former Heaven" era indicated the time between Creation and the first appearance of the Messiah, and the "Later Heaven" stage indicated the period spanning from the death of Christ until the Second Coming. This first stage, symbolized by the hexagram Tai (Peace; number 11), Bouvet called the age of the "Heavenly Way" (*Tiandao*), and the second stage, symbolized by the hexagram Pi (Obstruction; number 12) was designated the era of the "Earthly Way" (*Didao*). A third period of Redemption then followed, which Bouvet called the "Way of Man" (*Rendao*), symbolized again by the Tai hexagram. Later, Bouvet designated these three stages of human history as: (1) the *Xiantian weibian* (Former Heaven, not yet changed) period; (2) the *Xiantian yibian* (Former Heaven, already changed) period; and (3) the *Houtian bubian* (Later Heaven unchanged) period.[62] In this latter scheme, Bouvet ascribed 50 cycles to the *Xiantian* period, but only 49 to the *Houtian* era, since no one knew the exact date of the Second Coming of the Messiah.

Prior to 1711, Bouvet seems to have discussed his theories about the *Yijing* with the Kangxi emperor only sporadically and unsystematically. But in the spring and summer of that year, encouraged by the throne, he began to introduce his thoughts on the classic in a more methodical way, sending the emperor a flurry of charts, diagrams and written texts in rapid succession. It is difficult to know exactly which documents Bouvet conveyed to the Qing monarch for his perusal—in part because some of the essays and illustrations that have been attributed to him bear two different titles or exist in multiple versions, and also because comments concerning these documents in memorials and edicts of the period are often quite cryptic.

In any case, between June and July of 1711 Bouvet sent many dozens of documents to the throne, including one or more versions of the *Tianzun dibei tu* and a similar-looking diagram from Cheng Dawei's *Suanfa tongzong* (Systematic treatise on arithmetic, 1592), titled the "Chart of Extracting the Root" (*Kaifang qiulian tu*) or the "Chart of Methods of Transformation by Doubling" (*Jiabei bianfa tu*).[63] In fact, Bouvet used this latter document to help explain the mathematics of the former, as we can see clearly in his *Yixue waipian* (A study of the *Changes*, outer part).[64]

Although prohibited by his superiors from discussing anything other than physics and mathematics with the emperor,[65] Bouvet probably forwarded to the throne excerpts from at least some of the following Chinese manuscripts (most of them undated or unreliably dated), in addition to the *Yixue waipian*: the *Yikao* (aka *Yiyin*), the *Yiyin yuangao*, the *Yiyao*, the *Zhouyi yuanzhi tan* (An investigation into the original meaning of the *Zhou Changes*), the *Yijing zonglun gao* (Draft of a general discussion of the *Classic of Changes*; aka *Yijing zongshuo gao*), the *Taiji lueshuo* (A brief discussion of the Supreme Ultimate), and the *Shi Xiantian weibian shizhong zhi shu you tianzun dibei tu er sheng* (An explanation of how the numbers of the Former Heaven [sequence], not yet changed, from beginning to end, arise from the *Chart of Heavenly Superiority and Earthly Subordination*).[66]

Bouvet's broad-ranging investigations of the *Yijing*, and his study of various charts in the *Suanfa tongzong*, not only nourished the Kangxi emperor's already existing interest in the *Changes*; they also gave him a new-found appreciation for Cheng Dawei's book on mathematics.[67] The result was that the emperor arranged for the *Suanfa tongzong* to be republished immediately with additions and corrections. He also ordered his trusted Grand Secretary, Li Guangdi (1642–1718) to begin compiling a huge annotated edition of the *Yijing* which was published under the title *Zhouyi zhezhong* (A balanced [edition of the] *Zhou Changes*; 1715). As chief compiler of this latter work, Li not only took into account Bouvet's scholarship on the *Changes*, but he also shared his *Yijing*-related work with the French Jesuit.[68] In Han Qi's opinion, by helping the Kangxi emperor to understand the "mathematical mysteries" (*shuxue aomi*) contained within the *Yijing*, Bouvet played a significant role in sustaining the emperor's interest in the classic, and in so doing he contributed in a direct way to the general burst of scholarship on the *Yijing* in the Kangxi era and thereafter.[69]

This account of Bouvet's influence is perhaps a bit too sanguine. To be sure, at times the emperor complimented the French Jesuit for his hard work and the clarity of his numerological interpretations of the *Changes*.[70] But the emperor also periodically expressed frustration with his slow progress, and criticized Bouvet for being too independent and for failing to take into account relevant Chinese works. A succession of critical edicts in July and August called into question his piecemeal scholarship, describing his writings as prolix, repetitive and tedious, and stigmatizing some of his ideas as either strange, confused, redundant or unfathomable.[71] Although the emperor remained intrigued by Bouvet's effort to explain the "magic squares" of the *Hetu* and *Luoshu* by means of his *Tianzun dibei tu* and other diagrams, and although he clearly appreciated Bouvet's attempt to link Chinese musical notation with the *Yijing*, he felt that the French Jesuit needed additional assistance from both Western and Chinese scholars in order to complete his work.[72] Overall, it appeared to the Qing monarch that Bouvet, as an exponent of religious (as opposed to scientific) views, really had nothing more to offer than the fragmentary texts compiled by Buddhist monks, Daoist priests and Tibetan lamas.

Emboldened by the Kangxi emperor's criticisms, officials such as He Su and Wang Daohua, who had been serving as liaisons for Bouvet, quickly joined the chorus. For instance, in a memorial dated August 3, 1711, they wrote:

> After looking repeatedly at Bouvet's [studies of the] *Yijing* and his charts, we really don't understand [what he is trying to say]. The charts have things like ghosts and spirits, and things that [look] like flowers. Although we

> don't [claim to] know their secrets, they look ridiculous. In addition, [Bouvet] continually draws upon writings from Chinese books only to claim paradoxically that they belong to Western teachings.[73]

Complicating matters was Bouvet's involvement in factional disputes revolving around one of the Kangxi emperor's sons, Yinzhi, who was at that time responsible for overseeing the activities of the Jesuits and other Western "specialists" at court. Although the Kangxi emperor may have been amused or even slightly flattered by Bouvet's claim that together the Chinese ruler of "All under Heaven" and the French "Sun King" were destined to usher in a new era of world peace and universal redemption, Bouvet's religious reading of the *Yijing* seems to have made the Kangxi emperor increasingly uneasy. In a nutshell, he admired Bouvet's mathematical abilities but rejected his theology.

Meanwhile, the Figurist enterprise continued to endure vigorous attacks by other members of the Christian community. Charles Maigrot (1652–1730), for example—justly famous for alienating the Kangxi emperor with his ignorance of classical Chinese in a 1705 audience—assailed the *Yijing* as a "despicable medley of obscure nonsense," no better than the *Book of Nostradamus* or the *Prophecies of Merlin*.[74] And even Father de Prémare, who had nothing but contempt for the "deep ignorance and the intellectual poverty of Monsieur Maigrot," criticized the chaotic approach and excessive claims of Bouvet (who, for example, equated the ancient Chinese kingdom of Lu with Judea). Prémare wrote that he "would not dream" of comparing the *Classic of Changes*, the *Classic of History* or the *Classic of Songs* "with our Holy Writ."[75]

During the next few years Bouvet's fortunes waned, although the emperor never gave up completely on him.[76] But, in April of 1716, a delegation of Jesuit missionaries hostile to Figurism managed to convince the Qing monarch, whose enthusiasm for Western science had already diminished significantly, that Bouvet's ideas were not only eccentric but also dangerous. From this point onward, Bouvet fell increasingly out of imperial favor, although he continued to write on the *Yijing* until his death in 1730—convinced to the end that the astrologically grounded numerology of the *Changes* revealed the apocalyptic prophecies of the Book of Revelation, presaging the "Second Coming" of Christ.

Although the Figurism practiced by Bouvet and his followers has often been stigmatized—not only in its own time but also by later scholars—for its highly inventive etymologies, its numerological emphasis, its wide-ranging correlations and its far-reaching allegorical interpretations, we should remember that exegetical strategies of this sort were a part of the *Yi* tradition well before the arrival of the Jesuits in China. One prominent Qing intellectual whose ideas show certain clear affinities with those of Bouvet is Jiang Yong (1681–1762), discussed in Chapter 5 in this volume. Another was Jiao Xun (1763–1820), a gifted and innovative mathematician. Although at the heart of Jiao's theory of analogues or proportions (*bili*) was a sophisticated "computational logic," he sought to "use the mathematics of *bili* to seek the *bili* of the *Changes*," not the other way around.[77] In other words, his primary concern seems to have been finding a way to correlate the hexagrams, trigrams and line statements of the *Yijing* in order to reveal, and understand more fully, "the [moral] Way of Fuxi, King Wen, the Duke of Zhou, and Confucius." Indeed, according to Fang Chao-ying, Jiao's main contribution to the study of the *Changes* was "his application of the principles of mathematics to determine comparatively the amount of good fortune or calamity which … [according to the *Yijing*] ensued from various types of conduct."[78] It was the divinatory potential of the classic—specifically its value as a guide to moral behavior—that held the greatest attraction for Jiao, not its mathematical possibilities.

In short, *Yijing* exegesis of the sort indulged in by Bouvet, Jiang, Jiao and a great many other Qing scholars,[79] opened up a vast field of speculative endeavor in premodern China, as it does to this day. Although the *Changes*-based numerological, historical, and theological speculations of the Figurists may have been a bit farfetched in terms of traditional Chinese categories of concern, they were part of a long hermeneutical history of creative *Yijing* interpretations, textual transmissions and cultural transformations—one that not only preceded the Jesuits in East Asia by several centuries, but one that would also continue to manifest itself in various parts of the world for several centuries to come. Moreover, it is important to remember that the criticisms of Bouvet's work lodged by some Qing scholars and the throne in the eighteenth century were no more severe than the ones Chinese savants had been leveling against one another for nearly two thousand years before him. Charges of obscurity, deception, wrong-headedness, over-reading, superficiality and "forcing" an interpretation (*qiangci*) or "forcing a fit" (*qiangpei*) were common in the history of *Yijing* exegesis—particularly during the *kaozheng* debates of the early- and mid-Qing period.[80]

The *Yijing*'s journey to the West

In certain respects, the process by which the *Yijing* has moved to Europe and the Americas during the past four centuries parallels the earlier (and, in fact, on-going) process by which the document has traveled to East Asia.[81] Most significantly, in both cases, conscious efforts have continually been made by devotees of the *Changes* to "domesticate" it by various means. But the attempt by both missionaries and lay people to introduce the *Yijing* to Western audiences from the eighteenth century onward has involved unprecedented problems of translation and accommodation—not least, because the text has moved from a "mainstream" cultural environment in East Asia to a "counter-cultural" one in the West. And in this latter process, particularly during the twentieth century, market forces have come increasingly into play, complicating the already "complex dialectic" by which the words of the text have interacted with the intellectual concerns of its translators and commentators.[82]

Despite the unhappy fate of the Figurists in China, their writings captured the attention of several prominent European intellectuals in the late seventeenth and early eighteenth centuries—most notably the great mathematician Gottfried Wilhelm Leibniz (1646–1716).[83] Leibniz's interest in China had been inspired by,

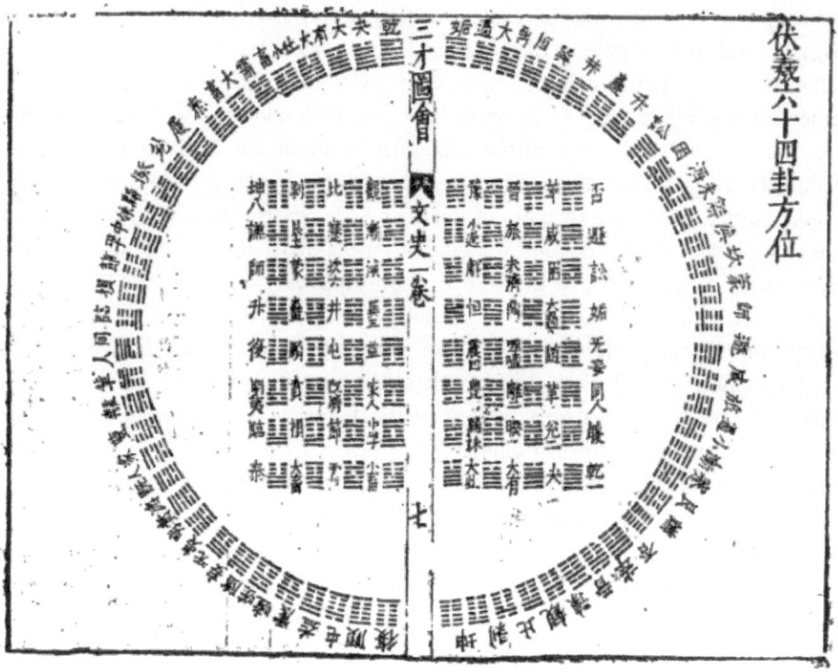

FIGURE 6.3 Shao Yong's Former Heaven chart from the Illustrated Compendium of the Three Powers

among other things, his search for a "Primitive Language"—one that existed before the Flood. In their correspondence, Bouvet convinced Leibniz that a systematic study of the *Yijing* could assist in this quest, and that it might ultimately contribute to the development of a comprehensive, scientific/mathematical language that Leibniz called the "Universal Characteristic." Such a language would make the act of thinking—like the act of calculation—a reflection of the binary structure of nature itself. In Bouvet's mind, Shao Yong's Former Heaven Chart (Figure 6.3) offered a mathematical point of entry: a depiction of line changes that expressed the same formal features as the binary system invented by Leibniz himself.[84]

When Bouvet sent a copy of Shao Yong's diagram to Leibniz, the latter was overjoyed to see cross-cultural confirmation of his binary system—a system that had religious and mystical significance because it suggested the idea that God (represented by the number one) had created everything out of nothing (0). But while there were certain similarities between the ideas and approaches of Leibniz and Shao Yong, there were also significant differences. In the first place, the numbers Shao employed in all his calculations were based on the decimal system; second, Shao was far more interested in correlative metaphysics than in the binary structure of the Former Heaven Chart *per se*. On the whole, Shao had little interest in quantitative and empirical methods, and he did not share Leibniz's optimistic belief in linear progress.[85]

The first complete translation of the *Changes* in a Western language (Latin) was undertaken by three missionary-scholars who were hostile to Shao Yong and extremely critical of Bouvet and his followers. This anti-Figurist group consisted of Jean-Baptiste Regis (1663–1738), Pierre-Vincent de Tartre (1669–1724), and Joseph Marie Anne de Moyriac de Mailla (1669–1748). All three men denied that the Chinese classics contained any truths of the Christian faith, and they all denounced the Figurists for producing what de Tartre disparagingly called the "Cabala [Kabbala] of the Enochists." Work began on the translation in 1707, but the preliminary draft was not completed until 1723. This final version was based on the imperially commissioned *A Balanced [Edition of the] Zhou Changes* and its official Manchu rendering. The Regis manuscript then languished for more than a decade in Paris, until a young sinologist named Julius Mohl (1800–1876) produced a two-volume printed version of several hundred pages in the 1830s titled *Y-King antiquissimus Sinarum liber quem ex latina interpretatione P. Regis aliorumque ex Soc Jesu P.P. edidit Julius Mohl* (*Yijing*, the most ancient book of the Chinese, edited by Julius Mohl based on the Latin translation of Father Regis and other Fathers of the Society of Jesus). Although this version attacked the Figurists as well as the theories of Shao Yong, it addressed most of the major issues of traditional Chinese *Yijing* scholarship in a systematic way, quoting from orthodox Neo-Confucian sources and citing the authority of the Church Fathers and Western philosophers for comparative purposes.[86]

Within decades, several additional translations of the *Changes* appeared in Europe, including Canon Thomas McClatchie's *A Translation of the Confucian Yi-king* (1876), Angelo Zottoli's 1880 rendering of the *Changes*, which appeared in volume 3 of his *Cursus literaturae sinicae neo-missionariis accomodatus* (1879–1882), James Legge's *The Yi King* (1882), P.L.F. Philastre's *Tscheou Yi* (1885–1893), and Charles de Harlez's *Le Yih-king: Texte primitif, rétabli, traduit et commenté* (1889).[87] These works, as Norman Girardot suggests, reflect a "scholarly vogue in European culture at this time concerned with the uncovering, and the rational and historical explanation, of all manner of apparent Oriental mysteries," including not only Buddhism and Daoism, but also theosophy, spiritualism, the Golden Dawn, and various occult novels.[88]

Like Bouvet, but with no acknowledgement of the Jesuit priest's pioneering scholarship, McClatchie believed that the *Yijing* had been brought to China by one of the sons of Noah after the Deluge. But whereas Bouvet tried to use the *Changes* to prove that the Chinese had knowledge of "the one true God," McClatchie believed that the work reflected a form of pagan materialism, "perfected by Nimrod and his Cushites before the dispersion from Babel." He identified Shangdi as the Baal of the Chaldeans. According to McClatchie's wild cross-cultural speculations, Baal/Shangdi was, among other things, a "Hermaphroditic Monad," personified in Chinese mythology as the ancient culture hero Shennong.[89]

In addition to offering a generally disparaged translation of the *Changes*, McClatchie published two articles in the *China Review* at about the same time—one titled "The Symbols of the *Yih-King*" and the other, "Phallic Worship." In these two works, particularly the latter, he identified the two primary hexagrams of the *Changes* with the male and female sexual organs:

> From the statements of the Yih King, and of Confucius in his Commentary, Kheen-khwan [Qian-Kun] or Shang-te [Shangdi; identified by McClatchie with both Baal and Shennong] is evidently the phallic God of Heathendom represented unmistakably by the usual symbols. Kheen or his Male portion is the membrum virile, and Khwan or his Female portion is the pudendum muliebre; and these two are enclosed in the circle or ring, or phallos, the "Great Extreme" [Taiji] or globe of Air [*qi*], from and by which, as the "Great Monad" [Taiyi], all things are generated.[90]

Scholars like Legge, and later, the eminent Russian Sinologist Iulian Shchutskii, ridiculed this decidedly sexual view (Shchutskii described it as an example of "pseudoscientific delirium"), but recent work by Edward Shaughnessy and others suggests its essential correctness—particularly in the light of research on the Mawangdui version of the *Changes*.[91]

Another individual with equally outspoken opinions was Albert Terrien de LaCouperie, whose study titled *The Oldest Book of the Chinese: The Yh-King and Its Authors* (1892) provides a systematic critique of Chinese and Western scholarship on the *Changes*. Terrien believed that the *Yijing* was originally "a handbook of state management … set forth under the sixty-four words [hexagram names]." Virulently critical not only of the Chinese commentarial tradition (the product of what he derisively described as "tortured minds" and "maddened brains") but also of most Western interpretations of the *Yijing* ("amusing enough to dispel the spleen"), he posited Near Eastern origins for the basic text of the classic but not the Ten Wings. According to Terrien, the *Changes* originated in the ancient kingdom of Akkad, which he believed to be Bactria. Following a great flood, the Bak people (*Baixing*) supposedly migrated eastwards to China and settled in the Yellow River valley around the year 2282 BCE.[92] De Harlez, for whom the acerbic Terrien seems to have had a certain respect, held a somewhat similar view of the *Yijing*, arguing that the *Changes* was originally the notebook of some ancient political figure, which had been transformed into a book of divination at a later date. Iulian Shchutskii points out, however, that something of the opposite process actually took place.[93]

James Legge, the object of some of Terrien's most vitriolic invective (he accuses Legge, for example, of offering an "unintelligible paraphrase" of each line of the *Changes*), began his translation in 1854, but for various reasons it was not completed for another twenty years or so.[94] Like Bouvet, Legge had a talented Chinese amanuensis to assist him with his translation—a comparatively well-read treaty-port scholar named Wang Tao (1828–1897). Legge believed that the Confucian classics were fundamentally compatible with Christian beliefs, but he was not a Figurist in any sense.[95] Nor was he enamored of most Western-language translations of the *Changes*. He was especially hostile to the work of McClatchie. In addition to denouncing him for focusing on the *Yijing*'s sexual imagery, Legge assailed

McClatchie for resorting to the methods of "Comparative Mythology." In Legge's words: "I have followed Canon McClatchie's translation from paragraph to paragraph and from sentence to sentence, but found nothing which I could employ with advantage in my own."[96]

Legge had no fondness for China nor any respect for the *Yijing*. Indeed, he described the work as "a farrago of emblematic representations." He admitted that the *Changes* was "an important monument of architecture," but he characterized it as "very bizarre in its conception and execution."[97] Legge's highly literal translation, published in 1882, followed the prevailing neo-Confucian orthodoxy of the Qing dynasty as reflected in the *Zhouyi zhezhong*, a work that probably encouraged Legge to separate the "basic text" from the Ten Wings. This intellectual orientation also informed the more famous and more influential rendering of the *Changes* by the German scholar Richard Wilhelm, which first appeared in 1924. Like Legge, Wilhelm had been a missionary in China before embarking on a more "scholarly" career, and like Legge he also had the assistance of a talented Chinese scholar, Lao Naixuan (1843–1921).

The standard English translation of Wilhelm's work was carried out by one of Carl Jung's students, Ms. Cary Baynes, and published in 1950. The standard comparison of the two works—a distortion in both directions—is that Legge's text indicates what the *Yijing* says while Wilhelm's conveys what it actually means.[98] The interesting point about Wilhelm's translation is that it bespeaks a man not only enamored of China, but also one who believed that the *Yijing* had something important to say to all of humanity. Like Bouvet, he considered the *Changes* to be a global property and a work of timeless wisdom, but unlike Bouvet he treated it solely as a Chinese document, with no genetic links with the ancient West or Near East.

Like many of his predecessors, both in East Asia and in the West, Wilhelm tried to "domesticate" the *Yijing* in various ways. One of his strategies was to call upon the authority of classical German philosophers and literary figures, like Kant and Goethe, to illustrate "parallel" ideas expressed in the *Changes*. Another was to cite the Bible in the same way. Yet another was to argue that the *Yijing* drew upon "some common foundations of humankind that all our cultures—unconsciously and unrecognizedly—are based." Wilhelm believed, in other words, that "East and West belong inseparably together and join hands in mutual completion." The West, he argued, had something important to learn from China.[99]

Wilhelm also tried to "demystify" the *Changes* by providing elaborate commentaries that paraphrased and explained the "spiritual" material that he felt might "confuse the European reader too much with the unusual." This strategy of "rationalization," as Michael Lackner points out, was somewhat similar to that of the Jesuit Figurists, "who frequently prepared second translations of certain texts because they claimed to know the intrinsic meaning of these texts: the prefiguration of Christian revelation."[100] In the case of the Figurists, this process often involved the willful misrepresentation (or at least the ignoring) of traditional commentaries in order to "dehistoricize" the "original" text. In Wilhelm's case, however, the impulse to explain away material that might be considered "superstitious" reflected the general climate of "rational" academic discourse in early twentieth-century Europe. Wilhelm remained a "missionary," so to speak, but a secular one, whose rendering of the *Changes* seemed to confirm Carl Jung's theories about archetypes and "synchronicity"—just as Bouvet's representations of the work had confirmed Leibniz's binary system and fed his speculations about a "Universal Characteristic" and a "Primitive Language."[101]

By contrast, Aleister Crowley (1875–1947), a wild-eyed Englishman who traveled to China during the first decade of the twentieth century, adopted a self-consciously mystical approach to the *Changes*—a harbinger of counter-cultural enthusiasm for the document that would peak worldwide in the 1960s. Upon his return from China, Crowley began the study of various Chinese texts, including the *Yijing*. At first, he relied heavily on Legge's translation, but found it unsatisfactory—in large measure because of the Scottish missionary-translator's hostility to the document ("what pitiable pedantic imbecility," Crowley once wrote of Legge's attitude). Eventually Crowley developed an approach to the classic that dispensed with the conventional attributes of some of the trigrams and tried to assimilate them, in the fashion of Bouvet, to the "Kabbalistic Tree of Life."

According to Crowley, the *Yijing* was "mathematical and philosophical in form," with a structure that was "cognate with that of the Qabalah." In fact, he maintained that "the identity [between the *Changes* and the Kabbala] is so intimate that the existence of two such superficially different systems is transcendent testimony to the truth of both." In Crowley's view, the Dao as expressed in the *Yijing* was "exactly equivalent to the Ain or Nothingness of our Qabalah," and the notions of *yang* and *yin* "correspond exactly with Lingam and Yoni." Furthermore, he equated *jing* (essence) with Nephesh, *qi* (material force) with Ruach, and *hun* (soul) with Neschamah. For Crowley, the Confucian virtues of *ren* (humaneness), *yi* (duty), *li* (ritual) and *zhi* (knowledge) suggested the principles of "Geburah, Chesed, Tiphareth, and Daath."[102]

In Crowley's decidedly sexual interpretation of the *Changes*, reminiscent of McClatchie's, the eight trigrams represented the male and female reproductive organs, the sun, the moon, and the four Greek elements—earth, air, fire and water. Thus he writes:

> In the place of Chesed, which is water in our Qabalah, we find Tui [Dui; "Lake"], which is water in the Chinese system. In Geburah, our Fire, is Kan [Zhen; "Quake" or "Thunder"]. In the place of Netzach, which is Earth in our Qabalah, there is Kan [Gen; "Restraint" or "Mountain"] on the Chinese plan. Finally, for the Sephira Hod, which in our system is Airy and Mercurial, we find Sun [Sun or Xun; "Compliance" or "Wind], the Chinese trigram of Air.[103]

Qian ["Pure *Yang*" or "Heaven"] is equated with Daath, Kun ["Pure *Yin*" or "Earth"] with Malkuth, Li ["Cohesion" or "Fire"] with Tiphareth, and Kan ["Sink Hole" or "Water"] with Yesod. With similar abandon, Crowley equates the four attributes of the judgment for the first hexagram, Qian—*yuan*, *heng*, *li* and *zhen*—with the four spheres of the Tree of Life and the four parts of the human soul, representing wisdom, intuition, reason and the animal soul.[104]

In more recent times, a great many books and articles (including website publications) have appeared that relate the *Yijing* to the conventional values of Christianity and that employ Figurist techniques and logic. The writings of Jung Young Lee are representative.[105] Lee asserts, for example, "God the Father is closely associated with the image of a dragon in *The Book of Change*." As the "hidden dragon," Lee writes, he is "the source of [all] creativity."[106] C. H. Kang and Ethel R. Nelson make the same kinds of connections in their 1979 book, *The Discovery of Genesis*. In it, they dissect several dozen characters in an effort to show that "the ancient picture writing of the Chinese language embodies memories of man's earliest days," and that "when broken down into component parts … [these characters] reflect elements of the story of God and man recorded in the early chapters of Genesis."[107] I. Mears and L.E. Mears, for their part, try to show in *Creative Energy* (1931) that "God-like qualities" can be found in the various symbols of the *Yijing*, from the characters of the text, to the trigrams, to the hexagrams.[108]

In a more strictly "Figurist" vein, but without an acknowledged debt to Bouvet or his followers, Joe E. McCaffree's *Bible and I Ching Relationships* (1982; first published in 1967) tries to show in 446 extraordinarily convoluted pages that: (1) certain Chinese characters were "specially designed" for correlative purposes in works such as the *Changes*; (2) the texts of the *Yijing* and the Hebrew Bible "follow the same sequence with respect to the order of events and their inherent characteristics" (for example, McCaffree maintains that the first fifty hexagrams of the *Yijing* correspond with the fifty chapters of the Book of Genesis); (3) various biblical personalities, including Joseph of Genesis, Moses and Solomon, are identified in the *Yijing*; (4) Biblical accounts have "a hexagrammic structure" which includes literal references to the trigrams; (5) the Zhou people were "probably" Israelites; (6) the *Yijing* "fulfills" biblical prophecy; and (7) the *Changes* were intended as a "study guide" to the Bible. In short, the *Yijing* unveils a "divine plan for the culture of man," and in so doing "consummates" what the Torah has "generated." In the process, McCaffree attempts to establish connections between the symbolism of the *Changes* and that of ancient Egypt, India and the Middle East.[109]

Ong Hean-Tatt's *The Chinese Pakua* (1991), one of the most recent and most comprehensive works of this sort, clearly draws a good deal of self-conscious inspiration from the Jesuit model, as well as other "Figurist" interpretations. Indeed, he even acknowledges Jesuit precedents explicitly on occasion. Moreover, he covers much the same ground that Bouvet and others covered in the sixteenth and seventeenth centuries: the "common origin" of Chinese-Hebrew "Kabalic [*sic*] Magic Squares," the "link between [the] Chinese and Hebrew languages," "similarities in [the] Structures of Chinese and Sumerian-Egyptian Characters," the identity between the twenty-two symbols of the Chinese stem-branch (*ganzhi*) system and the twenty-two letters of the Kabbala, and the appearance of "Middle East legends in Chinese ideograms" (for instance, he asserts that the *long* ["dragon"] is the winged "seraphim" of the Bible). Like the Figurists, he breaks down a number of Chinese characters into their constituent elements to reveal biblical "messages," arguing, for example, that a combination of the words for "hand" 手 (*shou*) and "spear" 戈 (*ge*) yields the character for "self" 我 (*wo*), which, when combined with the character for sheep or lamb 羊 (*yang*), produces the character *yi* 義 (*yi*; righteousness, duty, etc.). According to Ong, this depiction of a "hand using a spear to sacrifice the lamb" indicates that "the Chinese knew righteousness will come from [a] slain lamb."[110]

Some concluding remarks

In retrospect, the westward movement and eventual "globalization" of the *Yijing* are easy enough to comprehend. As indicated at the outset, the *Changes* was in many ways an ideal instrument for building bridges across cultures. Challenging, sophisticated, useful and adaptable, it was a potentially valuable commodity in the ever-widening marketplace of ideas. Simple only to the simple-minded, it encouraged inventive people to make creative connections of all sorts, linking in various ways not only the past, the present and the future, but also heavenly and earthly phenomena, numbers and images, moral principles and practical predictions. It generated innovative scholarship in a wide range of areas—from philology and philosophy to math and science—and it continues to do so to this day, not only in East Asia, but also in Europe and the Americas.[111] Attempts by contemporary scholars in China and the West to identify the "original meaning" of the *Changes*, like those of the Figurists two and a half centuries before them, proceed unabated, with greater or lesser degrees of methodological sophistication. That the *Yijing* has provoked much silliness and superficiality over the years is not the fault of the document.

Recognizing the *Yijing*'s unparalleled prestige, its rich and provocative symbolism, its deep philosophical content and its practical utility, Bouvet and his followers, facing East and West simultaneously, saw the document as a convenient means by which to link two seemingly different philosophical and religious traditions. The document was, in their eminently reasonable view, a capacious vehicle of cross-cultural understanding. Moreover, in using the *Yijing* to carry out their own evangelical agenda, the Figurists did nothing with the document that their predecessors in East Asia, and, of course, their successors on both sides of the Atlantic, did not also do.

What, then, went wrong? The failure of the Figurists in eighteenth-century China was clearly not a matter of will or intellect. It was simply, or perhaps not so simply, a product of politics. To be sure, the scholarship of Bouvet and his followers was flawed—at least by conventional Chinese standards of the time. But, as I have tried to suggest, the Figurists worked within interpretive traditions that virtually all Chinese scholars recognized and respected, and one suspects that if they had been allowed greater latitude in matters such as the choice of Chinese terms by which to render the Christian concept of "God," they would have made far more headway with the emperor and scholar-bureaucrats. The constraints imposed by the Church authorities on what Bouvet and others could say or write, however, had a deadening effect on Figurist discourse, and their embroilment in the factional politics of both the imperial court and the Catholic Church put them in an extraordinarily vulnerable position. After the failure of the Tournon Mission of 1706, the Jesuits fought an increasingly uphill battle, as Kangxi grew ever more distrustful of them.

In short, the Figurists in China had the right medium, but they lacked the freedom to articulate their cross-cultural message. The situation in Europe was considerably different—although there, too, politics, personalities and intellectual fashions affected the reception and interpretation of the *Yijing*. Complicating matters was the problem of language, for unlike the spread of the classic to Japan, Korea and Vietnam, where elites were completely comfortable with the classical Chinese script, in Europe and the Americas the *Changes* required translation.

In this and several other respects, the transmission of the *Changes* to the West parallels the process by which Buddhism and Daoism traveled westward. As Stephen Batchelor, James Coleman, J.J. Clark and others have indicated, in each case "missionaries" have played a role in the process of translation, and in each case there have been varied responses in the West over time, ranging from "blind indifference," to "rational knowledge", "romantic fantasy" and "existential engagement."[112] And as with the sophisticated texts of Buddhism and Daoism, the *Yijing* has been subjected to rigorous scholarly analysis as well as to superficial punditry. It has provoked fascinating speculations and idiotic drivel.

One of the most revealing examples of the way philosophical and religious systems intersect in the process of cross-cultural communication is the work of Ernst Lothar Hoffman, a German citizen who became a "homeless" lay Buddhist in the late 1920s and was given the name Lama Anagarika Govinda. In the 1980s, after many years of Buddhist study and practice, he wrote a book titled *The Inner Structure of the I Ching: The Book of Transformations* (1981). Replete with elaborate diagrams and illustrations, it claims to reveal "what the *I Ching* itself has to say," rather than what "various Chinese and European philosophers and scholars thought about this book."[113] The diagrams and text of this study, which draw upon Tibetan Buddhist traditions as well as Western astrology and Chinese Confucian and Daoist traditions, suggest the same sort of "grand synthesis" that Bouvet attempted more than 250 years before.

In all, the history of the *Changes* in the hands of Western interpreters serves as an especially vivid reminder that exegesis never occurs in a vacuum. It is always "motivated," and the historian's duty is to ferret out the motives and provide a context for them. As Daniel Gardner reminds us, "there simply is no one stable or definitive reading of a canonical text."[114]

For some two thousand years, the primary incentive for commenting on the *Yijing* in China was to clarify the meaning of the sages. This generally involved an effort to "fix" or "stabilize" the text in accordance with a particular philosophical or religious outlook, and it often involved an implicit or explicit reaction to previous interpretations. A related goal, also common to many cultural traditions, was to make a work that was diverse in origins appear coherent, consistent, and relevant to readers of a different time. This might be done to "legitimate" a text in the face of claims that the version in hand was not "authentic." And, of course, commentary allowed individuals to associate their own ideas with a classic, another strategy of "legitimation."[115] All of these motives have been evident in *Changes*' exegesis, both East and West, and they will continue, no doubt, to be.

NOTES

Introduction

1 P. Cohen 2003, 10. Cf. P. Cohen 1984, 161–72.
2 See the persuasive discussion in Johnson 2009, esp. 4–8.
3 Sadly, this last book was not completed until after Professor Fairbank's death.
4 See R. Smith 1988 and 1992. Cf. Morgan, 1980, N. Wang 1993 and J. Yeh, 2011.
5 Johnson 2009, 3–4.
6 Cited in Watson 1963, 89 and K.C. Liu 1978, 412.
7 It was not, of course, the only ordering device in China. Scholars such as Joseph Needham (SCC), Nathan Sivin and Benjamin Elman have identified longstanding Chinese traditions of science and technology, mathematics, medicine, linguistics and many other fields of knowledge that developed independently, and sometimes in spite of, the *Yijing*'s enormous epistemological influence.
8 Smith 2008a, 3.
9 Ibid., 3–4.
10 See M. Lewis 2006, 307.
11 See ibid., 284–303, esp. 297.
12 Ibid., 260–303 passim.
13 The theoretical literature on ritual is vast. See Bell 1997. Bell's own approach to ritual emphasizes its situational, relational, performative and strategic dimensions. Bell 1992.
14 M. Lewis 2006, 2. See also the essays in Kern, ed. 2005 and Elman and Kern, eds. 2010.
15 See Smith 1994, 6. Johnson 2009, 8 writes:

> Chinese culture was a performance culture: even in pre-Qin times Chinese philosophers were concerned more with how people should act, and what counted as good actions, than with using logic to prove propositions. Ritual was the highest form of action or performance; every significant life event, social, political, or religious, was embedded in and expressed through ritual. If we wish to understand how Chinese people thought and felt about the family, the community, the state, or the gods we must study the rituals by which those thoughts and feelings were expressed and shaped.

16 See Zito 1997, 4, 49–50 and 57ff.
17 Ibid., 58.
18 Smith 1991, xi–xii.
19 One of the most interesting and controversial approaches to questions of unity and diversity in late imperial China is James Watson's notion of orthopraxy. See the discussion of his work in the symposium titled "Reconsidering James L. Watson's Ideas: Ritual, Cultural Standardization, and Orthopraxy in China," in *Modern China* 3.3 (2007): 3–158.
20 See R. Smith 2012, 9–10.
21 Ibid.
22 Cohen 2003, 13–14.
23 Ibid.
24 Ibid.
25 See R. Smith 1996, 1–9.
26 See Qian, Fong and Smith (2008), 1–2.
27 See, for example, Hevia 1995, Hostetler 2001, 43–49, Wills 1995, and Wills, *et al.*, eds. 2011, 2–4.
28 See, in particular, Fairbank 1942 and 1953 and Fairbank and Teng 1939.
29 Hamashita 2008, 13. See also D. Kang 2010, 54–81. Cf. Rowe 2009, 133–38.
30 See, for example, Millward, *et al.*, eds, 2004, 3–4, Perdue 2005, 402–06, Millward 1998, 7–10, 48–49, 158–59, 199–203, Andre Schmid in Lary, ed. 2007, esp. 126–30, and Rowe 2009, 133–8.
31 See, for example, S. Yang 2007, Pines 2005, L. Wong 2010, Dabringhaus and Ptak, eds. 1997, Zhang Wanshou 2011, Abe 1946, Meserve 1982, Swope 2002, Brose 2006, Giersch 2006, Rowe 2001, esp. 406–26.
32 C.C. Huang 2007, citing the research of Wang Ermin.
33 Cited in C.C. Huang 2007, 407 modified. Cf. the discussions in S. Yang 2007, Pines 2005, M. Lewis 2006, 189–212, 252–60, 288–303, and Kang 2010, 29–33.
34 Cited in R. Smith 1994, 3. Cf. D. Kang 2010, 29ff.
35 R. Smith 1994, 138.
36 See the discussions in W. Fang 2001, Fiskesjö 1999, J.M. Lee 2008, Li Yuanquan 2004, Pines 2005, and Wong 2010. Lydia Liu's argument (2004, 31–69) that the term *yi* became simply a "supersign" that was manipulated by Westerners for their own purposes is too reductionist. For revealing Japanese, Korean and Vietnamese perspectives on the term *yi*, see Elman 1990, 216–18, Mizuno 2003, esp. 36–39 and J.M. Lee 2008, esp. 4–5, 18–20, 65–75. In the nineteenth century, and for hundreds of years before, the term *yi* was generally considered pejorative by most Asians and Westerners alike. It is true, however, that the Qianlong emperor famously declared that *yi* was not a disrespectful term, and that Mencius himself had described the great sage heroes Shun and King Wen as originally *yi*. Elman 1990, 218.
37 S. Yang (2007), 5–6.
38 P.T. Ho 1967, 191–92.
39 Rawski 1996, 131.
40 See Waley-Cohen 2004, 200.
41 Crossley 1990a and 1999; Elliott, 2001.
42 See Elliott 2009.
43 See Rawski 1998, 118–20.

44 P.T. Ho 1998.
45 Crossley, Siu and Sutton 2006, 6ff.
46 P. Huang 2011, 3–5.
47 Sahlins 1993, 15.

1 The languages of the *Yijing* and the representation of reality

1 This essay was originally written for non-specialists; hence, most of the footnotes—then and now—refer to relatively accessible Western-language sources.
2 R. Smith 2012, 4.
3 For a list of the 64 hexagrams in what became their received order, as well as various English translations of the hexagram names, see http://chaocenter.rice.edu/Content.aspx?id=601.
4 Archaeological excavations at sites such as Mawangdui (1973), Fuyang (aka Shuanggudui; 1977), Tianxingguan (1978), Baoshan (1987), Wangjiatai (1993), Guodian (1993), Ji'nan cheng (1993) and Jingmen (aka the Shanghai Museum version of the *Changes*; early 1990s) have unearthed several different versions of the *Changes* and/or *Changes*-related documents, some of which show affinities with earlier descriptions of the hexagram-based works known as the *Lianshan* ("Linked Mountains") and the *Guicang* ("Return to the Hidden"). For a general overview, see R. Smith 2008a, 48–56; for details, consult Shaughnessy 1996, 2001, 2002, and forthcoming.
5 For extensive discussions of the variant understandings of these cryptic formulas, see Kunst 1985, 150–211, 369–80 and 421–38. Cf. Shaughnessy 1983, 123–35, 175–287 and Fendos 1988, 18–134, passim.
6 These renderings are based on Kunst 1985.
7 Ibid.
8 Hon 2005b, 3.
9 The most complete written records of *Yijing*-based divination in the Zhou dynasty come from the *Zuozhuan* (Zuo commentary), a highly influential work dating from around the late fourth century BCE. The *Zuo Commentary* provides about two dozen fairly detailed examples of how the *Changes* came to be used during the so-called Spring and Autumn period (*c.* 770–476 BCE)—specifically, from 672 to 485 BCE. For examples, see K. Smith 1989 and Rutt 1996, 175–201.
10 *Ge* can be rendered "skin," "leather," "the wings of a bird," "to remove," "to shed [feathers or fur]," "to dismiss," "to change," "to flay," "the ends of reins," and "urgent [when pronounced *ji*]." This translation follows Kunst 1985, 337 and Rutt 1996, 272. For significantly different understandings of this text, see Lynn 1994, 444–49, R. Wilhelm 1967, 184–92 and Legge 1963, 167–68.
11 See discussions of commentaries in R. Smith, 1991 (esp. Chapter 3), 2008a (Chapters 3–7) and 2012 (Chapters 3–5).
12 For differing opinions, see R. Smith 2008a, 266, note 3.
13 R. Smith 2012, 55. For details on Chinese-style correlative thinking, see Henderson 1984, Graham 1986 and 1989, Hall and Ames 1995, and esp. the essays in Fiskesjö, ed. 2002.
14 For systematic discussions of these and many other cosmic variables, see R. Smith 1991 (Chapter 2), Feuchtwang 2002 (Chapters 2–4), and esp. Aylward 2007, passim. See also Chapters 3 and 5 on ritual and divination in this volume.

15 Sivin is cited in R. Smith 2012, 52.
16 See also Chapters 3 and 5 on ritual in this volume.
17 Cited in de Bary and Bloom, eds. 1999, 1: 338.
18 See the discussion in R. Smith 2012, 50–51; also R. Smith 1994, esp. 4–5, 121–24, 134–35, 147–48, 177–80, 211–12, 223–24, etc.
19 R. Smith 2012, 55–56. See also R. Smith 2008a, 33–36, 73–75, 116–18, 125–27.
20 See the detailed discussion in Aylward 2007, 38–53, 78–87, 112–22, etc. After the ten stems have been paired with the first ten branches, the initial stem is paired with the eleventh branch and the second stem is paired with the twelfth branch. Thereafter, the third is paired with the first branch, and so on. Ibid., 39.
21 For details, see ibid., 88–102, 131–47, 191.
22 The range and sophistication of these efforts are truly staggering. See, for example, Granet 1934, esp. Chapter 3, Henderson 1984, P.Y. Ho 2003, Aylward 2007, R. Smith 1991 and 2008a, etc.
23 For a few salient works, see A. Wang 2000, Pines 2002 and 2009, and Elman and Kern, eds. 2010.
24 R. Smith 2012, 29–30. For details, see R. Smith 2011.
25 R. Smith 2012, 58–60; see also R. Smith 2008a, 37–48, 89–91.
26 For an excellent overview of the language and purposes of the document, see Peterson 1982.
27 R. Smith 2012, 60.
28 Lynn 1994, 51 and 62 (modified); cf. R. Wilhelm 1967, 294 and 314.
29 R. Smith 2012, 61. On the concept of "knowing fate," see R. Smith 2008a, 36–38, 61–62, 102, 227, 232.
30 Lynn 1994, 56–57; cf. R. Wilhelm 1967, 304. See also Lynn 1994, 77; cf. R. Wilhelm 1967, 328.
31 R. Smith 2012, 61–62. See also the discussions of images in M. Gu 2004, 471–73, W. Zheng 2008, and R. Smith 2008a, esp. 8, 16, 37–48, 59–60, 86–88, 92–93, 95–98, 103–05, 115, 121–24, 129, 132, 147, 161–65, 168, 174, 176, 179, 187, 273–74, notes 35–41, 287, note 20, 296, note 48, 312, note 28, 315, note 69. For some Chinese-language sources on images, consult Li Zhongjun 1994, Liu Dajun, ed. 2003, Liu Yujian 1996, Wu Lin, ed. 1995, You Yi shufang zhuren 2003, and Zhang Qicheng 2003.
32 Lynn 1994, 49–50 (modified); cf. R. Wilhelm 1967, 287–88.
33 Lynn (1994), 50 (modified); cf. R. Wilhelm 1967, 290.
34 Lynn 1994, 119–26; cf. R. Wilhelm 1967, 262–79.
35 Lynn 1994, 444–51 (modified); cf. R. Wilhelm 1967, 290. Ge is number 49 in the received sequence of the hexagrams. Most translations of its hexagram name focus on the idea of change, but some emphasize more concrete meanings: skin, leather, a bridle. See note 10 above.
36 Lynn 1994, 444–51; cf. R. Wilhelm 1967, 189–92 and 635–40.
37 See R. Smith 2008a. See also Shilling 2009.
38 For valuable details on virtually all such schemes (organized alphabetically by Pinyin transliterations), consult Nielsen 2003.
39 Lynn 1994, 93–95; cf. R. Wilhelm, 1967, 352–55.
40 Lynn 1994, 54–61 and esp. 62–63 and 72–73 (notes 35–45); cf. R. Wilhelm, 1967, 300–12 and esp. 314–15.
41 Lynn 1994, 62–63 (modified); cf. R. Wilhelm, 1967, 314–15.

42 R. Smith, 2012, 77. See also Granet, 1934, esp. Chapter 3, and Fendos 1988, 148–57 and 264–67. For some Chinese works on images and numbers, see note 31 above.
43 See R. Smith 2008a, 59–66, 89, 103–06, 114, 125, 133, 136, 150, 173, 178, 202, 330 and 348. Wu Lin 1996 effectively compares and contrasts the two major schools of interpretation. See also Fendos 1988, 148–57 and 264–67 and 289–417, passim.
44 See, for example, R. Smith 2002 and 2008a, Chapters 3–7, passim.
45 See R. Smith 1991, esp. Chapter 2 and Aylward 2007, passim. Fendos 1988, 289–417 provides an enormous amount of detail on these and other Han-era schemes.
46 For details on these systems, see Nielsen 2003, 1–6 (eight palaces), 75–89 (hexagram breaths) and 180–84 (attached initial stems and attached initial notes). See also ibid., 7–9, 59–62, 67–69, 204–08 and 274–76 and Fendos 1988, 303–83, esp. 321–40.
47 For evidence that versions of these documents may have existed in Han times, see Nielsen 2003, 103–05, 169–71 and 236–37.
48 For the controversies surrounding the *Hetu* and *Luoshu*, see ibid., also R. Smith 2008a, 78–80, 134–35, 143, 151, 172–74, 177, 188, 193, 294, 312.
49 Significantly, this arrangement differs from the logic of the "Explanation of Trigrams" itself, in which the pairings are the same but the positions differ.
50 R. Smith 2012, 96. For details, see Nielsen, 2003, 263–68.
51 R. Smith 2012, 97. For details, see Nielsen, 2003, 107–10.
52 R. Smith 2012, 94–96. For details see Adler 2002, 3–14 and Nielsen 2003, 103–05.
53 R. Smith 2012, 96. For details, see Adler 2002, 3–14 and Nielsen 2003, 169–71.
54 See the discussion in R. Smith 2012, 97–98. See also Nielsen 2003, 103–05, 169–71, 236–37 and 254–56, and R. Smith 2008a, 117–20, 149, 180–81.
55 K. Smith, *et al.* 1990, For details on Shao Yong's *Huangji jingshi shu* (Supreme principles that rule the world), see, for example, Birdwhistell 1989, Wyatt 1996, Ryan 1996; on Lai Zhide's approach to the *Changes*, consult Schulz 1982.
56 For a discussion of this ordering system and its influence in the West and modern China, see Chapter 6 in this volume.
57 See R. Smith, 2008a, 160–65.
58 See R. Smith 1991, 70–74 and 89–91.
59 See the article on Qing divination in this volume; cf. R. Smith 2008b, 141, 184–86, 193, 209, 240 and 266.
60 See Lynn 1994, 76; cf. Wilhelm 1967, 283, 288–89, and esp. 326–27.
61 Divided lines are designated *zhai* (broken); solid lines, *dan* (single, or simple).
62 For detailed accounts of this process, see R. Smith 2008a, 226–32 and 2012, 109–23, passim.
63 These line positions are also sometimes correlated with moral values and even the body parts of both humans and animals. For some examples, see Nielsen 2003, 299–300, and Shchutskii 1979, 8.
64 For details, see Nielsen 2003, 294–99, 333, and R. Wilhelm 1967, 360–65.
65 When "holding together" occurs between the fifth and the sixth lines, it may be advantageous if the ruler is humble (a yielding line in the fifth place) and the sage is strong (a firm line in the sixth place). See Spence 1975, 147, 171, slightly modified. Cf. Wilhelm 1967, 129, 362–64 and 550.
66 See Nielsen 2003, 85–92 for other variables.
67 See R. Wilhelm 1967, 264, 289, 364–65 and 446–50.
68 Lynn 1994, 120.

69 See CC, *Mengzi*, 1: 559–60.
70 See the discussion in R. Smith 1991, 103.
71 For details and refinements, see Wilhelm 1967, 262 ff.
72 Lynn 1994, 242 (modified); cf. R. Wilhelm 1967, 72 and 473.
73 Ibid. (both sources).
74 R. Smith 2012, 81–83.
75 R. Wilhelm 1967, 709–13 discusses the implications of the relative positions of the nuclear trigrams for an understanding of the Jiji hexagram. Cf. Lynn 1994, 538–44.
76 What follows is a more or less conventional Song dynasty understanding of the Ge hexagram, based primarily on R. Wilhelm 1967, 635–40 and Zhuxi's *Zhouyi benyi* (Basic meaning of the *Zhou Changes*. Taibei: Xin wenfeng cuhuban gongshe, 1979), 169–72. See also Legge 1899, 167–70, and the extensive remarks by the Song dynasty commentator Cheng Yi, cited in Lynn 1994, 449–50, note 7.
77 See R. Wilhelm 1967, 635. Ibid., 636–38 devotes a good deal of attention to the calendrical correlations associated with this and other hexagrams.
78 This is how the Ge hexagram was generally understood. See, for example, Hon 2005, 86; W. Chan, trans. 1967, 189–90; R. Wilhelm 1967.
79 Discussed in Hon 2005b, 87.
80 See R. Smith 1991, 1998c, 2008a and 2012.
81 See the discussion in R. Smith 1991, 120, based on the stimulating work of Herbert Schneidau. Much of the following material on the cultural significance of the *Yijing* has been derived from ibid., 119–29.
82 For examples of such "inventories" in Chinese, see the collections by Tang Mingbang, *et al.*, eds. 1986; Zhu Bokun, ed. 1996; Zhang Qicheng, ed. 1996; etc. For a partial list of individual studies, see R. Smith 2008a, 218–19 and 253–55.
83 On the *Taixuan jing*, see Zheng Wan'geng 1997; also Sivin and Nylan 1995 and Ziporyn 1995–96. Cf. R. Smith 2008a, 73–77, 94, 108, 131–32, 144–45, 147, 156, 174, 190. For studies of the *Yilin* attributed to Jiao Yanshou, see Qian Shiming 1990. Cf. R. Smith 2008a, 71–73, 144, 152 and 281, note 51.
84 Lynn 1994, 129; cf. Wilhelm 1967, 3–4 and 369–70. On Yan's views, consult Y. Fung 1953, 2:636ff. Not surprisingly, the editors of the famous Qing compendium known as the *Siku quanshu* (Complete *Collection of the Four Treasuries*) criticize Yan for going "too far" in his opposition to Song Learning. SKQSZM, 97:9b–11b.
85 See, for example, R. Smith 2008a, 89–94, 101–03, 108–11, 130–31, 137–45, 165–69, 174, 179, 188, 190, 194, 219–22, 242–48 etc. Cf. Zhang Qicheng, ed. 1996, 1: 411–40 and 2: 377–390. On ethical thought in particular, see ibid., 1: 430–40.
86 Zhixu and Liu are both cited in R. Smith 1991, 121. Cf. R. Smith 2008a, 169–70 and 186–87.
87 See, for example, the views of the Qing scholar Hang Xingzhai in BRZ 7:27–29.
88 Cited in R. Smith 1991, 121. For discussions of the relationship between the *Yijing* and religion, see Zhang Qicheng, ed. 1996, 2: 377–90.
89 Lynn 1994, 368; cf. R. Wilhelm 1967, 147–48 and 573–75.
90 Lynn 1994, 87 (slightly modified); cf. R. Wilhelm 1967, 344–45.
91 Hall and Ames 1995 and 1998 and Fleming 2009 offer comparative reflections on Chinese and Western forms of logic. The philosopher Mozi (*c.* 470–391 BCE), did, however, develop the rhetorical strategies of "argumentation" (*bian*), category (*lei*) and cause (*gu*).
92 R. Smith 1994, 115.

93 R. Smith 2008a, 221–26. Cf. Zhang Qicheng, ed. 1996, 1: 452–83 and 2: 318–76.
94 Cited in R. Smith 1991, 122.
95 Cited in R. Smith 2008a, 222.
96 Cited in R. Smith 1991, 123.
97 Ibid., 123.
98 Cited in ibid, 123. R. Smith 1994 devotes a great deal of attention to the influence of *yin* and *yang* in all facets of traditional Chinese culture, from philosophy and religion to art, literature and social customs.
99 BRZ, 21:4–5.
100 R. Smith 1991, 124.
101 Ibid.
102 TSJC, *jingji*, 55:971–93.
103 Ibid., 55:986–95. A number of these poems have been translated and analyzed in R. Smith 2008a, 223–25. Other distinguished contributors to this collection of verse include the Tang scholars Meng Jiao and Zhu Qingyu, the Song scholar Qiu Cheng, the Yuan scholars Liu Yonglian and Huang Geng, and the Ming scholars Fang Xiaoru, Xue Xuan, Zhuang Chang, Hu Juren, Chen Xianzhang, Wang Ji, Yang Jue, and Chen Jiang.
104 Ibid., 55:993–95.
105 Lynn, 1994, 363–67; cf. R Wilhelm 1967, 143–47 and 569–73. For discussions of the *Yijing* and traditional Chinese political, social and military affairs, see Zhang Qicheng, ed. 1996, 1: 441–51 and 2: 284–98, 391–418.
106 P.Y. Ho 1972, 23–39 provides these identifications. Cf. SCC 2:315–21.
107 Cited in W. Chan, ed. 1967, 202 (slightly modified).
108 SKQSZM 1:3a. See also BRZ 7:20–21 and 13:3.
109 Ho Peng Yoke goes so far as to say: "If ... [the Chinese] were fully satisfied with an explanation they could find from the system of the *Book of Changes* they would not go further to look for mathematical formulations and experimental verifications in their scientific studies." P.Y. Ho, 1972, 38.
110 See R. Smith 2008a, 236–40 and 2012, 219–23. See also Zhang Qicheng, ed. 1996, 1: 484–501 and 2: 3–283.
111 SCC 2:292, 304–40; 3:56–59, 119–20, 464, 625; 4.1:14, 16; 4.2:143, 530; 4.3:125; 5.3:51–53; 60–66, 69–74, 128, 217, etc. See also P.Y. Ho 1972, esp. 30–38.
112 SCC 2:329–35.
113 Ibid., 2:337.
114 Ibid., 2:336.
115 Elman 2005 and 2006, Sivin 1990, 1995a, 1995b and 2005. Cf. R. Smith 2008a, 236–40 and 2012, 219–22.
116 See Elman 1984.
117 See Sivin 2005.
118 H. Wang 1995 and P.Y. Chu 1995, esp. 72–73. As both articles indicate, the interaction between evidential scholarship, "traditional" learning, and Western learning during Qing times was highly complex.
119 For a more thorough discussion, see R. Smith 2008a, 235–40.
120 See, for example, Chu 1995, 95; cf. Legge 1963, 38 and 329. This longstanding tendency to stress the Chinese origins of Western learning (*Xixue Zhongyuan*) sometimes reflected genuine conviction but it was also used as a rhetorical device.

121 Thereafter, Chinese scholars such as Guo Moruo, Gu Jiegang, Wen Yiduo, Yu Shengwu, Li Jingchi, Gao Heng and others examined the *Yijing* in a more critical and "scientific" spirit. See R. Smith 2008a, 199–201.
122 See Tang Mingbang, *et al.*, eds. 1986.
123 See Liu Zheng 1993.
124 The editors' "Introductory Note" to Tang Mingbang, *et al.*, eds. 1986, as well as a number of individual articles in this volume, underscore the theme of Chinese national and cultural pride. Regional differences are also discussed in this volume, however.
125 A large section of ibid. is devoted to this theme, and to the place of the *Yijing* in Chinese medicine. See esp. ibid., 423–552.
126 See Feng Youlan, "Congratulations in Absentia," in ibid., 7–9. See also the articles by Pan Yuting, Zhao Dingli and others in ibid., 423ff.
127 See C. J. Jung's "Foreword" to Wilhelm 1967, esp. xxxiii–xxxiv.
128 I have reviewed this book in the *Journal of Asian Studies*, 51.4 (November, 1992), 895–96.
129 Liu Zheng 1993.

2 Mapping China's world

1 David Woodward's analysis of medieval *mappaemundi* in the West has important implications for an understanding of Chinese maps of the world, since both tended to blend time and space and to project "historical events on a geographic framework." See Woodward 1985, esp. 514.
2 Nor are these relationships and distinctions always clear in Chinese discourse. The term *tianxia*, for example, often refers to the "Chinese empire," but there are contexts in which it clearly refers to the "five continents."
3 For one useful inventory of European collections of Chinese maps, see Li Xiaocong 1997.
4 The Beijing Library (*Beijing tushuguan*, now known as the National Library of China [*Zhongguo guojia tushuguan*]), the British Library and the Library of Congress have been particularly helpful in this regard. See also the useful website titled "Zhongguo gu ditu" (Ancient Chinese Maps) at: http://geog.hkbu.edu.hk/GEOG1150/Chinese/. Many of these images have been drawn from Chinese Academy of Surveying and Mapping, ed. 1998.
5 Among the most useful secondary works in Chinese that I have consulted are the writings of Cao Wanru, Chen Feiya, Chen Zhengxiang, Li Xiaocong, Lu Liangzhi, Que Weimin, Ren Jincheng, Yan Ping, *et al.* 1995, Zhao Rong and Yang Zhengtai; in Japanese, Akioka Takejiro and Unno Kazutaka; in Western languages, Walter Fuchs and Lothar Zogner (in German), Henri Bernard and Marcel Destombes (in French), and Sen-dou Chang, Kenneth Chen, Mark Elliott, Phillipe Foret, Funakoshi Akio, Laura Hostetler, Hsu Mei-Ling, D.R. Howland, Arthur Hummel, James Millward, Roderich Ptak, Emma Teng, Helen Wallis, David Woo, and esp. Cordell Yee (in English).
6 Farquhar and Hevia 1993, 492–94.
7 Ibid., 489.
8 I might add that my use of the term "traditional" in this chapter implies no pejorative judgment; it refers simply (or perhaps not so simply) to an inclination on the part of individuals and groups to focus primarily on inherited wisdom and indigenous experience as a guide to the present and future. I do not view "tradition" as bad and "modernity" (however defined) as good; nor do I accept the presumably related "Orientalist" dichotomies that artificially oppose "culture" to "reason" or "culture" to "history."

9 Crossley 1990a, 2.
10 Sahlins 1993.
11 Ibid., 15.
12 Ibid., 16.
13 Rundstrom 1991, 1–2.
14 Wood 1992, 1.
15 Boelhower 1988, 479.
16 See also the stimulating discussion in King 1996, 18–39, on "World Views," esp. 31–37.
17 Clarke 1988, 455.
18 See "Cartography as Art," by Arthur Robinson in Rhind and Taylor, eds. 1989.
19 Cited in Wood 1992, 78.
20 Harvey 1990, 226.
21 Wood 1992, 79.
22 Dathorne 1994, 1–7.
23 Boelhower, 1988, 478.
24 A 1761 inventory of the Qing dynasty's Map Archives (*Yutu fang*) in Beijing reveals the following cartographic categories: astronomical maps; regional maps; sea and coastal charts; river charts; military and strategic maps; route maps (particularly maps for imperial travel); maps and other views of well-known topographical sites; symbolic representations; pictures of life in foreign countries; maps and pictures connected with salt extraction; famous temples; imperial mausolea and geomantic maps. See Fuchs 1937, 22.
25 Zhang Yanyuan 1983, 130. Zhuang Tingfu uses the term "spiritual journey" (*shenyou*) in discussing the use of maps such as his *Da Qing tongshu zhigong wanguo jingwei diqiu shi* (Model of the myriad tributary states of the great Qing dynasty from around the globe; 1794).
26 It is true, however, that officials and the throne often recognized the special cartographic skills of particular individuals, including foreigners, and commissioned them to produce maps. For reasons of state security, the Chinese government relentlessly tried to control the production and circulation of strategically important maps. Note, for example, the Japanese appropriation in 1750 of Lü Junhan's strategically oriented *Lidai shiji tu* (Map of the traces of events over time; 1679). For additional information on cartography and state security in China, consult Yee in Harley and Woodward, eds. 1994, 83–84.
27 They did not always coincide, however. Thus, for example, certain powerful cultural symbols, notably the Great Wall, might appear prominently in Chinese maps even if they depicted times when the structure did not actually exist or existed only in fragments.
28 R. Smith 1996, 35–36.
29 For other discussions and depictions of Jambudvipa, consult Harley and Woodward, eds. 1994, 230, 368, 371–73, 375, 429, 619, 621–23, 632, 716, 722–26, 730–33, 735–36, 781. On "wheel maps," consult ibid., 174–75 and 262–63.
30 Smith 1991, 131ff. and 1996, 35ff.; cf. Yoon 1992.
31 See the commentary by Yang Wenheng on plates 201–02 in Cao, *et al.* 1990; also ibid., 33–34.
32 Foret 1995 and 2000.
33 See Foret 1995; also Yang Wenheng in Cao, *et al.* 1994; Harley and Woodward, eds. 1994, 59, 221–22; Yee, *et al.* 1996, 22, 67–68; S. Chang 1974, 7–10.
34 See Nathan Sivin in Harley and Woodward (1994), 30. For a more detailed discussion of the "flat-earth" issue, consult Cordell Yee in ibid., 117–24, 170, 171, 203, 225.
35 Yee 1992, 31ff.

36 See M. Lewis 1999.

37 De Weerdt 2009, 157–64 provides an excellent discussion of how Southern Song scholars who viewed them emphasized, sometimes in verse, the sense of territory that had been lost to foreign invaders.

38 For a discussion of this map, together with illustrations, see Cao, *et al.* 1990, plates 70–72 and commentary; SCC 3:550–51; see also De Weerdt 2009, and Yee in Harley and Woodward, eds. 1994, 47. On the textual emphasis of Chinese cartography, see Yee in Harley and Woodward, eds. 1994, 29–30, 64, 107, 158, 171, 228.

39 S. Chang 1974.

40 Jullien 1995, 103, citing Wang Fuzhi.

41 Yee 1992.

42 The most commonly used Chinese terms for foreigners in late imperial times (*yi*, *fan* and *hu*) implied a degree of cultural inferiority; hence, they are often translated by the English word "barbarian." This usage, derived from the ancient Greek distinction between themselves ("civilized") and non-Greeks (*barbaros*; "uncivilized"), basically parallels Chinese usage. Other terms for aliens, such as *yuanren* ("people from afar"), *wairen* ("outsiders") and *yangren* ("people from oceanic realms") were less pejorative. For an excellent discussion of "barbarians," their subdivision into categories of "raw" and "cooked," and their role in defining "China" and contributing to the Middle Kingdom's ongoing "civilizing project," see Fiskesjö 1999, Rowe 2001, esp. 406–26. Cf. W. Fang 2001, Hara 1998, Pines 2005, L. Wong 2010. Cf. R. Edwards 1987, Hinsch 2004, C. C. Huang 2007, Elliott 1996 and 2000.

43 For the importance of the Song period in the history of Chinese cartography, see De Weerdt 2009 and Lu Liangzhi 1984, 79ff. For useful background, consult Needham 1970, 500ff. and Cordell Yee's essays in Harley and Woodward, eds. 1994, esp. 37–46, 72–81. See also Cao, *et al.*, eds. 1990, 1–30.

44 This map is one of several dozen in a collection known as the *Lidai dili zhizhang tu* (Convenient geographical maps, chronologically organized; 1098–1100, supplemented in 1162). See Shanghai guji chuban she, ed. 1989. For a systematic comparison of the Song and Ming editions of this work, consult Cao, *et al.*, 1990, 32. Ibid., 32–34 summarizes the basic content; see also De Weerdt 2009, 157–58. For illustrations of several of these maps, including the *Gujin Huayi quyu zongyao tu*, see ibid., plates 94–101 and Yee in Harley and Woodward, eds. 1994, 58. For examples of Ming editions, consult ibid., 168; also M. Hsu 1997, 39.

45 Both maps may have been based on a prototype created by the legendary Tang cartographer, Jia Dan. See the full discussion in De Weerdt 2009, 152–57; see also Cao, *et al.* 1990, 31–34. For a systematic comparison between the *Huayi tu* and the *Gujin Huayi quyu zongyao tu*, consult ibid., 42–45. For illustrations of the *Huayi tu*, see ibid., plates 61–62; also SCC 3, 500 ff. and M. Hsu 1997, 32.

46 The *Huayi tu* probably inspired the *Yudi tu* (Territorial map; *c.* 1270), which is also heavily annotated in a similar manner. See Cao, *et al.* 1990, 56–60 and ibid., plates 82–83.

47 For illustrations of this map, see Cao, *et al.*, 1990, plates 54–56. Cf. the *Yuji tu* of 1142; ibid., plates 57–59. For other illustrations of the 1136 *Yuji tu*, consult SCC 3, plate LXXXI; Yee in Harley and Woodward, eds. 1994, 48 and Figure 3.15.

48 See also Cao, *et al.* 1990, 35–40 and plates 63–65 on the equally "modern-looking" *Jiuyu shouling tu* (Administrative Map of the Nine Regions; 1121).

49 For an abundantly illustrated overview, consult Fuchs 1937; see also Lu Liangzhi 1984, 100ff.; M. Hsu 1997, 41–46. Liu Gang 2007 speculates that Zhu Siben drew the world's first bi-hemispherical world map, which Luo Hongxian then rendered into rectangular projections. See esp. 187–88.

50 For information on the sailing charts used by Zheng He, consult Niu Zhongxun in Cao, *et al.* 1994; also Lu Liangzhi 1984, 163ff. and SCC 3:556ff. For other illustrations of this type of map, see Yee in Harley and Woodward, eds. 1994, 54–55 and Smith 1996, 34.

51 For an illustration of the former map, see Smith 1996, plate 1; also Fuchs 1935–1936, 388–89. For information on, and illustrations of, the *Huang Ming zhifang ditu*, see Cao, *et al.* 1994, plates 231–34 and the commentary by Ren Jincheng in ibid. Also Lu 1984, 105 ff.

52 For magnificent color illustrations of the *Da Ming hunyi tu*, consult Cao, *et al.* 1994, plates 1–5; see also the illuminating discussion by Shen Xiaoya in ibid. On the evolution of this type of map during the fifteenth and sixteenth centuries—focusing on the well-known works of Yang Ziqi, Wang Qingquan, and Wang Pan—consult Cao, *et al.* 1994, plates 12–15, 60–61, 94–96 and commentaries; also Funakoshi 1975, 152–53 and Figure 9, 168 and Unno 1977.

53 For detailed information on this Korean map, see http://www.henry-davis.com/MAPS/LMwebpages/236mono.html.

54 See the discussion in Ren 1983. A photograph of the original map appears in Harley and Woodward, eds. 1994, plate 1. Cf. the quite similar Korean map, dated 1666, ibid., 268.

55 For discussions of the *Shanhai jing* and its significance, see H. Cheng, *et al.* 1985, Fracasso 1993 and Birrell, trans. 1999. For examples of "wheel maps," consult Ledyard in Harley and Woodward, eds. 1994, 256–68; also Smith 1996, 17–19.

56 Both Zhuge Liang and Wen Tianxiang are mentioned several times in different parts of the map.

57 SCC 3:547ff. Cf. Hostetler 2001.

58 Elman 1981–83, 1–2; SCC 3:585–86; Lu Liangzhi 1984, 170ff.; Hostetler 2001. The secondary literature on the Jesuits in the history of Chinese cartography is vast. See the bibliography in Standaert, ed. 2001, 763–70. For a few relevant works in Chinese, consult Wang Mianhou, Cao Wanru and others in Cao, *et al.* 1994; Tan Qixiang 1993a and 1993b; Wang Qianjin 1995; Xiong Yuezhi 1994, 45ff. For a few relevant works in Western languages, consult K. Ch'en 1939; Fuchs 1935–36, 1938 and 1943; Elliott 1995; Hostetler 2001, esp. 63–64 and 71–80; Perdue 2005, esp. 443–61, Ptak n.d., and Cordell Yee's useful summary in Harley and Woodward, eds. 1994, 170ff.

59 Wang Qianjin 1995. Note the comments by the Kangxi emperor and his ministers cited in L. Fu 1966, 1: 127–28.

60 Elliott 1996 and 2000 offer provocative interpretations of the relationship between Jesuit cartography and Manchu (as opposed to Chinese) empire-building. See also Hostetler 2001, 51–80, Millward 1999, Elman 2003, and Perdue 2005, esp. 443–61.

61 Smith 1996, 42–59; cf. Ptak n.d. On the comparatively greater Japanese receptiveness to Western cartographic knowledge in the seventeenth century, see Funakoshi 1986; also Harley and Woodward, 2.2 1994, esp. 347, 376–86, 390–402, 404–10, 426–30, 432–42, and 454.

62 Perhaps for reasons of tact, Jesuit maps such as Ricci's also emphasized the tributary system and China's cultural grandeur. Smith 1996, 47.

63 For these and other criticisms, see K. Ch'en 1939, 348–50. Even admirers of the Jesuit cartographic enterprise saw their maps as targeting certain resources to be exploited by Westerners in various parts of the world, including China. See, for instance, the commentary by Zhuang Tingfu on his *Da Qing tongshu zhigong wanguo jingwei diqiu shi* (1794), which emphasizes the profit motive of the West.

64 See Chen Guansheng 1939. Chen attributes "the rapid decline of Ricci's influence" to Chinese complacency, growing hostility to Catholic missionary activity, the "underdeveloped state of science in China," and the carelessness with which Chinese scholars handled Ricci's cartographic materials. See also Ptak n.d.

65 It is true, however, that some Ming scholars were quite taken with Jesuit cartographic models. Cheng Boer and his scholarly associates, for instance, introduced renderings of the Eastern and Western hemispheres based on Ricci's maps, together with Ricci's written explanations of these maps and a six-*juan* supplement on the "Outer Barbarians" (*waiyi*) in their *Fangyu shenglüe* (Compendium on geography; 1612)—a work that was later banned during the Qing dynasty. For a picture of the two hemispheres, consult Yee in Harley and Woodward, eds. 1994, 178–79.

66 Destombes 1983, 57, suggests that the date of Liang Zhou's map should probably be 1603 or 1605 instead of 1593, the usual date given.

67 For illustrated discussions of Liang's map, see Cao, *et al.* 1994, plate 145 and commentary; also R. Smith 1996, 56. For other examples of the way Jesuit models were transmuted, see ibid., 57–59.

68 See John Henderson's essay in Harley and Woodward, eds. 1994, 225–27. Most Chinese continued to believe that the earth was flat until well into the nineteenth century.

69 K. Ch'en 1939, 350.

70 There is also a 1663 version of this map, with the same title, by Wang Junfu, and a similar work, dated 1679, with a somewhat different title by Lu Junhan. See Unno 1977, 208–09, 212; Fuchs 1935–36, 395–96.

71 For further information on Cao Junyi's map, consult Cao 1994, plate 146 and commentary by Ren Jincheng. See also Unno 1977, 208–09 and 212, and Fuchs 1935–1936, 395–395 on the versions of this work produced by Wang Junfu and Lu Junhan.

72 See R. Smith 1996, color plate 3.

73 For an interactive on-line version of this tripartite map with commentary, see http://www.rice.edu//fondren/erc/projects/jingban/.

74 Ball 1903, 418–19.

75 For a useful discussion of Huang's map, which includes several illustrations of the *Tianxia quantu* genre, consult Fuchs 1938, 208ff., especially the four plates preceding 209; also Fuchs 1935–36, 394–95. For other illustrations of such maps, see Smith 1996, color plates 12–14.

76 Reproduced in Zogner 1983.

77 By contrast, Yu Shi's *Gujin xingsheng zhi tu* (1555) and maps patterned after it—including Harvard University's *Lidai shiji tu* (Map of the traces of events over time; 1679), and the British Museum's strikingly beautiful but anonymous and untitled world map of 1743—devote a great deal of attention to historical figures and culture heroes.

78 Fuchs 1938, 210–11.

79 Dong Gao 1761.

80 See Qi Yunshi 1884.
81 Cited in R. Smith 1994, 283.
82 See Dong Gao 1761, 1: 23, 47ff., 61, etc.; cf. Fairbank 1953, 11ff.
83 For similar discussions of this sort of "confusion" in standard Qing reference works, consult Fairbank and Teng 1939, 187–90; see also Ptak n.d.
84 Another piece of misinformation that appears in these maps is attributed to what must have seemed to be unimpeachable foreign authority: the Dutch "ambassador's" claim that Holland's boundary is connected to that of Russia and that the languages of the two countries are mutually comprehensible. For the probable source of this confusion, see L. Fu 1966, 1: 88 and 2: 469, note 228.
85 See Harley and Woodward, eds. 1994, 254. Zhuang's personal name is misrendered in this account, but the map seems quite obviously to be his.
86 For another *qi*-oriented explanation of differences between the Chinese and foreigners, see R. Smith 1996, 9.
87 Hevia 1995, 25.
88 Ibid. Cf. W. Fang 2001, Fiskesjö 1999, J.M. Lee 2008, Li Yunquan 2004, Swope 2002 and 2009, D. Kang 2010, Hara 1998, Hamashita 2008, Iwami 2009, Meserve 1982, E. Ong 2004, Pines 2005, Ptak 1999, F. Zhang 2009 S. Zhang 2006, and Zhuang Wanshou 2011.
89 For recent scholarship on the economic importance of the tributary system, see Hamashita 2008, 12–26, Iwami 2009, and D. Kang 2010, passim. Cf. Rowe 2009, 133–38.
90 See example, Fairbank 1942, 1953, Pritchard 1943, Fairbank, ed. 1968 and Fairbank and Teng 1939. For more recent scholarship indicating the sources and/or persistence of this Sinocentric rhetoric, see Behr 2010, Bickers, ed. 1993, Cartier 1981, Duman 1981, Fang 2001, Fiskesjö 1999, Hinsch 2004, C. Huang 2007, Kelly 2003, J.M. Lee 2008, Li Zhongjun 1994, Pines 2005, Ptak 1999, R. Smith 1975, Wong 2010, S. Yang 1991, and Zhuang Wanshou 2011. Most of these studies, like those listed in note 92, make a sharp distinction between the "rhetoric" of the tributary system and the "reality." See esp. D. Kang 2010, J.M. Lee 2008, E. Ong 2004, S. Zhang 2006 and F. Zhang 2009. Kim 2009 and Poo, 2005, offer valuable comparative perspectives involving ancient Greece, Mesopotamia, Egypt and China.
91 See Mizuno 2003, esp. 36–39 and J.M. Lee 2008, esp. 4–5, 18–20, 65–75.
92 See, for example, Brose 2006, Crossley 1990b and 1999 and Crossley, Siu and Sutton, eds, 2006, Dabringhaus and Ptak, eds. 1997, Di Cosmo 1998 and 2010, Di Cosmo and Wyatt, eds. 2003, Elliott 1996, 2000, 2001 and 2009, Hamashita 2008, Hara 1998, Harrell 1995, Hostetler 2001, Howland 1996, Iwami 2009, Lary, ed. 2007, J.M. Lee 2008, Millward 1998 and 1999, Millward, *et al.*, eds. 2004, Mizuno 2003, E. Ong 2004, Perdue 2005, Swope 2002 and 2009, E. Teng 2004, Tikhvinsky and Perelomov, eds. 1981, Van Lieu 2009, Wills, *et al.*, eds. 2011, S. Yang, 1991, Yun 1998 and n.d., F. Zhang 2009, S. Zhang 2006.
93 As Wang Gung-wu has pointed out, at times the idea of China's superiority "was clearly a myth, a sustaining and comforting myth," but one that was contingent on China's strength; at other times, however, it was "a reality, a reality that nurtured cultural pride but also called for moral restraint." Cited in F. Zhang 2009, 570.
94 Hevia 1995, 17–18.

95 See esp. Hevia 1995, Chapter 5, in which Hevia downplays Chinese terms of condescension but accurately captures the idea that the Qing emperor possessed "exemplary power capable of moral ordering," and that he addressed "the imperial domain and the world at large by 'instructing, admonishing, cherishing, and rewarding his servants (*chen*).'" This imperial instruction, Hevia goes on to say "embodies aspects of sagely and virtuous kingly rule and, as such, has the power to reorient others, including lesser lords." Ibid., 119–20. For criticisms of Hevia's translations and/or political agenda, see Esherick 1998a and 1998b and Zhang Longxi 1998. For Hevia's defense, see Hevia 1998 as well as Elman and Huters, 1997.

96 As Pritchard 1943, 196–200 points out, although "there can be little doubt that the Chinese viewed their contacts with outside peoples as the relationship between superior and inferior," they did not think of this relationship as a humiliating one for the inferior; rather it was "part of the natural order of things from which mutual benefits flowed to both parties."

97 See the discussion in F. Zhang 2009, esp. 570ff.

98 Cited in R. Smith 1994, 138. Hevia 1995, 116–17, note 2 expresses the same idea in discussing the different editions of the *Da Qing tongli*. See also Chapter 3 on Qing ritual in this volume.

99 See Pritchard 1943, esp. 185ff.

100 Hevia 1995, 128; see also ibid., 164, 187.

101 I have discussed the strengths and weaknesses of Hevia's study at greater length in a joint review of this work and Bickers, ed. 1993, which appeared in the *Journal of the Royal Asiatic Society* third series, 7, 1997, 170–71. Some of the contributors to the Bickers volume are among the several Chinese scholars sarcastically and condescendingly critiqued by Hevia 1995, 239–48 for adopting, "wittingly or not," "the intellectual framework of the colonizer."

102 Wills, *et al.*, eds. 2011, 2–4.

103 Fu 1966 provides numerous illustrations from official documents. See, for example, vol. 1: 9, 11–12, 15–20, 46–47, 49–50, 64, 77, 81–83, 85–86, 98, 100, 134–35, 139, 142, etc.

104 Ibid., 4.

105 Duyvendak 1938, 36–37 (my translation of the Chinese text).

106 Ibid., 73–74; my translation.

107 R. Smith 1975, 124–32; Smith 1994, 282–84. See also Fairbank in Fairbank, ed. 1968, 204–231; Swisher 1953, passim; Fu 1966, passim and note 103 above. An excellent cartographic illustration of this point appears in a beautiful, six-scroll 1895 map, based on a late Ming *tianxia* model, but including some mythical countries from the *Shanhai jing*. See Dahlen 1997, 3–4.

108 Polachek 1992, 7. This basic argument, first articulated by Fairbank 1942 and 1953 and I. Hsu 1960, remains to be more fully debated and discussed. What, for example, do we make of the 1880 preface to the official Qing dynasty collection of documents on China's foreign relations known as the *Chouban yiwu shimo* (The management of China's barbarian affairs from beginning to end), which casts so many of the humiliations imposed upon China by the West in the period from 1842–1880—from demands for an audience with the emperor on terms of diplomatic equality, to the onerous unequal treaties, to forced Chinese missions of apology to Europe—in terms

of imperial condescension? See the quotation and discussion in R. Smith 1994, 283; cf. Wong 2010.

109 See R. Smith 1996, 59 and plates 7–11. Cf. Lü Junhan's *Lidai shiji tu* (Map of the traces of events over time; 1679) and the *Tianxia shijiu sheng* (The nineteen provinces of all under Heaven; 1895) reproduced in Dahlen 1997, pp. 3–4. For affinities between the 1743 map, Liang Zhou's map and another anonymous map of 1739, consult Destombes (1983), 62–65. Like many of its predecessors, the 1743 map includes a great number of inscriptions dealing with various foreign lands and peoples (including those mentioned in the *Shanhai jing*), Sino-foreign relations (including, of course, the tributary system), administrative changes within the Chinese empire, and miscellaneous facts about various landmarks. We are told, for instance, that when the Qin built the Great Wall, they extended the structure several *li* into the sea by dropping "raw iron" into the water to serve as a foundation. The *Lidai shiji tu* has similar (and sometimes identical) inscriptions.

110 See Smith 1996, 62. Later, Li produced a comprehensive map of the Qing empire with similar cartographic features. See Anonymous 1832, 33–34. For additional evidence of Western missionary influence on Chinese geography and cartography, consult Xiong 1994, 115ff.

111 See R. Smith 1996, plates 15 and 16.

112 Ibid., 60–75; also Yee in Harley and Woodward 1994, 106–08 (on Xu Jiyu and the *kaozheng* tradition of scholarship) and 192ff. (on Wei Yuan). For a view of Wei's *Haiguo tuzhi* that locates it more fully in the context of Qing factional politics, see Polachek 1992, 195ff. Polachek also discusses the works by Wei and Xu as part of a more general burst of Chinese interest in coastal defense, geography and "political economy." Ibid., 178.

113 For background and major developments, consult Amelung 2007; also Lu Liangzhi 1984, 191ff. and Yee in Harley and Woodward, eds. 1994, 195ff.

114 See R. Smith 1996, 74–75 and plate 17.

115 Xu Jiyu wrote, for example, "When ... [the Jesuits] say falsely that they come from a great distance of 90,000 *li*, it is because they want us to believe that they have no ulterior motives behind them, so as to prevent us from worrying about their aggressive purposes." Cited in Ch'en 1939, 349. See also the remarks by Shen Que in Harley and Woodward, eds. 1994, 171.

116 For a convenient (albeit tentative) typology of cartographic functions, consult Harley and Woodward 1994, 847–48. This typology includes the following categories: inventory, way-finding, intelligence, education, planning and engineering, administration, commemoration, divination, and worship. I would add "aesthetic" to this list.

117 See Cao, *et al.* 1994, plates 57–59 and 77–80; also ibid., plates 91–93, and 226; R. Smith 1996, 42–59.

118 Hightower 1970, 229ff. (modified). Cf. the discussion in De Weet 2009, 155–57 concerning the poems that were inspired by maps in the Tang period.

119 Hightower 1970, 230.

120 See R. Smith 1996, 72 and Dathorne 1994.

121 See Strassberg 1994, 3–4.

122 Woodward 1996, 5; see also Turnbull 1993–94 and 1996.

123 Woodward 1996, 5.

3 Ritual in Qing culture

1 A. Wright 1960, 234–35.
2 A. Wright 1963, 39.
3 L. Wieger 1913, 98 (Chinese text); cf. Ebrey 1981, 204. For a specific example of these general points, see HCJSWXB 61:8b.
4 See, for example, HCJSWB 68:1a–b. In this essay, Lu Longqi emphasizes the role of the state in educating the people and rectifying local customs. Despite his official support for Zhu Xi's orthodoxy, Lu argues in Legalist fashion for a system of rewards and punishments and for fixed rules and regulations. He also advocates a set system of capping, marriage, mourning, and sacrificial rites as prescriptions for proper housing, clothing, food, and drink. His stress on the transformative power of moral education (*jiaohua*) can be found in many other writings in the HCJSWB and its supplements. See, for example, HCJSWB 54:6a–8b; HCJSWXB 61:6a–b; HCJSWBXB 74:1a–8b.
5 For one of many Qing period discussions of the relationship between ritual and "order and disorder" *zhiluan*), consult HCJSWB 54:12a–b. JLHT 1:1a puts the matter simply: "With ritual there is peace; without ritual there is peril." See also JLDC, preface, 1; LDTWC 1:5b.
6 For an excellent overview of heterodoxy in late imperial China, see the essays in Liu and Shek, eds. 2004.
7 See, for example, the preface to DQTL, esp. 2a; see also HCJSWB 54:2b–4a, 6a, 7b–8a; 57:5b–6a; 68:1a–3a; HCJSWXB 61:3b–6b; HCJSWBXB 74:7a–8b. Cf. LDTWC 2:25a; 4:10b–11a.
8 See Smith 1994, 6–7; cf. Zito 1997, 57ff. For a comprehensive overview of the vast theoretical literature on ritual, consult Bell 1997. Bell's own approach to ritual emphasizes its situational, relational, performative and strategic dimensions. Bell 1992.
9 See ZWDCD, 10338–50. The wide range of topics covered in the subcategory on "Ritual and Etiquette" (*liyi*) in the TSJC and on "Ritual Administration" (*lizheng*) in the HCJSWB and its supplements suggests this breadth of meanings.
10 See, for example, Chen Guofu and Qiu Peihao 1964, Preface, 1. In some versions of this phase the *yi* denotes "etiquette" rather than "right behavior," reinforcing the primacy of ritual activity.
11 Balazs 1964, 139–40.
12 Hucker 1961, 67–68.
13 Zhou Lin'gen 1971, 128–29; Yang Shufan 1978, 2–3; QS 4:33–34, 83:1046. Ritual was particularly a preoccupation of the eighteenth century, and it served as a common denominator of Han learning throughout the Qing period. See Chow 1994.
14 Cited and discussed in J. Meyer 1991. HCXWXTK 155:9114, indicates that in fact the emperor did this several times during his reign.
15 QS 83:1046–47.
16 See DQLCSL, 575:2b–3a (July 11, 1907). For the fascinating proposals on ritual offered by the Bureau of Ritual Studies (*Lixue guan*), see Cao Yuanzhongng, ed., *Liyi* (Proposals on ritual) (1916).
17 See, for example, Xu Qianxue 1696, Fang Bao 1748, Jiang Yong 1810, Chen Baoquan 1820, Huang Yizhou 1893, Li Gong in CSJC, Sun Qifeng 1884, Li Fuyao 1891. See also the sources cited in WXL 4:6a and Kiong 1906, 85–91. See also C.Y. Li 2004 on Qin Huitian and Elman and Kern 2010 on the *Rites of Zhou*.

18 A summary of the main divisions of ritual law can be found in DQHD 53:3b.

19 On the authority of the DQHD and DQTL, consult WXL, preface, 1a–b; HCJSWXB 56:10a–b; 61:3b–5a; HCJSWBXB 74:20a–23b. For background, consult HCJSWB 54:5a–b.

20 QS 83:1046.

21 On the influence of this and other such works in the late Qing period, consult HCJSWXB 50:2a–4a, 6a. Several excerpts from the *Wuli tongkao* can be found in the HCJSWB section on "Ritual Administration" (*lizheng*).

22 Wu's careful study, based both on his scholarly investigations and on his practical administrative experience, draws heavily on the DQHD and DQTL as well as on a number of other official and private sources.

23 Such works included administrative guides (e.g., Wang Huizu's *Xuezhi yishuo* (Opinions on the study of government; 1793); and Wu Rongguang's WXL), handbooks on family ritual (for example, JLQJ, JLHT, JLDC, and JLTSJC), and compendia emphasizing proper deportment for students (notably Zhang Boxing's influential YZLB). See also the following note.

24 On ritual specialists, see James Hayes in Johnson, Nathan and Rawski, eds. 1985, esp. 100–03. I am deeply indebted to Dr. Hayes for his scholarly advice and friendly assistance in locating ritual handbooks and related materials over the years.

25 Cited in H. Giles 1964, 873–74; see also ZWDCD, 10339–40, 10345–46, 10348–49.

26 HCJSWB 54:7b–8a. Zhu Xi's *Jiali* (Family rituals) in particular inspired an enormous number of derivative works, modifications, supplements, commentaries, collections, and individual essays. See Ebrey 1991a and 1991b; also K. Ming 1986, esp. 317–19. Cf. HCJSWB 54:14b–10, which contains excerpts from Wang Mouhong's controversial study of the *Jiali*.

27 H. Liu 1959, 16–49.

28 Legge 1963, 1: 380–84, 388 (modified after consulting the Chinese original); WLTK, Preface, 1a; see also HCJSWXB 61:2b–3a; HCJSWXB 61:1a, 7a–b; JLHT, Preface, 1a; 1:2b; LDTWC 3: 30b–31a; ZWDCD, 10339.

29 See, for example, Legge 1963, 1: 78, 94, 119, 128, 181, 235, 238, 385, 428, 472; 2:51, 71, 84, 135–36, 156, 180, 223, 289, 295, 298, 320, 331, and esp. 349–51.

30 Legge 1963, 1: 375, 390 (modified). Cf. JLHT: 1:1a, 2a.

31 HCJSWB 54:5a. See also LDTWC 1:5b–6b; 2:25a; 3:31 a–b.

32 HCJSWB 54:10a. See also LDTWC 2:26a.

33 See, for example, Boulais, trans. 1924, 354–62. Cf. Jones, trans. 1994.

34 ZWDCD, 2368, 10345. See also Chen Daqi 1964, 144–52, 159–65; JLHT 1:1a.

35 See, for example, DQTL, Preface, 2a; WXL, Preface, lb; JLDC, Yongzheng Preface, 1; JLHT 1:1a; LDTWC 2:25a–26a.

36 HCJSWB 54:1a; 65:3a–b; cf. JLDC, Yongzheng Preface, 1; JLHT, Preface, 1a, 1:1b–2a.

37 See L. Yang 1957. On the relationship between *li* and *renqing*, consult ZWDCD, 10345 (*Li shun renqing*); JLTSJC, Preface, 1a–2a; HCJSWXB 50:5b; HCJSWBXB 61:2a; LDTWC 3:11a.

38 Cited in L.S. Yang 1957, 302–3. See also DQHD 36:1b, on sacrifice as a means of "exalting virtue and recompensing merit" (*chongde baogong*).

39 *Analects* 3.3.

40 W. Tu 1979, 3. See also W. Tu 1972. For a related view of ritualization, see Zito 1984, 1987, 1993 and 1997, esp. 49–50 and 57ff.

41 Legge 1963, 1: 63 (modified).

42 HCJSWB 54:1a–2b; LDTWC 2:25.
43 Cua 1979, esp. 380ff.
44 Plaks 1976, 46–47, 80–81, 178–79.
45 Legge 1963, 2: 98 (modified).
46 Ibid., 2: 99 (modified). Most Chinese dynastic histories—including the *Mingshi* and the *Qingshi gao*—contain full treatises (*zhi*) on music as well as ritual.
47 Legge 1963, 2: 99 (modified).
48 Although I have created a somewhat static model of Chinese ritual in order to indicate its principal structural elements, it is clear from even a cursory glance at sources such as the HCJSWXB and its supplements, the TSJC, the WLTK, the XWXTK, and the "Basic Annals" and "Treatises on Ritual" sections in the dynastic histories, that ritual practices changed continually throughout Chinese history, including the Qing period. This point is made explicitly in the preface to the WLTK and affirmed in the text. See also WXL, Preface, esp. 2a. For a detailed but somewhat ideological overview of Chinese ritual as it evolved over the centuries, consult Zhou Lin'gen 1971; cf. He Liankui 1979.
49 See R. Smith 1994, 1–7; also Sangren 1987b, 132ff.
50 HCJSWB 68:3a.
51 Miu Tianhua 1977, 649–50.
52 Ebrey 1981, 162. JLHT 1:2a offers a nearly identical admonition. Cf. LDTWC 3:11a.
53 Cited in Hugh Baker 1979, 39.
54 See, for example, JLHT 1:1aff. esp. 6b; LDTWC 3:11a. For additional remarks on the relationship between wealth and ritual activity, consult HCJSWB 54:8b–9, 63:1a–b, 66:7a–b, 68:1b.
55 See the discussion in Ebrey 1991a, esp. 216–18.
56 Most major Qing works on ritual are organized according to these categories. See, for example, the DQHD, DQTL, and WLTK. The DQHD 26:2a–3a identifies 129 separate *jili*, 74 *jiali*, 18 *junli*, 20 *binli*, and 15 *xiongli*. This relative emphasis is reflected, more or less, in the number of chapters (*juan*) devoted to each category in the DQTL, WHTK, and the "Treatise on Ritual" in the QS.
57 See R. Smith 1994, 44–47 and 157–64; cf. Zito 1984 and 1997.
58 The standard primary sources are the DQHD, DQTL, LBZL, and WXL. See also S. Wu 1970.
59 For lists of official rank and appropriate terms of address, see WXL 7:1a–9a; also JLDC, 71–78, and JLHT 1:10a–11b.
60 See, for example, Bodde and Morris 1967, 275–78.
61 WXL, Introduction: 4b; 8:1a–11b; 12:1a–10b; 13:1a–10b; 14:1a–8b. Kiong 1906 provides a detailed and abundantly illustrated account of late Qing ceremonial usage, including official dress, salutations, gift-giving, etc.
62 See R. Smith 1994, 82–83 and 165–67; also QS 116:1383–84.
63 K. Hsiao 1967, Chapter 6, esp. 226, 229–31, 275–81 and notes. See also the articles in Sutton, ed. 2007c, esp. Sutton 2007b and Szonyi 2007.
64 Feuchtwang 1977, 596ff.
65 R. Smith 1994, 95–98. For some examples of the highly refined vocabulary of personal relations based on such ties, consult JLHT 1:18b–19a; JLDC, 41–42; JLTSJC 1:13b–17a. For a more comprehensive but less detailed view of traditional networks of affiliation, see the appropriate sections in the TSJC's subcategory on Social Intercourse (*jiaoyi*).
66 For details on Qing efforts to exert control over non-kin corporate activities, including festivals and pilgrimages, consult K. Hsiao 1967, 224–35, 275–81, 313–16.

67 Ibid., Chapter 8, esp. 348–57, 361, 370. See also HCJSWB, *juan* 58–60; HCJSWXB *juan* 55–56; HCJSWBXB *juan* 67–68.

68 Boulais, trans. 1924, 393, 396–400. Cf. W. Jones, trans. 1994.

69 The importance of proper mourning ritual (*sangli*) is suggested by the extraordinary amount of space devoted to the subject in both official sources (e.g., DQTL, *juan* 45–50) and private writings (e.g., WXL, which devotes five of its twenty-four *juan* to *sangli*). See also the sources cited in note 70. For exceptions to the practice of *dingyou* in times of exigency, consult HCJSWXB 58:8a.

70 See, for example, JLDC, I:7–20, 24–78, 82–98; JLTSJC, 1:1a–17b, 20a–22b, 38b–39a; 2:1a–8b; JLHT 1:4a–6a, 7a–b, 10a–22b; 2:1a–6b, 27a–31b; 4:27a–29b; YZLB 1:1–3; 3:7–25; see also JLQJ, *juan* 1–2, 4; LYBL, *juan* 1–2, 4. See also Ebrey 1991b and Kiong 1906.

71 See, for example, JLHT 3:15a–19a; JLQJ 1:10a, 4:44b, 48a; JLDC, 251–53, 275–79; JLTSJC 1:21b–22b, 3:32b–34b, 4:26a; WXL 3:4a, 14:6a–7b; 14:9b–13a, 18a; ZZJL 1:38b–59a, 2:29a–32a, 3:34b–38a, 4:34b–68a. See also Kiong 1906 and the many illustrations in sources such as the LBZL, the subcategory on "Ritual and Ceremony" (*liyi*) in the TSJC, and Guiliang 1883.

72 J. H. Gray 1878, 1: 347. See also W. Martin 1897, 323–24.

73 A.H. Smith 1899, 240. Note also the observations of R. F. Johnston and others, cited in R. Smith 1994, 6–7 and 266–68.

74 In the minds of many Qing officials, ritual provided the means by which to unite the family and the state. See, for example, HCJSWB 54:14a. See also Rowe 1998.

75 On the evolution of these institutions, see QS 86:1081, 115:1362–64. For details, consult DQHD, *juan* 27–40.

76 See QS 116:1376–83. For details, consult DQHD, *juan* 41–42, 71–73, 75. Ibid., 34:1a–2b discusses the major types of official seals.

77 See R. Smith 1988, 123–45.

78 DQHD, *juan* 77–80, esp. 77:1a–b, 80: 2b–3a.

79 R. Smith 1992; cf. Hayes 1985, 82–83.

80 R. Smith 1988. See also Chapter 5 on divination in this volume.

81 Feuchtwang 1977, 585–87 provides two convenient charts of the three main levels of state sacrifice. See also QS 83:1046–53; WXL, *juan* 9–11. For additional details, consult DQHD, *juan* 26–40, esp. 35 and 36; also E. Williams 1913 and Zito 1984, 1987 and esp. 1997.

82 Sources such as the XWXTK (*juan* 167–69) indicate that an astonishing number of individuals were so honored. Even foreign military employees could be "deified." See R. Smith 1975, 127–28.

83 See Taylor 1990, 126–57; also the discussion in Feuchtwang 1977.

84 Feuchtwang 1977, 588–92 and 596ff.; see also Zito 1987, 356–58.

85 R. Smith 1994, 174–85.

86 Legge 1963, 2:411–12, 416–20 (modified); see also W. Chan 1967, 202; YZLB, 13–14; JLHT 1:1b–2a. JLDC, 23, declares that the well-ordered state begins with the well-ordered family and that "for ordering the family nothing precedes ritual." See also Rowe 1998, esp. 381ff.

87 JLDC, 24–55, 226–28, and JLHT 1:3a–b, 6b–7a, 2:32a–34b. The contrast between Chinese and Western conceptions of the family as a ritual unit is highlighted by Tang Junyi 1979, 198–205, 253–58 and Wei Zhengtong 1981, 54–58, 302–08. On the subordination of women to men in most domestic rituals, see Ebrey 1991a, esp. 227–28.

88 Ebrey 1991a and 1991b.

89 JLDC, 238–41; JLHT 4:24a–26b; Doré 1914–33, 4:417ff. See also the critical discussion of *wuzi pai* in HCJSWB 66:10b–11a.

90 Ebrey 1991a, 21–23, 26–27, 46–67, 53–56, 62–63, 128–29 and esp. 207–08.

91 For a convenient overview, consult P. Chao 1983, 101–31. See also WXL, *juan* 14; JLQJ, *juan* 4; JLTSJC, *juan* 4, and esp. JLHT 1:4a, 6a; 4:14a–23b, 27a–29b. There were, of course, many variations in individual practice.

92 See, for example, JLHT 1:2b–3b, 6b–7a, and esp. 27a–44b. Cf. Ebrey 1991a, 81–84, 197–98 and esp. 211.

93 See the various essays in Watson and Rawski, eds. 1988; also Rowe 1998, esp. 380, Szonyi 2002, esp. Chapters 4 and 5, and R. Smith 1994, 88–93 and 179–81.

94 See Baker 1979, Chapter 4 and Aijmer 1978.

95 C. Yang 1961, 43. These elements are reflected in the language of prayers, such as those appended to JLTSJC 4:33a–40b. A few such prayers have been translated in "The Worship of Ancestors in China," *Chinese Repository* (July 1849).

96 R. Smith, 1991, passim. Jiang Yong's *Lishu gangmu* provides additional examples of the pervasiveness of divination, not only in life-cycle ritual, but also in other realms of ritual—including military affairs.

97 See HCJSWXB 63:1a–b, 8b–9b, 10b–11a; WXL 19:3a–7b; Doré 1914–33, 4:402ff. For details on *fengshui* theories and practices, consult Feuchtwang 2002 and Bruun 2002.

98 R. Smith 1994, 98–99.

99 Wieger 1913, 112 (Chinese text).

100 Cited in C.K. Yang 1961, 194.

101 Ibid., 192–98, 204–10, 218–43; see also HCJSWB 68:20a–b, 69:5b–9a; HCJSWBXB 74:5a–8b, 48a–b.

102 See Boulais, trans. 1924, 360–95, 469–74; cf. W. Jones 1994. See also HCJSWB 68: 8a–12b.

103 See, for example, Yang 1961, 206.

104 See J. de Groot 1903, 183ff.

105 See, for example, HCJSWB 58:12a–b; 68:12a–b; HCJSWXB 61:8b. HCJSWB 69:6b–7b provides an excellent example of the transformation of a Christian church (*Tianzhu tang*) into a Heavenly Consort temple (*Tianhou gong*).

106 See, for example, Bohr 2004 and Bohr in Hao and Wei, eds. 1998, 2: 907–53; cf. Michael 1971, passim.

107 Philip Kuhn 1980, iii–ix, Chapter 5; Shih 1976, Chapter 8; Liu and Shek 2004, 1–25, passim.

108 Ling Shanqing 1969, 5:6, 6:1ff.

109 V. Shih 1976, 188, 192, 214–66, esp. 243.

110 Ibid. See also Ling Shanqing 1969, 5:18, and Michael 1971, 2:25, 53, 115–18, 147, 162, 260, 296, 324ff., 415, 442, 452, 474–75.

111 For a discussion of the calendars of the Taipings and Ming loyalists, see Smith 1988.

112 See, for example, HCJSWXB 50:1a–2a; 61:1a–8b, esp. 6b–7b; LDTWC 2:25a; see also HCJSWBXB 74:1a–62a, passim.

113 See Watson's "Standardizing the Gods" in Johnson, Nathan and Rawski, eds. 1985, esp. 322–24.

114 See the symposium titled "Reconsidering James L. Watson's Ideas: Ritual, Cultural Standardization, and Orthopraxy in China," in *Modern China* 3.3 (2007): 3–158. Cf. an

earlier "Symposium on Hegemony and Chinese Folk Ideologies," in *Modern China* 13.1 (January 1987): 1–110 and 13.3 (July 1987): 259–371.

115 See the articles by Donald Sutton, Mellissa Brown, James Pomeranz and Michael Szonyi in Sutton, ed. 2007c.

116 See Rawski's "A Historian's Approach to Death Ritual," in Watson and Rawski, eds. 1988, 20–34; also Ebrey 1991a. Watson's critics have pointed out, however, that local diversity was sometimes "disguised" by elites "who wrote one thing [about orthodox ritual behavior] and did or permitted something else." See Sutton 2007a, 7.

117 See esp. Szonyi 2007, esp. 62–63 and note 10.

118 See Sutton 2007b, 139.

119 See ibid., 146–47.

120 Ebrey 1991a, 204.

121 Ibid., 229.

122 I have developed this argument at some length in Smith 1991a and 1994. See also Weller 1987 and Sangren 1987a and 1987b.

123 See Elman 2000, 311–61, esp. 312.

124 R. Smith 1994, 174–85, Sangren 1987b, Chapters 8, 9, and 11. See also JLHT 4:24a–26b, HCJSWB 66:1a–b, HCJSWXB 61:1a–3b.

125 For an inventory, see Feuchtwang 2002, 35–153. For examples from the heterodox tradition (in addition to the Taiping examples already noted), consult Naquin 1976, 17–18, 33–35, 53–54, etc.

126 Wolf, "Gods, Ghosts, and Ancestors," in Wolf, ed. 1974b, 145.

127 Ahern 1981, 92.

128 R. Smith 1994, 60–63, 228; cf. WXL, Introduction, 4b, 8:1a–11b, 12:1a–10b, 13:1a–10b, 4b, 14:1a–8b.

129 R. Smith 1994, 270–73; cf. JLHT 4:27a–29b, and JLDC, 324–32.

130 Baker 179, 148–52, 219–25.

131 See, for example, Hsiao 1967, 229–34, 352–55; S. Wu 1979, 57–58.

132 For a recent work on this topic, see Poon 2011, and Chapter 4 in this volume.

4 The teachings of ritual and the rectification of customs

1 Metzger 1990, 287.

2 Ibid. See esp. 271–72, 275, 276, 28, 284–85, 289–90. Cf. H. Chang 1987, 17ff. and 181ff., esp. 190.

3 The most direct source of scholarly inspiration for this essay is Professor Liu's masterful essay "Socioethics as Orthodoxy: A Perspective," in K.C. Liu, ed. 1990a.

4 K.C. Liu's "Introduction," in ibid.; also K. Chow 1994, 9–14.

5 See Chapter 3 in this volume; also R. Smith 1994, 6–7, and CWDCD, 10346.

6 See Chow 1994, 50–52; also CWDCD, 10339, Zito 1984 and 1997, Hevia 1995, esp. 15–25 and 121–25.

7 Cited in Chow 1994, 189. I have modified this translation somewhat. See also Chai and Chai, eds. 1967, 1: 380–84, 388 and 2: 100–01.

8 Cited in K.C. Liu 1978, 412.

9 For an excellent description of the process from an imperial perspective, see Hevia 1995, 121ff.; cf. Zito 1997, 49–101, passim.

10 Chai and Chai, eds. 1967, 2: 319–20 (modified). See also C.C. *Mengzi*, 455–56 and Metzger 1977, 67, 120–26, 157–58, 247–49.
11 CWDCD, 10898; cf. Gardner, trans. 1990, 109.
12 Cited in R. Smith 2012, 93.
13 Cited by O. Ng in Smith and Kwok, eds. 1993, 43 (modified). Gardner, trans. 1990, 109.
14 See, for example, Qian Mu 1977, C. Chang 1977, Tang Junyi 1979, Wei Zhengtong 1974 and 1981.
15 CC, *Mengzi*, 7.1:13.3, 455, HCJSWB, 54:1a–2b.
16 Chai and Chai, 1967, 2: 412 (modified).
17 B. Watson 1963, 25.
18 See Mair 1985, 325–59 for background; also K. Hsiao 1967, 184ff.
19 See Mair 1985, 325–26. A convenient (but incomplete) version of one Chinese text appears in Wieger 1913, 133–34. Cf. Milne, 1870.
20 Mair 1985, 336–37.
21 Ibid., esp. 337ff.
22 Although various texts differ in particulars, their general line of argument is much the same. Cf. Wieger 1913.
23 See, for example, JLHT, 1:3a–b.
24 Wieger 1913, 22–23.
25 Ibid., 98–100.
26 Ibid., 100–03.
27 Ibid.
28 Ibid., 106–09.
29 The *xiangyue* system originated in the Song dynasty as a voluntaristic "rural community-action and mutual-aid institution," but by Ming and especially Qing times it had become a propaganda tool of the bureaucratic managerial state. K. Hsiao 1967, 201–05.
30 For details, consult ibid., 185–201.
31 Ibid., 202–05.
32 Ibid., 194–201.
33 Ibid., 195–201. Cf. Anonymous, "Reading the Sacred Edict," *Chinese Repository*, 17 (November, 1848), 586–88.
34 Mair 1985, 355.
35 See note 22 above.
36 K. Hsiao 1967, 197–201.
37 H. Chang 1987, 5–7, 99–100, 181 ff, esp. 184–87.
38 Teng and Fairbank, eds. 1979, 239–45; see also *Xin qingnian* (The new youth), 3: 239–45 and 6: 578–80.
39 Teng and Fairbank, eds. 1979, 232; cf. Wei Zhengtong 1981, 1–2 and 11ff.
40 R. Smith 1994, 289–91.
41 See Duara 1991, Nedostup 2010, and Poon 2011, esp. 67–91 and 147–55. Poon's study reveals in fascinating detail the strategies used by devotees of stigmatized religious cults and practices to circumvent government policies that were designed to suppress them.
42 Cited in M. Wright 1962, 304.
43 See Ferlanti 2010, esp. 977–81.
44 For several valuable Western-language accounts of the New Life Movement, see Dirlik 1975, S. Chu 1980, Averill 1981, Oldstone-Moore 2000 and Ferlanti 2010.

45 YLLB, part 1, 2.
46 See XSHJY, part 1, 1.
47 XSHGY, 5; see also XSHXZ, 26–29.
48 K. Chiang 1974, 115. See also XSHJY, part 1, 6; XSHHB, part 1, 4–6; XSHXZ, 46–47, 56–57, etc.
49 XSHXZ, 17, 20, 47–48; XSHJY, part 1, 1.
50 Cited in S. Hsiung 1948, 363.
51 YLLB, part 1, 19, 33, etc.
52 M. Wright 1962, 305–08.
53 See M. Wright 1962, 99.
54 XSHHB, 3: 29; J. Li 1986, 23. Cf. XSHXZ, 18.
55 XSHGY, 9.
56 This "necessary knowledge" also appeared as an appendix to the "Official Outline" of the Movement (XSHGY). Chiang took pains to point out that his use of the term *xing* included both its narrow sense of walking (*xingzou*) and its broader sense of "movement" or "action" (*xingdong*). XSHGY, 11.
57 XSHGY, 19–22.
58 See, for example, Madame Chiang, ed. 1937. 63. On the importance of women in the New Life Movement, see Ferlanti, 989–91 and Yen 2005, passim; cf. XSHHB, 3:141.
59 XSHGY, 16–18.
60 Ibid., 15.
61 Eastman 1990, 67.
62 Averill 1981, 599–616.
63 YLLB, part 1, 92.
64 See, for instance, XSHGY, 18.
65 Ibid., 20–21.
66 XSHXZ, 10, 19, etc.
67 Tong, 1938, 72–73; J. Li 1986, 24–25; Ferlanti 2010, 972–77.
68 XSHGY, 5, Chen, "New Life Movement," 193–94, Madame Chiang, ed., 1937, 68ff.
69 XSHGY, 16.
70 Ferlanti 2010, 981–86 places special emphasis on the themes of "state building" and "hygienic modernity."
71 W. Chen 1936, esp. 544ff.
72 The XSHXZ's front material includes illustrations and descriptions of the flag, the shield, as well as the lyrics and music to the song.
73 For details, see W. Chen 1936.
74 J. Li 1986, 28.
75 W. Chen 1936, 208.
76 Ibid., 201, 204; J. Li 1986, 72. See also Feranti 2010, 990 and H. Yen 2005, passim.
77 Averill 1981, 604 ff; cf. W. Chen 1936, 209–10.
78 W. Chen 1936, 195–96.
79 J. Li 1986, 29; Dirlik 1975, 970.
80 YLLB, part 1, 43. See also XSHJY, part 1, 45–46.
81 See Averill 1981. On Liang Shuming's efforts to promote "rural reconstruction" and "Confucian modernization" during this period, consult Alito 1978, 192ff.
82 Averill 1981, 600ff., esp. 614; S. Chu 1980, 56ff.

83 S. Chu 1980, 58 ff; see also W. Chen 1936, 192, 195–96.
84 Tong, 1938, 2: 560–61.
85 Madame Chiang, ed. 1937, 59.
86 Peck 1967, 96. For other critical evaluations of the New Life Movement, see Averill 1981, 595–96 and 610ff., Wright 1962, 300–12, J. 1986, 30–37, Eastman 1990, 66–68, Wakeman 1994, 234–35. See also the divergent points of view expressed in XSHQK, 11, 21, etc. Ferlanti offers a relatively positive view of the movement, arguing that it had a lasting impact on Chinese society, and that it contributed to the shaping of Chinese "citizenship and national identity."
87 For more sanguine appraisals, based on short-term achievements, consult W. Chen 1936, Madame Chiang, ed., 1937, and Tong 1938.
88 See W. Wei 1985, 80, Dirlik 1975, 961, Tozer 1970, S. Wang 1987–88.
89 See, for example, Joachim 1992, and Jacobs 1982.
90 Dirlik 1975, 976.
91 R. Smith 1994, 292–94 discusses Mao's complex and somewhat ambivalent response to the Chinese past.
92 Ibid., 294–95.
93 See R. Smith 1989. A very selective and unauthorized version of this article appeared in Chinese as "*Zhongguo wenhua di weilai* (The future of Chinese culture)," in *Faxian zazhi* (Discovery Magazine) (Spring, 1990).
94 Whyte 1974. Cf. the "unification in groups" (*tuanti hua*) mentality of the New Life Movement.
95 Ibid.
96 See Dynon 2008 for a recent overview of "Post-Mao socialist ideology."
97 WJSMSC, 2–3 and 7–8. On the "crisis of faith," see Falkenheim 1982, Dirlik 1982, and Gold 1984. Cf. RMRB, March 13, 1980 and February 14, 1981.
98 WJSMSC, 3.
99 Ibid., 3–4.
100 Ibid., 7–8.
101 *Beijing Review*, March 22, 1982.
102 Anagost 1992, 179.
103 WJSMSC, 11–12.
104 Ibid., 2; see also ibid., 99–100.
105 *Beijing Review*, March 22, 1982.
106 J. Li 1986, 54. See also Dirlik 1982
107 RMRB, March 1, 1982.
108 Ibid., May 6, 1982 and February 27, 1983; *Beijing Review*, March 22 and April 5, 1982.
109 WJSMSC, 8; also 100–01.
110 Ibid., 11.
111 Ibid. See also ibid., 71 ff, esp. 73.
112 See, for example, RMRB, April 1, 1983; also Anagost 1992, esp. 187–88; cf. WJSMSC, 73–74.
113 Gold 1984; cf. Baum 1995.
114 See, for example, RMRB, April 1, 1983.
115 Gold 1984, 947 and 950.
116 Ibid., 957.

117 WJSMSC, passim.
118 Popular almanacs included similar kinds of information. See R. Smith 1992, 53ff.
119 WJSMSC, 247–53.
120 Ibid., 6 and 8.
121 Ibid., 69–73.
122 See, for example, RMRB, July 29, 1984 and January 12, 1985.
123 J. Li 1986, 107, note 124. On gift-giving in contemporary China, see M. Yang 1994, Y. Yan, 1996.
124 RMRB, February 26, 1985.
125 J. Li 1986, 58 and 106, notes 84 and 85.
126 RMRB, March 11, 1983.
127 See also J. Li 1986, 67.
128 See, for example, WJSMSC, 40–41, also RMRB, March 11, 1983.
129 WJSMSC, 27–31, 41, etc.
130 Ibid., 218.
131 Cf. H. Chang 1987, 190.
132 See R. Smith 1992, 40ff. See also Chapter 5 in this volume.
133 See Metzger 1990, 211–15. Thus, the goal of the New Life Movement and the Five Emphases campaign to "conquer material substance with spirit" (*yi jingshen zhansheng wuzhi*) seemed fruitless.
134 See Dicks 1989 and Anagost 1992; cf. Dynon 2008.
135 See R. Smith 1997.
136 Reported in *Newsweek*, September 13, 1993.
137 Yang Zong, *et al.*, eds. 1991, esp. 754.

5 Divination in the Qing

1 For a Chinese version of this essay, see R. Smith forthcoming. R. Smith 1991 explores in detail the issues discussed below. An online version of this out-of-print book is available at http://www.questia.com, but this version lacks illustrations, notes, bibliography and glossaries.
2 Quan's tract is reprinted in HCJSWB 69:15b–16a.
3 See CC: Lunyu, II.4; VI.2; IX.1, 5–6; XI.6; XII.5; XVI.8; XX.1, 3; Mengzi II.1.iv; IV.1.iv, viii; V.1.v–viii; VI.1.vii; VI. 2.xv; VII. 1.i–ii. Cf. Ma Xulun 1941, 15–34, Wei Zhengtong 1974, 181–84.
4 On the ambivalence of Qing scholars, see HCJSWB 69:15b–16a; also CC 3:338–39, note 31. For indications of Chinese attitudes toward fate over time, consult Y. Fung 1952, 1:30–31, 86, 129, 181, 224–25, 237, 285, 293, 313, 370, 374–75, 384–85 and 2:30, 46, 60, 62, 71–75, 125–30, 162, 191–93, 195, 417–18, 421, 446, 453, 466, 494, 514, 516, 527, 531, 592, 631, 650; also the essays in Lupke, ed. 2005.
5 Chai and Chai, eds. 1967, 1:94 (modified). For other *Liji* references to divination, consult ibid., 1:92, 94, 103, 119, 235, 259, 297–98, 380–81, 367, 385–86, 428–29, 472; 2:4, 51, 53–54, 71, 74, 84, 135–36, 156, 223, 233, 289, 295, 349, 363.
6 SKQSZM 146:8b. On the issue of Confucian rationalism, see, for example, Fu-kuan Hsü's essay in W. Chan, ed. 1986, 52–55; for a Qing period illustration, see the illuminating two-part essay by Zhou Yuanding titled "An Essay on Ghosts and Spirits (*guishen pian*)," in HCJSWXB 61:2a–3a; also Henderson 1984, Chapters 6–9, passim.

7 On commoners, see Liu Zhiwan, 1974, Ma Xisha and Han Bingfang 1992, and Eberhard, "Fatalism in the Life of the Common Man in Non-Communist China," in W. Eberhard, ed. 1971. On the relationship between women and divination, see R. Smith 1991, esp. 4, 9, 15, 18, 96, 107, 125, 149, 164–65, 175, 178, 181, 189, 198–99, 200, 206, 207, 215, 230–33, 236, 244–45, 254–55, 261–62, 286, 318, note 80, 331, note 142, 311, note 68, 329, notes 83 and 100, 325, note 25, 339, notes 127 and 128, 342, note 12, 347, note 97, 348–49, notes 132 and 33.

8 See BRZ. There is now an easily accessible Taiwan reprint (1998) and a PDF version of this work.

9 In addition to his *Zhongguo lidai buren zhuan*, see also Yuan Shushan 1919, 1928, 1947, 1948b, 1961a, 1961b and n.d. For a brief discussion of Chinese and Western scholarship on divination following publication of this book, see the "Addendum" to the essay in hand.

10 See de Bary, *et al.*, eds. 1964, 2: 286. The second edition of this two-volume anthology of translated writings (de Bary and Lufrano, eds. 2000) eschews such language, and gives much greater attention to "popular" tracts.

11 Scarborough and Allan, eds. 1926, 179.

12 The introductory sections on technicians (*fangji*) in local gazetteers often refer to Zixia's famous remark on *xiaodao* in the *Analects*. See also Lu Yao's essay, "Technical Arts," in HCJSWB 69:13b–14a.

13 See, for example, the Preface to Ouyang Chun, *Fengshui yishu* (Geomancy, book 1; 1814), 7a.

14 See Wang Fuzhi, *Zhouyi neizhuan fali* (Introduction to the inner commentary of the *Zhou Changes*), reprinted in YJJC 141:3b.

15 These variables are discussed at some length in the "Languages of the *Yijing*" essay that appears in this volume (Chapter 1). For more extensive discussions of these variables, consult R. Smith 1991, 51–90, and Feuchtwang 2002, 35–153.

16 See Zhang Yaowen n.d., esp. 2–28.

17 On different Chinese conceptions of fate, see R. Smith 1991, 10, 14, 24, 33, 35–36, 42, 95, 173–74 and 177ff.; also Lupke, ed., 2005.

18 Cited in R. Smith, 1994, 149.

19 See R. Smith, 1991, 173–74.

20 For an illustration of the elite's disdain for commoners who are "not content with right behavior and fate" and who "flatter spirits to seek blessings," consult WXL 9:10a.

21 See Henderson 1984, 108, 168, 159, 178–97, 207–25; also Ropp, 1981, 161–65, 171–79, 183–87, 189, 232, and Elman 1981–83 and 1984.

22 Chapter 2 of R. Smith 1991 develops this argument at considerable length. See also Smith 2008a, 192–94.

23 TSJC 47:5681–7135; 48:7136–854. This does not include material on the *Yijing* in the section on classics (*jingji dian*), on extraterrestrial phenomena in the section on Heaven (*qianxiang dian*), on spirits in the section devoted to the supernatural (*shenyi dian*), and on local divination practices, recorded in gazetteers, in the section on political divisions (*zhifang dian*). For examples of regional variation in customs related to divination, consult Hu Pu'an, comp. 1968. Generally speaking, southerners—the Cantonese in particular—were viewed as "most inclined to believe in spirits and ghosts." See, for example, HCJSWXB 61:8b.

24 The QBLC's sections on technicians (*fangji*) and superstitions (*mixin*) provide a vast amount of material related to divination. On the collected statutes and other official sources, consult R. Smith 1988.

25 Most of the major novels of the Ming-Qing period include references to divination. For a concrete example from the late Qing, see S. Liu, trans. 1975, 107–11 (on "spirit-writing").

26 All of these authors are quoted and evaluated in R. Smith 1991, 5–6; citations appear on 289, note 3. For some other observations of foreigners about Chinese divination, see ibid., 75–76, 88–90, 117–19, 132, 137–38, 158–59, 165–66, 169, 175–76, 185, 187, 205–07, 218–19, 222–23, 226, 234, 244, 256, 299, note 40, 311–12, note 86, 315, notes 10 and13, 320–21, note 132, 326, note 37, 330, note 116, 331, note 14, 336, note 61.

27 See, for example, R. Smith 2008a, 173–77, 183–84 and 192–94.

28 W. Chan, trans. 1967, vii, 108.

29 Lynn 1994, 120 (modified); cf. Wilhelm 1967, 262.

30 Lynn 1994, 66; cf. Wilhelm 1967, 320. Although Mao Qiling and other Qing scholars had conclusively demonstrated that existing versions of the *Hetu* and *Luoshu* were added to the *Yijing* only in the Song period, the diagrams continued to be linked to the classic by many individuals, commoners and scholars alike. See R. Smith 2008a, 79–80, 114–19, 134–6, 143, 172–77.

31 See SKQSZM 109:18a–20b; also ibid., 109:43b–45b; 111:15b–24a; 111: 41a–49b.

32 On traditional correlative systems involving these variables, consult Henderson 1884, 7–20, 23ff; cf. R. Smith 1991, 49–91 and R. Smith 2012, 49–58 and 76ff.

33 For details see P.Y. Ho 2003.

34 For details on these and other divination systems, see R. Smith 1991. For useful summaries in Chinese, consult Zhang Yaowen n.d., Yuan Shushan 1926, and the sources cited in the "Addendum" of this chapter.

35 See TSJC 47:5919ff., esp. the Ming work *Sanming tonghui* (Comprehensive compilation on the three fates), 6175–517. For an overview of "fate extrapolation" in English, see R. Smith, 1991, 174–87.

36 For a full discussion, consult R. Smith 1991, 173–87.

37 See the critique in ibid. 1991, 186–87.

38 Chen Gaoyong 1939 provides a convenient chronology from Qin times onward. For some indications of the importance of omens to the imperial government in the Qing period, see QS, *benji* 1:32; *zaiyi zhi* 1:655–723; etc. See also note 42, below.

39 Metzger 1977, 51. For detail on omens and portents, see R. Smith 1991, 21, 25–26, 39, 50., 65ff., 233ff., 247.

40 Spence 1975, 52; see also 150 and 174.

41 See, for example, *Da Qing shichao shengxun* 1880, Kangxi, *jingtian* 10:7a, 10:1aff., esp. 8a–b. Cf. ibid., Tongzhi, *jingtian* 11:1aff. Local gazetteers invariably recorded portents, and many included discussions of *fenye* (allocated field) systems that linked specific terrestrial areas with their celestial counterparts. A typical entry, which refers to the cosmological authority of the *Yijing* and emphasizes the principle of a "resonance [*ganying*] between Heaven and Man," can be found in the *Huangpo xianzhi* (Gazetteer of Huangpo County) (Hupei, 1871), *tianwen zhi* 1:9a–b.

42 Almanacs often included information on interpreting portents such as birdcalls and body sensations. See R. Smith 1992. Cf. R. Smith 1991, 233–34.

43 See R. Smith 1991, 233–45, passim; QBLC, *mixin*, 46ff.

44 For details on dream divination, see R. Smith 1991, 245–57. For an excellent translation of one of the most famous Chinese works on dream divination, see Strassberg 2008.

45 For details on this work, see R. Smith 1991, 252–56. The earliest complete version of the *Zhougong jiemeng* that I have seen, which corresponds closely with comprehensive contemporary editions of the work, has almost a thousand seven-character dream interpretations under a total of twenty-seven overlapping categories. It is contained in a fragment of the British Museum's *Yuxi chi guangji* (Comprehensive version of the jade casket record) (c. 1700). For a discussion of dream divination sections in almanacs, consult R. Smith 1992, 28–30, 72, 77, 81.

46 Doolittle 1865, 2:130.

47 QBLC, *mixin*, 75, 79, 86, 87, 95, 96, 106, 119, 120.

48 For details on Qing approaches to physiognomy, see R. Smith 1991, 187–201.

49 Gray 1878, 2: 3. The primary text for physiognomy was the fifteenth-century Ming work *Shenxiang quanbian* (Complete guide to spiritual observation), compiled by Yuan Zhongzhe and reprinted in the TSJC.

50 On these and other notables, see BRZ 3:19; 14:27–28; 18:10–11; 18:21; 30:21–22; 33:8–9; etc. On the role played by physiognomers in administrative affairs, consult ibid., 14:25; 14:27–28; 27:10; 33:22–23; also QBLC, *fangji*, 79, 84–85, 118ff., esp. 121, 123, 124, 125–26, 131.

51 Gray 1878, 2: 2.

52 See R. Smith 1991, 200.

53 CSJCXB 25:113–15; see also Topley, ed. 1967, 61.

54 For details, see R. Smith 1991, 201–04.

55 BRZ 9:20–21.

56 See Xu Dishan 1940, 32–60, passim; also QBLC *fangji*, 15–19, 21–22, 26.

57 See R. Smith 1991, 204–05. On the importance of *zhanhou* predictions to the Kangxi emperor, see *Da Qing shichao shengxun*, Kangxi, *jing tian* 10:3a. For some representative works, consult CSJCXB 25:5ff. On the analysis of seals, see Zhang Yaowen, n.d., 356ff.; cf. QBLC, *fangji*, 118.

58 The best single book on this subject is Feuchtwang 2002. See also R. Smith 1991, 131–71.

59 See BRZ, 1:16; 1:17; 1:22–23; 1:29; 2:2; 2:11; 2:16; 3:6; 3:11; 4:16–17; 5:7; 5:8; 5:9; 5:9–10; 5: 10; 5:18–19; 5:25–26; 6:7–8; 8:9–10; 8:10–11; 8:14; 10:13–14; 11:3; 12:3; 12:7; 12:14; 12:22–23; 12:30–31; 13:14; 13:32 (two entries); 14:31; 15:27; 15:31; 16:25–26; 16:34 (two entries); 17:19–20; 17:26; 17:28–29; 18:22f; 18:35; 18:40; 20:2; 20:12; 20:20; 21:25; 21:31; 21:39; 22:10; 23:3; 23:11; 25:13; 25:14; 25:19–20; 25:21–22; 26:10; 29:18; 29:29; 30:30; 32:3–4; 32:5; 32:9; 32:10; 33:5; 33:8; 33:11–12; 33:25; 35:6–7; 35:9; 36:3; 36:5; 36:7; 36:12; 36:22; 36:26–27; 36:28–29; 37:6–7; 37:11; 37:14–15; 37:19; 37:29. For the use of medical analogies in geomantic writings, see, for example, the preface to the *Huangdi zhaijing* (The Yellow Emperor's siting classic) in TSJC, yishu, 47: 6724.

60 For exquisite detail on the *luopan*, see Feuchtwang 2002, 37–58.

61 CSJCXB, *kanyu*, 25: 350, 360.

62 TSJC, *yishu*, 47: 6724.

63 See the discussion in R. Smith 1991, esp. 161–71.

64 For details, see ibid., 148–59. Of the 1200 or so biographies of diviners I examined in the course of doing research for ibid., about one-third of them explicitly mention expertise in geomancy (including the selection of auspicious days).

65 R. Smith 1991, 148ff.

66 BRZ, 1:18–19; 13:23; 18:14; 33:23.

67 The standard work in Chinese is Xu Dishan 1940; in English, consult Jordan and Overmyer 1986. For a summary of spirit-writing practices, consult R. Smith 1991, 221–30.

68 Xu Dishan 1940, 32; see also QBLC, *fangji*, 28.

69 Xu Dishan 1940, 32ff.; see also QBLC, *fangji*, 19, 22. Jordan and Overmyer 1986, 45–63, discuss the relationship between *fuji* divination and morality books.

70 See, for example, Xu Dishan 1940, 44; QBLC, *fangji*, 15, 17, 19; Jordan and Overmyer 1986, 41–44.

71 On women and shamanism, see R. Smith 1991, 230–33 and Wang Guiyuan 1991. Significantly, Yuan Shushan's *Zhongguo lidai buren zhuan* (BRZ) ignores spirit writing entirely as a category of divination, and the TSJC devotes a mere fifteen pages to shamans (48:8453–68), as against 2172 pages on divination.

72 Xu Dishan 1940, 36; see also QBLC, *fangji*, 16; Sutton 1981, 43; HCJSWB 68:1aff., esp. 8b–10a.

73 For a discussion of these techniques, see R. Smith 1991, 1–2, 234–44. Cf. QBLC, *mixin*, 9.

74 Rong Zhaozu, 1969, 43–44; see also QBLC, *mixin*, 10–13.

75 See the Qianlong emperor's preface to the *Qinding xieji bianfang shu* (Imperially approved treatise on harmonizing the times and distinguishing the directions), translated in Aylward 2007, 56–58.

76 See Wakeman 1986, 1:245, 258, 262, 281, 289, 312, 504. Significantly, the ascension of the Shunzhi emperor in 1644 was delayed until the first day (*jiazi*) of a new sexagenary cycle, to assure an auspicious start for the Qing dynasty. Ibid., 2:857.

77 See R. Smith 1991, 74–83; also R. Smith 1988, 123–45. For a discussion of basic calendrical terms and concepts, consult Morgan 1980, esp. 158–244.

78 R. Smith 1991, 74.

79 See ibid., 83–91; also R. Smith 1992a and N. Wang 1993, esp. 30–53.

80 Morgan 1980, 92–113.

81 R. Smith 1992, 10–39.

82 DQHD, 80:2bff.

83 R. Smith 1992, 13–16.

84 R. Smith 1992, 13–16; R. Smith 1991, 82

85 On the popularity of almanacs and calendars in late imperial times, see the essays by Evelyn Rawski and James Hayes in David Johnson, *et al.*, eds. 1985, esp. 23–24, 82–83 and N. Wang 1993, 89–108 and 118–28. See also R. Smith 1992, 8–9. Cf. J. Yeh (2011).

86 These and other observations by foreigners are cited in Smith 1991, 89–90.

87 Ibid., 89. See also R. Smith 1988, 128–29. The new calendars (*xinli*) of the Taipings are especially interesting as countercultural documents. In certain ways they looked like their imperial counterparts; for instance, they had yellow covers adorned with double dragons and double phoenixes, and they used a few conventional calendrical markers for time and space (the sexegenary system, the twenty-four solar terms and the twenty-eight lunar lodge). But Taiping calendars contained no references to auspicious or inauspicious days. Indeed, the preface indicates that the express purpose of the new "Heavenly Calendar" was to eradicate the heretical doctrines (*xieshuo*) of previous calendars, including the Qing dynasty's *Shixian shu*.

88 R. Smith 1991, 89. J. J. M. de Groot 1903, 458–59.
89 See Naquin 1976, 12–18; Overmyer 1976, 146–50, 172; C.K. Yang 1961, 230–43.
90 XJBFS, Preface, 1a. Cf. Aylward 2007, 56.
91 HCJSWB 69:15b–25a; HCJSWBXB, 74:1a–8b.
92 HCJSWB 69:14b–16a, 23a–b, 23b–24a, 24a–25a; ibid., 63:1a–b, 8b–9b, 10b–11a; WXL 9:10a; 19:3a–7b.
93 Morality and sincerity were constant themes in many forms of divination during Qing times. See R. Smith 1991, 4, 22, 24, 74, 93, 95, 153–54, 165, 201, 208, 227–28, 229, 241–42, 245, 260, 265, 294, note 88, 295, note 106, 345, note 45. On the self-confirming character of popular beliefs generally, see K. Thomas 1971, 642–43.
94 See R. Smith 1991, 118–19.
95 See Thomas 1971, cited and discussed in R. Smith 1991, 271–72.
96 Zuesse 1987, 4:375.
97 Sivin 1990, 169–70. See also the discussions in R. Smith 1991, 265–66 and 283–84, R. Smith 2008a, 61–62, 192–94, 220–21, 235–40 and R. Smith 2012, 49–58, 214–21.
98 I have discussed these features of Chinese divination at length in R. Smith 1991, passim.
99 Ibid., esp. 1–4, 205–15, 222–33, 232, 236–37, 244–45, 266.
100 Ibid., 22, 26, 51, 82, 87, 127, 131–32, 148, 152, 182, 188, 193–95, 222–30, 236, 240–42, 248, 266, 281. See also Unschuld 1985, esp. 197.
101 See the discussion in Smith 1991, 194 and 266.
102 On Zeng, see Yuan Shushan 1928, 5:14a; on Guo, see BRZ 18:11. For relevant proverbs on fate, consult Plopper 1926, 291–315, esp. 294.
103 See Sivin 1990, 169–70; cf. Thomas 1971, 657.
104 Sivin 1990, 170, slightly modified.
105 Konings 1990.
106 See the discussion in R. Smith 2008a, 212, 235–40 and R. Smith 2012, 219–21.
107 See Yates, 2005; also P.Y. Ho 2003, 10, 64–66, 68, 102–05, 135–6; R. Smith 1991, 213–14.
108 See R. Smith 1991, 157–58, 268.
109 On the therapeutic effects of divination, see Smith 1991, 204, 206–11, 232–33, 241–44, 245, 268–69, 345, note 57.
110 Zuesse, "Divination," 380.
111 See, for example, R. Smith 1991, 108, 154; BRZ, 1:21–22, 5:23, 8:37, 12:14, 13:28, 13:32 15:10, 36:23.
112 Jordan 1972, 85–86.
113 BRZ, 28: 2; cf. CC 2:363 and 370.
114 See, e.g., BRZ 1:24; 3:8–9; 5:13; 5:22; 7: 17–18; 7:22–23; 10:9; 12:13; 14:12–13; 15:8–9; 18:13–14; 18:19–20; 18:24–25; 20:18; 20:27; 21:19–20; 23:16–17; 26:5, etc.
115 See Jordan and Overmyer 1986, passim.
116 See R. Smith 1991, 154–55, 210–16.
117 Foulks 2011 insightfully discusses Ouyi Zhixu's divinatory theories and practices. For evidence concerning other clerics, see BRZ, 20:8–9 and 33, 38: 39–43; also QBLC, *fangji*, 68 and 83.
118 On cosmological kingship, see H. Chang 1987, 5–7, 99–100, 181ff., esp. 184–87.
119 See R. Smith 1988, 40–42.
120 Henderson 1984, 173.

121 Cited and discussed in R. Smith 1991, 274.
122 Cited and discussed in ibid., 259–60.
123 Ibid., 274–75, R. Smith 1992, 41–50, cf. Nedostup 2010.
124 J. Yeh 2011.
125 See, for example, R. Smith 1991, 276–81, R. Smith 1992, 50–80, passim, R. Smith 2008a, 205–08, N. Wang, 1993, 103–08 and 118–28.
126 See the discussion in R. Smith 1991, 278–81.
127 Tambiah 1990, 18–21, 106–07, 113; see also R. Smith 1991, 283–85.
128 For a select bibliography of Chinese and Western works on the *Yijing* published in the period from 1985 to 2008, see R. Smith 2009. There is no space here to discuss any of the vast amount of recent scholarship on the *Changes*.
129 For a few of the many broad-ranging studies that have both drawn upon and contributed to what Chinese scholars often refer to as China's "magical" or "mystical" culture (*wushu wenhua*, *shenmi wenhua*, etc.), see Chemla, *et al.* 1999, Field 2008, Gao Guofan 1999, Hu Wenhui 2000, Hu Xinsheng 1998, Jiang Xiaoyuan 1992, Kalinowski 2002, Li Ling (2000a, 2000b), Liang Zhaotao 1989, Nickerson 1994, Wang Yude 1994a, Wang Yude, *et al.*, eds. 1993, Wei Shaosheng 1992, Yu Xiaoqun 1994, Zhang Ronghua 1991, Zhang Rongming 2000, Zheng Xiaojiang 1993, etc.
130 Shen Zhi'an 1995, Su Ai 1996, Weng Naijian 1993, etc.
131 See, for example, Bruun 2002, Meyer 1991, Tsu 1997, Wang Yude 1994a and 1994b, Yi Ding, *et al.* 1996, etc. Wang Yude 1994a is particularly useful, because of the wide range of viewpoints it presents, including both advocates and critics.
132 See, for example, Clart 1997 and 2003, Mori 2002, Zeitlin 1998, Kleeman 1994.
133 See, for example, Cao and Zhang 1995, Huang Yinong 1996, Jiang Xiaoyuan 1992, Kalinowski 1991, Robinet 1995, Yu Xiaoqun 1994, P.Y. Ho 2003.
134 See, for example, Jiang Xiaoyuan 1992, Li Shenglong, ed. 1993, Liu and Zhou 1994, Xiong 2002, etc.
135 See esp. Yates 2001. This type of divination is generally known as *bingzhan* in Chinese.
136 See, for example, Chen Meiying, *et al.* 1995, Huang Yinong 1998, F. Lin 1995, Liu Wenying 2000, 2003, Yao Weijun 1994.
137 See, for example, Fuehrer 2001, Lippiello 2000, Yang and Song 1992.
138 See, for example, Fuehrer 2001, Wang Bugui 1993, Xie Gui'an 1998, etc. These texts are generally known in Chinese as *chenyao*, *chenwei*, etc.
139 Many of the works on divination published in the 1990s appeared under thematic series titles, such as *Zhongguo fangshu daquan* (Complete collection of the predictive and prescriptive arts of China), *Zhonghua fangshu wenhua congshu* (Collectanea on the culture of the predictive and prescriptive arts of China), *Zhongguo shenmi wenhua yanjiu congshu* (Collectanea on research into the culture of mystery of China), *Zhongguo chuantong wenhua toushi* (Perspectives on traditional Chinese culture), *Minsu wenhua congshu* (Collectanea on popular culture), *Zhonghua bentu wenhua congshu* (Collectanea on Chinese local culture), *Wenhua tanmi* (Investigations into the mysteries of [Chinese] culture), *Zhongguo ren congshu* (Collectanea on the Chinese people), etc. Some of these collections consist of "modern" monographic studies, others feature reprinted "traditional" works. For an overview of developments from a Taiwan perspective, see the Preface to Wang Yude 1994a, 1–2; cf. the introductory discussion to the first of two *zhanxing* volumes edited by Li and Liu 1993, 1: 2–3.

140 Some such works include Gu Jianqing, 1991, Jin Liangnian 1994, Liu and Wang, eds. 1991, Lü Shaogang, ed. 1992, Shi Wei and Qiu Xiaobo, eds. 1994, Wu Hua 1993, Xiao Yuan, *et al.*, eds. 1992, Yin Xiaolin, ed. 1993, Zhang Guiguang 1994, Zhang Shanwen, ed. 1992, and Zhang Yongtang, ed. 1992a.

141 For just a few examples of recently edited collections of source materials, see Li and Liu, eds. 1993, Liu Yongming, ed. 1995, etc. For useful indices, consult Zhang Yongtang, ed. 1992a and 1992b. See also the collections mentioned above and below.

142 See in particular Part 3 on "Technical Arts," which includes articles and appendices on early Chinese cosmology, numerology and various forms of divination by authorities such as Christopher Cullen, Mark Kalinowski, Li Jian, Vivienne Lo and Michael Nylan.

143 See, for example, Lin Meirong 1997, Liu and Wang, eds. 1991, Ma and Han 1992, Zhang Ronghua 1991, etc.

144 See, for example, the following Chinese- and Western-language works that illustrate the close relationship between Chinese medicine and mantic practices, past and present: Chang 2001, Farquhar 1996, Harper 1999a, 1999b, 2001, Raphals 2001, Strickmann 2002, etc. For an appeal to a "popular" awareness of such connections, consult the preliminary "to the reader" (*zhi duzhe*) remarks in Jiang Yong 1989, not paginated.

145 See in particular R. Smith 1991.

146 Marc Kalinowski's continued efforts to refine the discourse and elevate the study of Chinese divination over the past decade or so are particularly admirable. For just a few of these efforts, see Kalinowski 1990, 1991, 1998, 1998–99, 2002, 2004, 2011 and Kalinowski, ed. 2003.

147 The literature on these topics is too vast to allow a systematic summary, even of the most superficial sort. For a few of the many recent studies that have discussed and analyzed these ancient materials, see Cook 1998, Giele 1998–1999, Harper 1999a, 1999b, and 2001, Hu Wenhui 2000, Kalinowski 1998–1999, Loewe 1994, Li Ling 2000a and 2000b, Pankenier 1995, Poo 1998, Raphals 2005, Shaughnessy 1998, Wang Aihe 2000, Ying Dingcheng 1990, etc. See also the illuminating set of articles on early Chinese cosmology in *Bulletin of the Museum of Far Eastern Antiquities* (2002).

148 See, for example, the various materials in collections such as the *Dunhuang wenxian fenlei lujiao congkan* series, notably Deng 1998.

149 This project, located at the École pratique de Hautes Études in Paris, has been supported by the Chiang Ching-kuo Foundation since 1996. The list of collaborators includes many of the most recognizable names in Chinese divination studies, including not only Kalinowski, but also Alain Arrault, Deng Wenkuan, Catherine Despeaux, Jean-Pierre Drege, Dimitri Drettas, Donald Harper, Liu Lexian, Jean-Claude Martzloff, Christine Mollier and Carole Morgan.

150 The literature on these new discoveries is vast in both Chinese and Western languages. For a useful introduction, see Giele 1998–99.

151 Xiong 2002 offers an illuminating perspective from the "center."

152 Davis 2001. See also Kleeman 1994 and H. Liao 2005. As I have indicated earlier, recent *Yijing* scholarship on the Song and later periods in both Asian and Western languages is almost unmanageably abundant.

153 For my preliminary effort to explore the relationship between divination and the examination system, see R. Smith 1991, 6–7, 41, 49–50, 128, 146, 149, 152, 154, 157, 199–200, 201, 210, 228ff. 272, 273, 278.

154 Elman 2000, 311.
155 Ibid., 311–61, esp. 312.
156 See, for example, R. Smith 2002 and Brook 2002.
157 This is one of the main arguments of R. Smith 2008a.
158 Elman 2005, 111, 115–17, 121.
159 See the initial chapter in this book; also the discussion in Zhang Qicheng 1996, 1: 67–83; cf. Zhu Bokun 1995, 3: 349–66, and 384–426.
160 Zhu Bokun 1995, 3: 336–514. 3: 336–45. For a summary of the disputes surrounding these documents, see the essay by Sun Guozhong in Jiang Yong 1989, 445–71; cf. Henderson 1984, 159–63, 181–99 and 218–24.
161 Zhu Bokun 1995, 3: 336–45.
162 Ibid., 3: 384.
163 I am grateful to Dr. Chu Pingyi (see P. Chu 1995) for alerting me to the complexity of Jiang's worldview.
164 P. Chu 1995, 23. Yuan 1948a, 16: 25 indicates that Jiang also had an interest in medicine. His preface to the *He Luo jingyun* is included in the "Confucian Scholars" section (*Rulin*) of the Draft History of the Qing (*Qingshi gao*). Ibid.
165 The other book was titled the *Buyi yuanji* (The perfect mechanics of *Changes* divination).
166 For information on Bouvet, see Chapter 6 in this volume on Jesuit approaches to the *Yijing*.
167 Jiang Yong 1989, personal preface, 16.
168 Jiang Yong 1989, 337–39.
169 See the discussion in P. Chu 1995.
170 Henderson 1984, 188, 243–44, 248, 250, 252.
171 For detail on *fenye* correlations and calculations, see the discussion in R. Smith 1991, 34, 67–71, 136, 152, 177, 199.
172 See Jiang Yong 1936, 10ff.
173 E. Wang 1993. On Li Fu's cosmological views, see C. S. Huang 1995, 77ff.
174 Jiang Yong 1989, personal preface, 16–17. Jiang discusses these technical terms at length in the *Hetu jingyuan*. On the *chengfang* system, for example, see ibid., 292ff.
175 See P. Chu 1995, 95ff.
176 Ibid., 101–02.
177 See, for example, BRZ, 1:19; 2:7; 5: 2; 5:25–26; 6:17; 8:11; 8:16; 9:6; 9:17–18; 12:10; 14:14; 15:27; 15:29; 16:25; 17:13; 18:7; 18:15; 18:25; 18:26–27; 19:34; 22:7; 22:15; 22:16–17; 22:21; 22:24; 22:25; 24:7; 24:19; 26:4–5; 28:2; 28:15–16; 28:20; 29:36–37; 31:18; 31:25; 31:26; 32:14; 33:25–26; 34:8; 34:9; 34:10; 34:22–23; 35:11–22; 35:12–13; 36:16; 36:18; 36:26; 37:31. Note also the large number (155) of Chinese-language works dealing with the *Hetu* and *Luoshu* in Duong and Lê 1999, 748–60.
178 R. Smith 1991 281ff.
179 See, for example, Chapter 2 of Sawada 2004, "Divination as Cultivation," which makes a number of points about the social role of divination in Tokugawa Japan that conform closely to the Chinese case, especially regarding the perceived relationship between morality and "fate." Cf. R. Smith 1991, 4, 22, 24, 95, 153–54, 165, 201, 208, 227–29, 241–42, 245, 260, 265, etc.
180 See, for example, Kalinowski 1990, F. Lin 1995, Nickerson 1994, Robinet 1995, Sakade 2000.

181 See also Foulks 2008 and 2011 and Fukui and Fussman 1994. One hopes for rapid publication of the several papers delivered in the two-part panel on "Buddhism and Divination" for the Sixteenth Congress of the International Association of Buddhist Studies, Jinshan, New Taipei City, Taiwan, June 20–25, 2011.
182 See, for example, Luo Yiqun 1993. Cf. R.J. Smith 1991, 260–61. QBLC 1928, *fangji*, 34–35 and 61–62 devotes some attention to "minority" divination.
183 I have discussed these themes briefly in R. Smith 1991, 209–10, 232, 244–45, 255–56 and 266.
184 See Yuan 1919, 1928, 1947, 1948a, 1948b, 1961a, 1961b and n.d. For a valuable preliminary study of Yuan, see Lackner 2012.

6 Jesuit interpretations of the *Yijing* in global perspective

1 For a useful discussion of the concept of "globalization" in an Asian context, consult Evans 2002, 147–62. For one of several useful Chinese language studies relating globalization specifically to the *Yijing*, consult Yang Hongsheng 1995. Other valuable studies on the transnational circulations of texts and ideas include Batchelor 1994, Clark 2000, Coleman 2001, Baumann and Prebish 2002, and Hofmeyer 2004.
2 Figurism, to be described more fully below, refers to an effort on the part of certain Jesuit missionaries to find "signs" (*figurae*) in the Chinese classics that indicated God's revelations.
3 See Hacker, Moore and Patsco, eds. 2002.
4 Legge 1963, 154–57, modified. Cf. R. Wilhelm 171–73, Lynn 1994, 410–16, Kunst 1985, 326–27.
5 See the discussion in R. Smith 1991, 96–97.
6 For an early understanding of this line statement, see Kunst 1985, 246–47. Wang Bi's commentary is translated in Lynn 1994, 161–62. Cheng Yi's commentary appears in ibid., 163–64.
7 See the discussions in R. Smith 2010, 114, Nielsen 2003, 146–47, and M. Lewis 1999, 199, 206–09, and esp. 272–74.
8 See the discussion in R. Smith 1991, 201–04, 217, 227–29, 249–51; also R. Smith 1994, 106–10.
9 See, for example, D. White, 1976, esp. 153ff., Foulks 2008, Y. Lo 2008, Cleary, trans. 1987.
10 Cited in R. Smith 2008a, 170.
11 Ibid., 221.
12 Zurcher 1994, 35–36. Similarly, a late Qing treatise traces Islam back to Fuxi. Ibid., 37.
13 See the discussion by Chengguan (738–838) in White 1976, 58–60.
14 W. Chan, trans. 1967, xxi, 6, 11–12 and 26–27, 38, 53, etc.
15 See the discussion in the Introduction to this volume. For a more complete inventory, see R. Smith 2008a, 218–40, and Zhang Qicheng, ed. 1996, 1: 401–501 and 2: 3–418.
16 See R. Smith 2012, 129–30.
17 W. Ng, 2000, 96–132. I am deeply indebted to Professor Ng for his *Yijing*-related scholarship, not only on Japan but also on Korea and Vietnam.
18 Ibid., 47–51, 78–95, 135–205.
19 Ibid., 107. I have modified this translation somewhat, and patched two disconnected but related passages together.

20 Ibid., 109–10.
21 See R. Smith 2012, 148–50.
22 See http://diendan.lyhocdongphuong.org.vn/danh-muc/52-the-ancient-oriental-learning/. According to some accounts, the Lac Viet people founded a state known as Van Lang around 3000 BCE.
23 R. Smith 2012, 155–57.
24 Ibid., 137–40, 141–2.
25 This is one of the principal arguments of R. Smith 2008a. See also R. Smith 2012, 48, 93, 94, 104, 108, 132, 146, 154, 161, 169, 182, 197.
26 R. Smith 2012, 160–67.
27 Ibid., 171. Malatesta 1997 provides a convenient overview of the Jesuit mission in China. See also Mungello 2009. For bibliographies of Western and Asian materials relating to the Jesuits, consult the relevant sections of Standaert, ed. 2001 and Standaert 1997, 569–613.
28 See Zürcher 1994. Significantly, the Jesuits did not have the same respectful attitude toward all alien peoples. See, for example, Lackner 2009.
29 Chinese converts such as Yang Tingyun (1557–1627) became similarly adept, comparing, for instance, the "virgin birth" of Jesus to the births of Laozi and the Buddha. See von Collani 1998, 34–48, esp. 47–48.
30 See, for example, the many stimulating essays in O'Malley, *et al.*, eds. 1999.
31 See Rule 1987, 158–159, 165, and 284, notes 46–48.
32 On Bouvet's life and work, see von Collani 1985.
33 Ibid., 156. Bouvet's position, summarized neatly by Michael Puett 2001, 4–5, was that there was a single figure—known by different names in different places (Fuxi in China, Hermes Trismegistus in Egypt and Greece, Enoch among the Hebrews, and Zoroaster in Persia)—who transmitted God's laws to all ancient peoples.
34 Rule 1987, 154ff.
35 See the excellent descriptions of Figurism by Claudia von Collani in Standaert, ed. 2001, 668–79 and von Collani 2007, 239ff. The literature on the Figurists in Western languages is quite substantial, and much of it has been produced by Dr. von Collani. For two useful works in Chinese, consult Zhuo Xinping 1998, 292–317 and Yang Hongsheng 1995, 179ff.
36 For other examples see Standaert, ed. 2001, 675, von Collani 1985, 133ff. and 173ff.
37 R. Smith 2012, 172–74.
38 In Bouvet's mind, Fuxi was Hermes Trismegistus and a "figure" of Christ (as was the Yellow Emperor). See Lundbaek 1991, 128–29. Another Jesuit, Niccolo Longobardo (1565–1655) claimed Fuxi was Zoroaster, who came to China and whose magic power was expressed in the trigrams.
39 The Jesuits used this phrase and this general argument to justify their own exegetical efforts. See, for example, Bouvet's *Yiyao zixu* in VL, Borg. Cin. 317 (2), 2a–b. Cf. Lynn, *The Classic of Changes*, 87.
40 Han Qi 1998, 185–202.
41 Huang Yinong 1994, 163–66.
42 On Zhou Zhi, consult Han Qi 1998, 190, note 19. Zhou was the author of the 1678 work on the *Yijing* titled *Du Yi ji* (Record of reading the *Changes*). This 36-page tract refers repeatedly to concepts such as "original sin" and "the Trinity," and uses several

standard interpretive techniques involving line and trigram relationships to analyze a total of nine hexagrams, drawing in particular on the rich symbolism of the first hexagram, Qian. See VL, Bor. Cin. 357 9 (C) under the name "Jacques Tcheou Tche." Of course, the enemies of Christianity also drew upon the *Changes* for authority. See, for example, J. Young 1975, 155–86, esp. 178–80.

43 VL, Borg. Cin. 439 A (h). Cf. Liao Mingchun 1991, 458.

44 VL, Borg. Cin. 317 (2), 1b. *Yiyao zixu.*

45 VL, Borg. Cin. 317 (4), 1a–b.

46 VL, Borg. Cin. 317 (6), 49–50. See also Han Qi 1998, 190, note 18.

47 See, for example, Secret 1979, 35–53, von Collani 2000b, 527–558, Javary 1991, 113–22.

48 Cited in Secret 1979, 40. A convenient Western-language exposition of Bouvet's basic ideas about the *Changes* is his "Idea generalis doctrinae libri ye kim" (General idea of the doctrinal book *Yijing*). For a discussion of the text (including the Latin original and a German translation), see von Collani 2007, 277–88 and 360–65. Ibid., 288–313 provides the Latin and French texts and English summaries for some of Bouvet's letters that deal with the *Changes*.

49 Ibid., 39. "Sap." refers here to the "apocryphal" work known as *Liber Sapientiae* or "Book of Wisdom."

50 Bouvet's subsequent explanation of the relationship between *li*, *shu*, *xiang*, and *tu* suggests an understanding derived not only from the orthodoxy of Zhu Xi but also from the less orthodox speculations of Shao Yong. See, for example, VL, Borg. Cin. 439 A (C) 1. For a useful analysis of various Jesuit responses, including Bouvet's, to Shao's ideas, see Lundbaek 1993.

51 I do not know for certain which of the "magic squares" Bouvet borrowed from Cheng Dawei, but it is clear that he drew upon Cheng's version of "Pascal's Triangle." See VL, Borg. Cin. 317 (3) loose diagram, and Borg. Cin. 361 (2) 73a.

52 See the illuminating discussion in Mungello 1989, 312–28. For evidence of Bouvet's use of Cheng Dawei's mathematics to explicate the *Yijing*, see Han Qi 1998, 193–94.

53 See von Collani 1985 frontispiece and 168ff. R. Smith 2012, 176 reproduces one of these diagrams.

54 A number of these diagrams can be found in VL, Borg. Cin. 361 (1) C (IV), Borg. Cin. 361; (4) II, and Borg. Cin. 361 (6) III; also Borg. Cin. 361 (1), 1a–74a, passim, Borg. Cin. 361 (10), 57ff., and Bibliothèque Nationale, Ms. n.a. lat. 1173, f. 136. See also the illustration in Javary 1991, 112.

55 On the Chinese version of "Pascal's Triangle," see SCC, 3: 133ff., esp. 136 and Li Yan 1984, 109ff. Cf. Jiang Yong 1989, 292–94.

56 To both Shao and Bouvet (and to most Chinese scholars as well), the *Yijing* had the capacity to reveal the nature of "ten thousand affairs and ten thousand transformations, from beginning to end." See Bouvet's *Yixue zongshuo hui* (A Classified General Discussion of *Changes* Scholarship), VL, Borg. Cin. 317 (8).

57 VL, Borg. Cin. 317 (8–12), esp. (8), 32aff. and (10), (*Yixue waipian*, section 7), 57aff. and (11), 8aff.

58 *Shi Xiantian weibian shizhong zhi shi you tianzun dibei tu er sheng* (An explanation of the numbers of the Former Heaven [sequence], not yet changed from beginning to end, according to the *Chart of Heavenly Superiority and Earthly Subordination*), VL, Borg. Cin. (11) 8a.

59 See Landry-Deron 1995.

60 Rule 1987, 163 tells us that right triangles represented to Bouvet the "hypostatical union of God and man," but most of the triangular diagrams drawn by Bouvet that I recall are not right-angled. See the many illustrations and related discussions in VL, Borg. Cin. 361 (14) and VL, Borg. Cin. 317 (8–11), es317 (8), 32aff. Bouvet asserts on more than one occasion that the triangle is "the beginning of heavenly images and earthly forms" (*tianxiang dixing zhi shi*).

61 For details, see von Collani 1985, 169ff. Lundbaek 1993, 180 indicates, correctly I think, that in a certain sense Bouvet viewed Shao Yong as a "rival."

62 For Bouvet's written account of these three periods, see his *Zhouyi yuanzhi tan* (An investigation into the original meaning of the *Zhou Changes*), VL, Borg. Cin. 312 (L) 1: 1a–4a. It would appear that inspiration for some of Bouvet's ideas may have come from the writings of Zhou Zhi, discussed above.

63 For the various memorials and edicts dealing with Bouvet's essays, charts and diagrams on the *Yijing* during 1712, see Guan Xiaolian and Qu Liusheng, eds. 1996, 718ff., esp. documents 1716, 1719, 1724, 1725, 1731, 1734, 1738, 1741, 1752, 1755, 1759, 1760, 1761, 1764, 1767, 1768, etc. One replica of the *Jiabei bianfa tu* appears as a loose item in VL, Borg. Cin. 317 (3); cf. VL, Borg. Cin. 361 (2) 73a, which has the same item under a different title. This diagram is also included in the *Qimeng fulun* (Appendix to the *Yixue qimeng*) in Li Guangdi, ed. 1998, 1205–06.

64 See VL Borg, Cin. 317 (10), esp. 57a–67b. Cf. Li Guangdi's memorial cited in Han Qi, 1998, 194.

65 See Witek 1982, 166ff., esp. 176–77, and von Collani 2007, esp. 245–48.

66 References to these and other writings by Bouvet can be found in D. Yu 1996 and Takata, ed. 1995.

67 Han Qi 1998. See also Luo Lida, 1997. For some of the many links between the *Yijing* and Chinese mathematics, consult SCC 3: 40, 56, 57, 59, 69, 119, 140, 199, 287, 291, 301, 464, 470, 625,

68 Bouvet's influence can be detected in the *Qimeng fulun*, discussed above. See Li Guangdi, ed. 1998, 1168–206; also von Collani 2007, 253–56.

69 Significantly, each man claimed to admire the contributions of the other. Han Qi, 1998, 193.

70 Ibid., 194–95.

71 Guan and Qu, eds. 1996, 722ff., esp. documents 1724, 1734, 1738, 1752, 1755, 1760, 1764, and 1768; see also VL, Borg. Cin.439 A (c) 1 and Han Qi 1998, 193. Cf. von Collani 2007, 250–56.

72 As a result, Fouquet was summoned from the provinces to assist Bouvet. See von Collani 2007, esp. 245ff.

73 Guan and Qu, eds. 1996, 734, document 1760 and 735, document 1764.

74 Lundbaeck 1994, 129–48, esp. 136.

75 Ibid., 138. Lundbaek 1991, 117 quotes Prémare in April of 1716 as saying that the Kangxi emperor "got tired" of Bouvet "more than a year ago and believes this poor old man has gone mad from meditating on the *Yi Jing*." Cf. von Collani 2007, 251–52.

76 See Paul Rule 1995, 303–31, esp. 330.

77 See the preface to the *Yi tongshi*, 2a, in Jiao Xun 1970. For details on Jiao's theories, consult Lai Guisan 1994.

78 Cited in R. Smith 2002a 7–8.

79 See, for example, Yuan Shushan 1948, 1:19; 2:7; 5:2; 5:25–26; 6:17; 8:11; 8:16; 9:6; 9:17–18; 12:10; 14:14; 15:27; 15:29; 16:25; 17:13; 18:7; 18:15; 18:25; 18:26–27; 19:34; 22:7; 22: 15;22:16–17; 22:21; 22:24; 22:25; 24:7; 24:19; 26:4–5; 28:2; 28:15–16; 28:20; 29:36–37; 31:18; 31:25; 31:26; 32:14; 33:25–26; 34:8; 34:9; 34:10; 34: 22–23; 35:11; 35:11–22; 35:12–13; 36:16; 36:18; 36:26; 37:31. Note also the large number of Qing dynasty works (more than 100) in the bibliographical section on the *Hetu* and *Luoshu* in Duong and Lê 1999, 748–60.

80 See, for example, Henderson 1984, 178ff. and Smith, 1991, 70ff.

81 For a general account of the process by which the *Yijing* traveled westward, consult Smith 2012, 177–94.

82 For an insightful analysis of this dialectic, see Gardner 1998, esp. 416–18.

83 See Mungello 1977, Li and Hans Poser, eds. 2000, von Collani 1989 and 2007, esp. 241–43, Cook and Rosemont, 1994. To this day, Chinese scholars continue to celebrate the connection between Leibniz (and other Western scientists) and the *Yijing*. See, for example, Zhang Qicheng, ed. 1996, 1: 5.

84 R. Smith 2012, 177–78.

85 Ibid. See also von Collani 2007, 241ff.

86 This summary is drawn primarily from R. Smith 2012, 180–81. For further background and a great deal of valuable detail, consult von Collani 2007, 256–76.

87 R. Smith 2012, 182ff.

88 Girardot 2002, 371–72.

89 See Shchutskii 1979, 23–24.

90 Cited in R. Smith 2012, 183–84.

91 See Shaughnessy 1996, 17.

92 R. Smith 2012, 186–88. For a severe critique of Terrien's scholarship, see Shchutskii 1979, 24–27.

93 Ibid., 27.

94 On Legge's translation of the *Changes*, and the debates that surrounded it, see Girardot 2002, esp. 366–374.

95 See the discussion in ibid., 372–73.

96 Legge 1963, xvii.

97 Ibid., xiv–xv, 10, 17, 25–26, 38, etc.

98 See Gerald Swanson's introduction to Shchutskii 1979, xi–xii. For an excellent, historically sensitive analysis of these two works, see Hon 2005a.

99 R. Smith 2012, 189–91.

100 See Michael Lackner 1998.

101 See R. Smith 2012, 177–78 and 190–91.

102 Cornelius and Cornelius 1998, 19ff.

103 Ibid., 21.

104 For a similar effort to link the *Yijing* to ancient mysticism—including not only the Kabbalah, but also various forms of astrology and Tarot card reading—see Higgins 1997.

105 Lee 1994.

106 Ibid., 136–50.

107 Kang and Nelson 1979.

108 Mears and Mears 1931.

109 For a convenient summary of his basic argument, see McCaffree 1982, 428–32. At various points in the book, McCaffree links the Li hexagram (number 30) with (A) the Torah as a whole, (B) the thirtieth chapter of Genesis, (C) Jesus at age thirty, (D) Jacob, (E) the "Kabbalistic Tree of Life," (F) the fleur-de-lis, and (G) the unicorn (Chinese: *qilin*).

110 Ong 1991. For dozens of Ong's Figurist-style etymologies, see ibid., 166ff.

111 For some examples, see R. Smith 1998c, 2003, 2008a, esp. 208–11 and 2012, esp. 221–23.

112 See Batchelor 1994, xi; also Coleman 2001, Baumann and Prebish, eds. 2002, Clark 2000, and R. Smith 2012, esp. 127–28 and 170.

113 Govinda 1981, xi. For details on Govinda, consult Batchelor 1994, 312–14, 320–21, 327–29.

114 Gardner 1998, 416–17.

115 Henderson 1991 provides an illuminating comparative study of several commentarial traditions that display common characteristics of the sort discussed in this paragraph. See also R. Smith 2008a, 241–49.

BIBLIOGRAPHY

In the interest of clarity, I have followed the following bibliographical conventions in the notes: In citing the author of a work written in Chinese or Japanese I have observed the East Asian practice of placing the family name first, followed by the entire personal name. In citing the author of a Western-language work, even if he or she is of East Asian descent, I give only the family name, unless two or more authors have the same surname, e.g., Kidder Smith and Richard Smith or Shiming Zhang and Weiwen Zhang. In these instances, I add the first (and sometimes, if necessary, the second) initial of the personal name; hence K. Smith, S. Zhang, and so on. In the case of individuals who write in both Asian and Western languages, I have followed the rules indicated above.

Abbreviations

BRZ Yuan, Shushan. 1948. *Zhongguo lidai buren zhuan* (Biographies of diviners in China by dynastic periods). Shanghai: Rude shuju, 1948. A new 1998 edition was published in Taibei by the Xingwenfeng chuban gongsi.

CC Legge, James, trans. 1893–1895. *The Chinese Classics*. Oxford: Clarendon Press. Reprinted by the Hong Kong University Press in 1979.

CNA *China News Analysis*.

CSJC Congshu jicheng, ed. 1965–1966. *Congshu jicheng: chubian* (Complete collection of collectanea; first series). Reprinted Taibei: Taiwan shangwu yinshu guan. Also Beijing: Zhonghua shuju, 1985–1991.

CSJCXB Xinwen feng chuban gongsi bianji bu, comp. 1986. *Congshu jicheng xinbian* (New edition of the complete collectanea). Taibei: Xinwen feng chuban gongsi.

DQHD *Qinding Da Qing huidian* (Imperially approved collected statutes of the Qing dynasty, 1899).

DQLCSL *Qinding lichao shilu* (Veritable records of the successive reigns of the Qing dynasty). 1964. Taibei: Hualian reprint of a 1937–1938 edition.

DQTL *Qinding Da Qing tongli* (Imperially approved comprehensive guide to Qing dynasty rituals, 1759). Reprinted Taibei: Commercial Press, *c.* 1978.

HCJSWB He, Changling, ed. 1826. *Huangchao jingshi wenbian* (A collection of Qing dynasty writings on statecraft). Beijing, n.p.

HCJSWBXB Sheng, Kang, ed. 1897. *Huangchao jingshi wenbian xubian* (Additional supplement to the collection of Qing dynasty writings on statecraft). Wujin, Jiangsu: Sheng shi sibu lou.

HCJSWXB Ge, Shijun, ed. 1888. *Huangchao jingshi wen xubian* (Supplement to the collection of Qing dynasty writings on statecraft). Shanghai: Tushu jicheng.

HCXWXTK Liu, Jinzao, comp. 1935. *Huangchao xu wenxian tongkao* (Supplement to the complete analysis of the literary and historical records of the Qing dynasty). Reprinted Shanghai: Shanghai shangwu shuju.

IWSM Wenqing, *et al.*, eds. 2002. *Chouban yiwu shimo* (The management of China's barbarian affairs from beginning to end). Available online at the Chinese Text Project: http://ctext.org/library.pl?if=en&res=1905.

JJZS Li, Daoping, ed. 1994. *Zhouyi jijie zuanshu* (A compilation of collected subcommentaries on the *Collected Explanations of the Zhou Changes*). Reprinted Beijing: Zhonghua shuju.

JLDC Lü, Zuzhen. 1975. *Jiali dacheng* (A complete book of family rituals). Eighteenth-century. Reprinted Tainan: Xibei.

JLHT Zhang, Rucheng, comp. 1723. *Jiali huitong* (A general collection of guides to family rituals). Changzhou: Jixin tang.

JLQJ Liang, Jie, ed. 1895. *Jiali quanji* (A complete collection of guides to family rituals). Shanghai: n.p.

JLTSJC Chen, Mingsheng. 1842. *Jiali tieshi jicheng* (A complete collection of documentary forms regarding family rituals). Youwen tang.

LBZL *Qinding Libu zeli* (Imperially approved collection of statutes and precedents regarding the Board of Rites). 1966. Reprinted Taibei: Chengwen.

LDTWC Yixin (Prince Gong). 1867. *Ledao tang wenchao* (Collected prose from the Hall of Delighting in the Way). Beijing.

LYBL Zhou, Liang. 1911. *Liyi bianlan* (Handbook on rituals and etiquette). n.p.

QBLC Xu, Ke, ed. 1928. *Qingbai leichao* (Classified anecdotes from the Qing). Shanghai: Shangwu. Original edition 1916.

QS Zhang, Qiyun, *et al.*, eds. 1961. *Qingshi* (History of the Qing dynasty). Taibei: Guofang yanjiu yuan.

RMRB *Renmin ribao* (People's Daily).

SCC Needham, Joseph, *et al.* 1956–present. *Science and Civilisation in China*. Cambridge: Cambridge University Press.

SKQS Ji, Yun, *et al.*, eds. 1983–1986. *Qinding Siku quanshu* (Imperial edition of the *Complete Collection of the Four Treasuries*). Reprinted Taibei: Shangwu yinshu guan, under the title *Yingyin Wenyuan ge Siku quanshu*.

SKQSZM Ji, Yun, *Qinding siku quanshu zongmu* (Reviews from the *Complete Collection of the Four Treasuries*). Reprinted Taibei: Shangwu yinshu guan, 1983. Cf. ZMTY below.

TSJC Chen, Menglei, *et al.*, eds. 1977. *Qinding gujin tushu jicheng* (Imperial edition of the complete collection of illustrations and writings, past and present, 1726). Reprinted Taibei: Dingwen shuju.

VL Vatican Library (Biblioteca Apostolica Vaticana). Rome, Italy.

WJSMSC *Wujiang simei shouce* (Handbook of the five emphases and the four beautifications). 1983. Beijing: n.p.

WLTK Qin, Huidian. 1880. *Wuli tongkao* (A comprehensive analysis of the five categories of ritual). Shanghai: Jiangsu shuju.

WXL Wu, Rongguang. 1832. *Wuxue lu* (Record of my studies). Minting huijing tang.

XBZZJC Shijie shuju bianji bu, ed. 1974. *Xinbian zhuzi jicheng* (New edition of the compendium on philosophers). Taibei.

XJBFS Yunlu, *Qinding xieji bianfang shu* (Imperially approved treatise on harmonizing the times and distinguishing the directions). Beijing, 1724. Reprinted Taibei, 1974.

XSHJY Xin shenghuo yundong cujin zonghui, ed. 1936. *Xin shenghuo yundong jiyao* (A collection of important documents on the New Life Movement.) Guangzhou: Xin shenghuo yundong cujin zonghui.

XSHXZ Xin shenghuo congshu she, ed. 1935. *Xin shenghuo yundong xuzhi* (Necessary information on the New Life Movement). Shanghai: Xin shenghuo congshu she.

YJJC Yen Ling-feng, ed., *Yijing jicheng* (Compendium on the *Classic of Changes*). Taibei, 1975.

YLLB Zhongyang xuanzhuang, ed. 1938. *Jiang Weiyuanzhang yanlun zhang leibian* (Categorized collection of the speeches of Chiang Kai-shek). Nanjing: Zhengzhong shuju.

YZLB Zhang Boxing. 1868. *Yangzheng leibian* (Classified primers for moral rectification). Fuzhou: Zhengyi shuyuan.

ZMTY Siku quanshu zongmu tiyao bianwei hui, ed. 1999. *Siku quanshu zongmu tiyao* (Annotated general catalogue of the Complete Collection of the Four Treasuries). Haikou: Hainan chubanshe.

ZWDCD Zhang Qiyun *et al.*, eds. 1968. *Zhongwen da cidian* (Encyclopedic dictionary of the Chinese language). Taibei: Zhongguo wenhua yanjiu so.

ZZJL Qiu, Jun, ed. 1701. *Zhuzi jiali* (Zhu Xi's family rituals). Guangzhai tang. There are a great many editions of this work, which are edited by different scholars and go by different names (e.g. *Wengong jiali*).

References

Abe, Takeo. 1946. "Shincho to Ka i shisho" (Thinking about Chinese and barbarians in the Qing dynasty). *Jimbun kagaku*, 1.3 (December): 1–23.

Adler, Joseph. 2002. *Introduction to the Study of the Classic of Change (I-hsüeh ch'i-meng)*. New York: Global Scholarly Publications.

Ahern, Emily. 1981. *Chinese Ritual and Politics*. Cambridge: Cambridge University Press.

Aijmer, Goram. 1978. "Ancestors in the Spring: The Qingming Festival in Central China." *Journal of the Hong Kong Branch of the Royal Asiatic Society* 18: 59–82.

Alito, Guy S. 1978. *The Last Confucian: Liang Shuming and the Chinese Dilemma of Modernity*. Berkeley, CA: University of California Press.

Alleton, Viviane and Michael Lackner, eds. 1998. *De l'un au multiple. La traduction du chinois dans les langues européennes*. Paris: Maison des Science de l'Homme.

Amelung, Iwo. 2007. "New Maps for the Modernizing State: Western Cartographic Knowledge and its Application in 19th and 20th Century China." In Bray, Lichtman and Metailie, eds. 2007: 685–726.

Amitai, R. and Biran, M. eds. 2005. *Mongols, Turks and Others: Eurasian Nomads and the Sedentary World*. Leiden: Brill.

Anagost, Ann. 1992. "Socialist Ethics and the Legal System." In Wasserstrom and Perry, eds. 1992: 177–205.

Anonymous. 1832. Review of *Ta Tsing Wan-neen Yih-tung King-wei Yu-too* [*Da Qing wannian yitong jingwei yutu*] (A map, with latitude and longitude, of the eternally unified great Qing empire) by Lee Mingche [Li Mingche; aka Li Qinglai]. *Chinese Repository* 1.2 (June), 1.4 (August) and 1.5 (September).

——. 1848. "Reading the Sacred Edict." *Chinese Respository* 17 (November).

Atwill, David. 2005. *The Chinese Sultinate: Islam, Ethnicity, and the Panthay Rebellion in Southwest China, 1856–1873*. Stanford, CA: Stanford University Press.

Averill, Stephen. 1981. "The New Life in Action: The Nationalist Government in Jiangxi, 1934–1937." *China Quarterly* 88 (December): 594–628.

Aylward, Thomas F. 2007. *The Imperial Guide to Fengshui and Chinese Astrology*. London: Watkins Publishing.

Baker, Hugh. 1979. *Chinese Family and Kinship*. New York: Columbia University Press.

Balazs, Etienne. 1964. *Chinese Civilization and Bureaucracy*. New Haven, CT: Yale University Press.

Ball, J. Dyer. 1903. *Things Chinese: Being Notes on Various Subjects Connected with China*. New York: Kelly and Walsh.

Batchelor, Stephen. 1994. *The Awakening of the West: The Encounter of Buddhism and Western Culture*. Berkeley, CA: Parallax Press.

Baum, Richard. 1995. "Deng Liqun and the Struggle against 'Bourgeois Liberalism.'" *China Information* 9.4 (Spring): 1–35.

Baumann, Martin and Charles Prebish, eds. 2002. *Westward Dharma: Buddhism beyond Asia*. Berkeley, CA: University of California Press.

Behr, Wolfgang. 2010. "Role of Language in Early Chinese Constructions of Ethnic Identity." *Journal of Chinese Philosophy* 37.4: 567–87.

Bell, Catherine. 1992. *Ritual Theory, Ritual Practice*. Oxford: Oxford University Press.

——. 1997. *Ritual: Perspectives and Dimensions*. Oxford: Oxford University Press.

Bickers, Robert A., ed. 1993. *Ritual and Diplomacy: The Macartney Mission to China 1792–1794*. London: Wellsweep Press.

Birdwhistell, Anne. 1989. *Transition to Neo-Confucianism: Shao Yung on Knowledge and Symbols of Reality*. Stanford, CA: Stanford University Press.

Birrell, Anne, trans. 1999. *The Classic of Mountains and Seas*. London: Penguin Classics.

Bodde, Derk and Clarence Morris. 1967. *Law in Imperial China*. Cambridge, MA: Harvard University Press.

Boelhower, William. 1988. "Inventing America: A Model of Cartographic Semiosis." *Word and Image* 4.2 (April–June): 475–97.

Bohr, P. Richard. 2004. "The Taipings in Chinese Sectarian Perspective." In Liu and Shek, eds., 2004: 393–430.

Boulais, Guy. 1924. *Manuel du code chinois*. Shanghai: Imprimerie de la mission catholique.

Bray, Francesca, Vera Lichtman and Georges Metailie, eds. 2007. *Graphics and Text in the Production of Technical Knowledge in China: The Warp and the Weft*. Leiden: Brill.

Brook, Timothy. 2002. "Is Smoking Chinese?" *Ex/Change* (February). Available at: http://www.cityu.edu.hk/cityu/about/exchange.htm.

Brose, Michael C. 2006. "Realism and Idealism in the *Yuanshi* Chapters on Foreign Relations." *Asia Major*, 3rd series 19.1–2: 327–47.

Bruun, Ole. 2002. *Fengshui in China: Geomantic Divination between State Orthodoxy and Popular Religion*. Richmond, Surrey: Curzon Press.

Cadonna. A. and E. Bianchi, eds. 2002. *Facets of Tibetan Religious Tradition and Contact with Neighboring Cultural Areas (Orientalia Venetia)*. Casa Editrice Leo S. Olschki.

Cao, Fujing and Zhang Yueming. 1995. *Da Liuren jingjie* (Essential explanation of the Great Six Yang Waters). Harbin: Heilongjiang renmin chuban she.

Cao, Wanru, *et al.* 1990. *Zhongguo gudai ditu ji: Zhanguo zhi Yuan* (A collection of ancient Chinese maps: from the Warring States period to the Yuan dynasty). Beijing: Wenwu.

——. 1994. *Zhongguo gudai ditu ji: Mingdai* (A collection of ancient Chinese maps: the Ming dynasty). Beijing: Wenwu.

——. 1997. *Zhongguo gudai ditu ji: Qingdai* (A collection of ancient Chinese maps: the Qing dynasty). Beijing: Wenwu.

Cao Weilin. 1873. *Yixue shijing* (Historical mirror of *Changes* studies). Reprinted Taibei: Xin wenfeng chuban she, 1980.

Cao, Yuanzhong. 1916. *Liyi* (Proposals on ritual). n. p.

Cartier, Michel. 1981. "Barbarians through Chinese Eyes: The Emergence of an Anthropological Approach to Ethnic Differences." *Comparative Civilizations Review* 6: 1–14.

Chai, Chu and Winberg Chai, eds. 1967. *Li Chi: Book of Rites*. New York: University Books.

Chan, Gilbert, ed. 1980. *China at the Crossroads: Nationalists and Communists, 1927–1949*. Boulder, CO: Westview Press.

Chan, Wing-tsit, trans. 1967. *Reflections on Things at Hand*. New York: Columbia University Press.

——, ed. 1986. *Chu Hsi and Neo-Confucianism*. Honolulu: University of Hawaii Press.

Chang, Che-chia. 2001. "Medicine and Astrology: Disorders in Chinese Astrological Cases." Paper for the Conference on Divination in honor of Professor Ho Peng Yoke, Needham Research Institute, Cambridge University. December 7–8.

Chang, Chi-yun. 1977. *The Essence of Chinese Culture*. Taibei: China Academy.

Chang, Hao. 1987. *Chinese Intellectuals in Crisis: Search for Order and Meaning (1890–1911)*. Berkeley, CA: University of California Press.

Chang, Sen-dou. 1974. "Manuscript Maps in Late Imperial China." *Cartographia* 11.1 (June): 1–14.

Chao, Paul. 1983. *Chinese Kinship*. London: Kegan Paul.

Chemla, Karine, Donald Harper and Marc Kalinowski (eds.). 1999. *Divination et rationalité en Chine ancienne*. (Vol. 21 of Extrême-Orient/Extrême-Occident: Cahiers de recherches comparatives). Saint-Denis: Presses Universitaires de Vincennes.

Chen, Baoquan. 1820. *Lishu fulu* (Appendices to the ritual classics). n.p.

Chen, Daqi. 1982. *Kongzi xueshuo* (On Confucian learning). Taibei: Zhengzhong.

Chen, Gaoyong. 1939. *Zhongguo lidai tianzai renhuo biao* (Chart of heavenly calamities and human misfortunes in China by dynastic period). Shanghai: n.p.

Chen, Guansheng [Kenneth Ch'en]. 1939. "Li Maodou dui Zhongguo dili xue zhi gongxian ji qi yinxiang" (Matteo Ricci's contribution to and influence on geographical knowledge in China). *Yugong* (Chinese Historical Geography) 5.3–4 (April), 1–50.

Chen, Guofu and Qiu Peihao 1964. *Tongli xinbian* (A new comprehensive guide to rituals). Taibei: Zhengzhong . Reprint of 1950 edition.

Chen, Jiujin and Yang Yi. 1994. *Zhongguo gudai di tianwen yu lifa* (Ancient Chinese astronomy and calendrical methods). Taibei: Taiwan shangwu yinshu guan.

Ch'en, Kenneth [Chen Guansheng]. 1939. "Matteo Ricci's Contribution to and Influence on Geographical Knowledge in China." *Journal of the American Oriental Society* 59: 325–59.

Chen, Meiying, *et al.* 1995. *Zhonghua zhanmeng shu* (The art of Chinese dream divination). Taibei: Wenjin chuban she youxian gongsi.

Chen, Walter Hanming. 1936. "The New Life Movement," *Information Bulletin of the Council of International Affairs* 2.11 (December 21).

Chen, Yongzheng, ed. 1991. *Zhongguo fangshu da cidian* (An encyclopedia of the predictive and prescriptive arts in China). Guangzhou: Zhongshan daxue chuban she.

Chen, Yun'gen. 1989. *Shushu pipan* (A critique of the predictive and prescriptive arts). Taibei: Tiandi tushu.

Cheng, Hsiao-Chieh, *et al.* 1985. *Shan Hai Ching: Legendary Geography and Wonders of Ancient China*. Taipei: National Institute for Compilation and Translation.

Chiang, Kai-shek. 1974. *Aphorisms of President Chiang Kai-shek*. Taipei: Government Information Office.

Chiang [Kai-shek], Madame (Song Meiling), ed. 1937. *General Chiang Kai-shek*. New York: Book League of America.

Chinese Academy of Surveying and Mapping, ed. 1998. *Treasures of Maps: A Collection of Maps in Ancient China*. Harbin: Cartographic Publishing House.

Chow, Kai-Wing. 1994. *The Rise of Confucian Ritualism in Late Imperial China: Ethics, Classics, and Lineage Discourse*. Stanford, CA: Stanford University Press.

Chu, Pingyi. 1995. "Ch'eng-Chu Orthodoxy, Evidential Studies and Correlative Cosmology: Chiang Yung and Western Astronomy." *Philosophy and the History of Science*, 4.2 (October): 71–108.

Chu, Samuel. 1980. "The New Life Movement before the Sino-Japanese Conflict: A Reflection of Kuomintang Limitations in Thought and Action." In Gilbert Chan, ed., 1980.

Clark, J.J. 2000. *The Tao of the West: Western Transformations of Taoist Thought*. London: Routledge.

Clarke, G. N. G. 1988. "Taking Possession: The Cartouche as Cultural Text in Eighteenth-Century American Maps." *Word and Image* 4.2 (April–June): 455–74.

Clart, Philip. 1997. "The Phoenix and the Mother: The Interaction of Spirit Writing Cults and Popular Sects in Taiwan." *Journal of Chinese Religions* 25: 1–32.

——. 2003. "Moral Mediums: Spirit-Writing and the Cultural Construction of Chinese Spirit-Mediumship." *Ethnologies* 25.1: 153–89.

Cleary, Thomas, trans. 1986. *The Taoist I Ching*. Boston: Shambhala Press.

——, trans. 1987. *The Buddhist I Ching*, Boston: Shambhala Press.

Cohen, Paul. 1984. *Discovering History in China: American Historical Writing on the Recent Chinese Past*. New York: Columbia University Press.

——. 2003. *China Unbound: Evolving Perspectives on the Chinese Past*. New York: Routledge.

—— and Merle Goldman, eds. *Ideas across Cultures: Essays on Chinese Thought in Honor of Benjamin Schwartz*. Cambridge, MA: Harvard University Press.

Coleman, James W. 2001. *The New Buddhism: The Western Transformation of an Ancient Tradition*. New York: Oxford University Press.

Cook, Constance A. 1998. "Myth and Fragments of a Qin *Yi* Text: A Research Note and Translation." *Journal of Chinese Religions* 26: 135–43.

Cook, Daniel J. and Henry Rosemont, Jr., eds. 1994. *Gottfried Wilhelm Leibniz: Writings on China*. Chicago: Open Court.

Cornelius, J. Edward and Marlene Cornelius. 1998. "*Yi King*: A Beastly Book of Changes." *Red Flame: A Thelemic Research Journal* 5: 1–226.

Crossley, Pamela K. 1990a. "Thinking about Ethnicity in Early Modern China." *Late Imperial China* 11.2 (June): 1–36.

——. 1990b. *Orphan Warriors: Three Manchu Generations and the End of the Qing World*. Princeton, NJ: Princeton University Press.

——. 1999. *A Translucent Mirror: History and Identity in Qing Imperial Ideology*. Berkeley, CA: University of California Press.

——, Helen Siu and Donald Sutton, eds. 2006. *Empire at the Margins: Culture, Ethnicity, and Frontier in Early Modern China*. Berkeley, CA: University of California Press.

Cua, A.S. 1979. "Dimensions of *Li* (Propriety): Reflections on an Aspect of Hsün-tzu's Ethics." *Philosophy East and West* 29.4: 373–94.

Dabringhaus, Sabine and Roderich Ptak, eds. 1997. *China and Her Neighbors: Borders, Visions of the Other, Foreign Policy 10th–19th Century*. Wiesbaden: Harrassowitz.

Dahlen, Martha. 1997. *The Vanishing Border: A Retrospective of China and Hong Kong through Maps*. Hong Kong: Credit Lyonnais Securities Asia.

Da Qing shichao shengxun (Sacred instructions for ten reigns of the great Qing). 1880. Beijing.

Dathorne, O.R. 1994. *Imagining the World: Mythical Belief Versus Reality in Global Encounters*. Westport, CT: Bergin and Garvey.

Davis, Edward L. 2001. *Society and the Supernatural in Song China*. Honolulu: University of Hawaii Press.

De Bary, Wm. Theodore, *et al.*, eds. 1964. *Sources of Chinese Tradition*. Two vols. New York: Columbia University Press.

—— and Irene Bloom, eds. 1999. *Sources of Chinese Tradition*. Second edition, vol. 1. New York: Columbia University Press.

—— and Richard Lufrano, eds. 2000. *Sources of Chinese Tradition*. Second edition, vol 2. New York: Columbia University Press.

De Groot, J. J. M. 1903. *Sectarianism and Religious Persecution in China*. Leiden: J. Miller.

De Meyer, Jan A.M. and Peter M. Engelfriet, eds. 2000. *Linked Faiths: Essays on Chinese Religion and Traditional Culture in Honour of Kristofer Schipper*. Leiden: E.J. Brill.

Deng, Wenkuan. 1999. *Dunhuang tianwen lifa wenxian jijiao* (Collected collations of the astronomical texts and calendars from Dunhuang). Nanjing: Jiangsu guji chuban she.

Destombes, Marcel. 1983. "Wang P'an, Liang Chou et Matteo Ricci: Essai sur la cartographie chinoise de 1593 à 1603." In *Actes du IIIe Colloque internationale de Sinologie*. Paris: 47–65.

De Weerdt, Hilde. 2009. "Maps and Memory: Readings of Cartography in Twelfth- and Thirteenth-Century Song China." *Imago Mundi* 61.2: 145–67.

De Woskin, Kenneth. 1983. *Doctors, Diviners and Magicians of Ancient China*. New York: Columbia University Press.

Dicks, Anthony. 1989. "The Chinese Legal System: Reforms in the Balance." *China Quarterly* 119 (September): 540–76.

Di Cosmo, Nicola. 1998. "Qing Colonial Administration in Inner Asia." *The International History Review*, 20.2 (June): 287–309.

——. 2003. "Kirghiz Nomads on the Qing Frontier: Tribute, Trade, or Gift-Exchange?" In Di Cosmo and Wyatt (2003): 351–72.

——, ed. 2009. *Military Culture in Imperial China*. Cambridge, MA: Harvard University Press.

——. 2010. "Ethnography of the Nomads and 'Barbarian' History in Han China." In Foxhall, *et al.*, eds. 2010: 229–325.

—— and Don J. Wyatt, eds. 2003 *Political Frontiers, Ethnic Boundaries, and Human Geographies in Chinese History*. London: Routledge Curzon.

Dirlik, Arif. 1975. "The Ideological Foundations of the New Life Movement: A Study in Counterrevolution." *Journal of Asian Studies* 34.4 (August): 945–80.

——. 1982. "Spiritual Solutions to Material Problems: The Socialist Education and Courtesy Month in China." *South Atlantic Quarterly* 81 (Autumn): 359–75.

Dong, Gao. 1761. *Huang Qing zhigong tu* (Illustrations of the tribute-bearing people of the imperial Qing). Taibei: Huawen shuju (1968 reprint).

Dong, Guangbi. 1995. *Zhongguo jindai kexue jishu shi* (History of modern Chinese science and technology). Changsha: Hunan jiaoyu chubanshe.

Doolittle, Justus. 1865. *Social Life of the Chinese*. New York: Harper and Brothers.

Doré, Henri. 1914–1933. *Researches into Chinese Superstitions*. Trans. M. Kennelly. Shanghai: Tusewei Printing Press.

Drake, Fred W. 1975. *China Charts the World: Hsu Chi-yü and His Geography of 1848*. Cambridge, MA: Harvard East Asian Monographs.

Duara, Prasenjit. 1991. "Knowledge and Power in the Discourse of Modernity: The Campaigns against Popular Religion in Early Twentieth Century China." *Journal of Asian Studies* 50.1 (February): 67–83.

Duman, L.I. 1981. "Ancient Chinese Foreign Policy and the Origins of the Tribute System." In Tikhvinsky and Perelomov, eds. 1981: 17–42.

Duong, Ngoc Dung and Lê Anh Minh. 1999. *Kinh dich va cau hinh tu tuong Trung Quoc* (The *Classic of Changes* and the structure of Chinese thought). Hanoi: Nha xuat ban Khoa hoc xa hoi.

Duyvendak, J. J. L. 1938. "The Last Dutch Embassy to the Chinese Court (1794–1795)." *T'oung Pao* 34.1–5: 1–137.

Dynon, Nicholas. 2008. "'Four Civilizations' and the Evolution of Post-Mao Chinese Socialist Ideology." *The China Journal* 60 (July): 83–109.

Eastman, Lloyd. 1990. *The Abortive Revolution: China under Nationalist Rule, 1927–1937.* Cambridge, MA: Harvard University Press.

Eberhard, Wolfram. 1971. *Moral and Social Values of the Chinese.* Taibei: Ch'eng-wen.

Ebrey, Patricia B., ed. 1981. *Chinese Civilization and Society.* New York: Free Press.

—— 1991a. *Confucianism and Family Rituals in Imperial China: A Social History of Writing about Rites.* Princeton, NJ: Princeton University Press.

——, trans. 1991b. *Chu Hsi's Family Rituals: A Twelfth-century Manual for the Performance of Cappings, Weddings, Funerals, and Ancestral Rites.* Princeton, NJ: Princeton University Press.

Edwards, R. Randall. 1987. "Imperial China's Border Control." *Journal of Asian Law* 1.1 (Fall): 1–36.

Elliott, Mark C. 1995. "Mapping the Qing State: All under Heaven?" Paper for the Association for Asian Studies annual meeting, Washington, D.C., April 6–9.

——. 1996. "Manchu (Re)Definitions of the Nation in the Early Qing." *Indiana East Asian Working Papers Series on Language and Politics in Modern China* 7: 47–78.

——. 2000. "The Limits of Tartary: Manchuria in Imperial and National Geographies." *The Journal of Asian Studies* 59.3 (August): 603–46.

——. 2001. *The Manchu Way: The Eight Banners and Ethnic Identity in Late Imperial China.* Stanford, CA: Stanford University Press.

——. 2009. *Emperor Qianlong: Son of Heaven, Man of the World.* New York: Pearson-Longman.

Elman, Benjamin. 1981–83. "Geographical Research in the Ming-Ch'ing Period." *Monumenta Serica* 35: 1–18.

——. 1984. *From Philosophy to Philology: Intellectual and Social Aspects of Change in Late Imperial China.* Cambridge, MA: Harvard University Press.

——. 1990. *Classicism, Politics, and Kinship: The Ch'ang-chou School of New Text Confucianism in Late Imperial China.* Berkeley CA: University of California Press.

——. 2000. *A Cultural History of Civil Examinations in Late Imperial China.* Berkeley, CA: University of California Press.

——. 2003. "The Jesuit Role as 'Experts' in High Qing Cartography and Technology." *Taida lishi xuebao* 31 (June): 223–50.

——. 2005. *On Their Own Terms: Science in China, 1550–1900.* Cambridge, MA: Harvard University Press.

——. 2006. *A Cultural History of Modern Science in China.* Cambridge, MA: Harvard University Press.

—— and Theodore Huters. 1997. "Maga'erni shituan houxiandai zhuyi yu jindai Zhongguo shi" (The Macartney embassy, postmodernism, and modern Chinese history). *Ershiyi shiji* 44 (December): 122–28.

—— and Martin Kern, eds. 2010. *Statecraft and Classical Learning: The Rituals of Zhou.* Leiden: Brill.

Esherick, Joseph. 1998a. "Tradutore, Traditore: A Reply to James Hevia." *Modern China* 24.2 (July): 328–32.

——. 1998b. "Cherishing Sources from Afar." *Modern China*, 24.2 (April): 135–61.

Evans, Grant. 2002. "Between the Global and the Local There Are Regions, Culture Areas, and National States: A Review Article." *Journal of Southeast Asian Studies* 33.1 (February): 147–62.

Fairbank, John K. 1942. "Tributary Trade and China's Relations with the West." *Far Eastern Quarterly* 1.2 (February): 129–49.

——. 1953. *Trade and Diplomacy on the China Coast*. Cambridge, MA: Harvard University Press.

——, ed. 1957. *Chinese Thought and Institutions*. Chicago: University of Chicago Press.

——, ed. 1968. *The Chinese World Order*. Cambridge, MA: Harvard University Press.

——, ed. 1978. *The Cambridge History of China*, vol. 10, part 1. Cambridge: Cambridge University Press.

—— and S.Y. Teng. 1939. "On the Ch'ing Tributary System." *Harvard Journal of Asiatic Studies* 4.12: 135–246.

Falkenheim, Victor. 1982. "Popular Values and Political Reform: The 'Crisis of Faith' in Contemporary China." In Sidney Greenblatt, *et al.*, eds. 1982: 237–51.

Fang, Bao. 1748. *Sanli yishu* (Commentaries on the three ritual classics). n.p.

Fang, Weigui. 2001. "Yi, Yang, Xi, Wai and Other Terms: The Transition from 'Barbarian' to 'Foreigner' in Nineteenth-Century China." In Lackner, Amelung and Kurtz, eds. 2001: 95–124.

Farquhar, Judith. 1996. "'Medicine and the *Changes* Are One': An Essay on Divination Healing with Commentary." *Chinese Science* 13: 107–34.

—— and James Hevia. 1993. "Culture and Post-War American Historiography of China." *positions* 1.2: 486–525.

Fendos, Paul G. 1988. "Fei Chih's Place in the Development of 'I-Ching' Studies." Ph.D. dissertation, University of Wisconsin, Madison.

Ferlanti, Federica. 2010. "The New Life Movement in Jiangxi Province, 1934–1938." *Modern Asian Studies* 44.5: 961–1000.

Feuchtwang, Stephan. 1977. "School Temple and City God." In Skinner, ed. 1977: 581–608.

——. 2002. *An Anthropological Analysis of Chinese Geomancy*. Bangkok, Thailand: White Lotus Press.

Field, Stephen L. 2008. *Ancient Chinese Divination*. Honolulu: University of Hawaii Press.

Fiskesjö, Magnus. 1999. "On the 'Raw' and 'Cooked' Barbarians of Imperial China." *Inner Asia* 1.2 (1999): 135–68.

——, ed. 2002. "Reconsidering the Correlative Cosmology of Ancient China." *Bulletin of the Museum of Far Eastern Antiquities*, Special Issue. 72 (November): 1–196.

Fleming, Jesse. 2009. "A Set Theory Analysis of the Logic of the *Yijing*." *Journal of Chinese Philosophy* 36 (supplement): 37–47.

Foret, Philippe. 1995. "Mapping the Empire's Future: The Qing Definition of Maps as Prognostic Material." Paper for the Association for Asian Studies annual meeting, Washington, DC, April 6–9.

——, ed. 2000. *Mapping Chengde: The Qing Landscape Enterprise*. Honolulu: University of Hawaii Press.

Foulks, Beverly. 2008. "Duplicitous Thieves: Ouyi Zhixu's Criticism of Jesuit Missionaries in Late Imperial China." *Chung-Hwa Buddhist Journal* 21: 55–76.

——. 2011. "Divination as a Karmic Diagnostic: The Divination Texts of Ouyi Zhixu." Paper for the Sixteenth Congress of the International Association of Buddhist Studies, Jinshan, New Taipei City, Taiwan, June, 20–25, 2011.

Foxhall, Lin, Hans-Joachim Gehrke, and Nino Luraghi, eds. 2010. *Intentional History: Spinning Time in Ancient Greece*. Stuttgart: Franz Steiner Verlag.

Fracasso, Ricardo. 1993. "Shan Hai Ching." In Loewe, ed. 1993: 357–67.

Fu, Lo-shu. 1966. *A Documentary Chronicle of Sino-Western Relations (1644–1820)*. 2 vols. Tucson, AZ: Association for Asian Studies.

Fuchs, Walter. 1935–36. "Materialen zur Kartographie der Mandju-Zeit," part 1, *Monumenta Serica* 1: 386–427.

——. 1937. "The Peking Map Collection." *Imago Mundi* 2: 21–22.
——. 1939. "Materialen zur Kartographie der Mandju-Zeit." *Monumenta Serica* 3: 189–249.
——. 1943. *Der Jesuiten-Atlas der Kanghsi-Zeit: China und die Au enländer*. Peiping: Fu Jen University Press.
Fuehrer, Bernhard. 2001. "Seers and Jesters: Predicting the Future and Punning by Graph Analysis." Paper for the Conference on Divination in honor of Professor Ho Peng Yoke, Needham Research Institute, Cambridge University. December 7–8.
——. 2006. "Seers and Jesters: Predicting the Future and Punning by Graph Analysis." *East Asian Science, Technology and Medicine* 25: 47–68.
Fukui, Fumimasa and Gérard Fussman. 1994. *Bouddhisme et cultures locales: quelques cas de réciproques adaptations*. Paris: Ecole française d'Extrême-Orient.
Funakoshi, Akio. 1975. "Some New Lights on the History of Chinese Cartography." *Annual Report of Studies in the Humanities and Social Science* (Nara Women's University) 19: 147–70.
Fung, Yu-lan. 1952. *A History of Chinese Philosophy*. Trans. Derk Bodde. Two vols. Princeton, NJ: Princeton University Press.
Gao, Guofan. 1999. *Zhongguo wushu shi* (A history of the Chinese magical arts). Shanghai: Shanghai sanlian shudian.
Gao, Huaping and Cao Haidong. 1995. *Zhonghua wushu* (The magical arts of China). Taibei: Wenjin chuban she youxian gongsi.
Gardner, Daniel, trans. 1990. *Learning to Be a Sage: Selections from the Conversations of Master Chu, Arranged Topically*. Berkeley, CA: University of California Press.
——. 1998. "Confucian Commentary and Chinese Intellectual History." *Journal of Asian Studies* 57.2 (May): 397–422.
Giele, Enno. 1998–1999. "New Sources of Early Chinese History: An Introduction to the Reading of Inscriptions and Manuscripts." *Early China* 23–24, 247–336. Supplemented by Giele's "Database of Early Chinese Manuscripts" (updated Feb. 19, 2001). Available at: http://humanities.uchicago.edu/depts/easian/earlychina/research_resources/databases/early_chinese_manuscripts/preface2.htm.
Giersch, C. Patterson. 2006. *Asian Borderlands: The Transformation of Qing China's Yunnan Frontier*. Cambridge, MA: Harvard University Press.
Giles, Herbert. 1964. *Chinese–English Dictionary*. New York: Paragon. Reprint of the 1912 edition.
Girardot, Norman J. 2002. *The Victorian Translation of China: James Legge's Oriental Pilgrimage*. Berkeley, CA: University of California Press.
Gold, Thomas. 1984. "Just in Time!: China Battles Spiritual Pollution on the Eve of 1984." *Asian Survey* 24.9 (September): 947–74.
Gong, Yi. 1992. *Zhongguo tusi zhidu* (The native official system of China). Kunming: Yunnan minzu chubanshe shuju.
Gottschalk, Louis. ed. 1963. *Generalization in the Writing of History*. Chicago: University of Chicago Press.
Govinda, Lama Anagarika Brahmacari. 1981. *The Inner Structure of the I Ching, the Book of Transformations*, Tokyo and New York: Weatherhill.
Graham, A.C. 1986. *Yin-Yang and the Nature of Correlative Thinking*. Singapore: Institute of East Asian Philosophies, National University of Singapore.
——. 1989. *Disputers of the Tao: Philosophical Argument in Ancient China*. Chicago: Open Court.
Granet, Marcel. 1934. *La pensée chinoise*. Paris: La Renaissance du livre.
Gray, John H. 1878. *China: A History of the Laws, Manners, and Customs of the People*. London: Macmillan.
Greenblatt, Sidney, *et al.*, eds. 1982. *Social Interaction in Chinese Society*. New York: Praeger.

Gu, Jianqing. 1991. *Zhongguo fangshu da cidian* (Dictionary of China's predictive and prescriptive arts). Guangzhou: Zhongshan daxue chuban she.

Gu, Ming Dong. 2004. "Elucidation of Images in the *Book of Changes*: Ancient Insights into Modern Language Philosophy and Hermeneutics." *Journal of Chinese Philosophy* 31.4: 469–88.

Guan, Xiaolian and Qu Liusheng, eds. 1996. *Kangxi chao Manwen zhupi zhouzhe quanyi* (Complete translation of the Manchu language vermillion rescripts on memorials of the Kangxi reign). Beijing: Zhongguo shehui kexue chubanshe.

Guiliang. 1883. *Liqi yueqi quantu* (Complete illustrations of ritual equipment and musical instruments). n.p.

Guy, R. Kent. 2010. "Ideology and Organization in the Qing Empire." *Journal of Early Modern History*. 14.4: 355–77.

Hacker, Edward, Steve Moore and Lorraine Patsco, eds. 2002. *I Ching: An Annotated Bibliography*. London: Taylor & Francis.

Hall, David L. and Roger T. Ames. 1995. *Anticipating China: Thinking Through the Narratives of Chinese and Western Culture*. Albany, NY: State University of New York Press.

——. 1998. *Thinking from the Han: Self, Truth, and Transcendence in Chinese and Western Culture*. Albany, NY: State University of New York Press.

Hamashita, Takeshi. 2008. *China, East Asia and the Global Economy: Regional and Historical Perspectives*. Edited by Linda Grove and Mark Selden. New York. Routledge.

Han, Qi. 1998. "Bai Jin de *Yijing* yanjiu he Kangxi de shidai de 'Xi xue Zhong yuan' shuo" (Joachim Bouvet's research on the *Classic of Changes* and the idea of the 'Chinese Origin of Western Learning' during the Kangxi period). *Hanxue yanjiu* 16.1: 185–202.

Hao, Yen-p'ing and Wei Hsiu-mei, eds. 1998. *Between Tradition and Change: Studies in Modern Chinese History*. Nangang: Academia Sinica. Two vols.

Hara, Takemichi. 1998. "Korea, China, and Western Barbarians: Diplomacy in Early Nineteenth-Century Korea." *Modern Asian Studies* 32.2: 389–430.

Harley, J.B. and David Woodward, eds. 1994. *The History of Cartography*, vol. 2, Book 2: *Cartography in the Traditional East and Southeast Asian Societies*. Chicago: University of Chicago Press.

Harper, Donald. 1999a. "Warring States, Natural Philosophy and Occult Thought." In Loewe and Shaughnessy, eds. 1999.

——. 1999b. "Physicians and Diviners: The Relation of Divination to the Medicine of the *Huangdi neijing* (Inner Canon of the Yellow Thearch)." *Extrême-Orient, Extrême-Occident* 21: 91–110.

_______. 2001. "Iatromancy, Diagnosis, and Prognosis in Early Chinese Medicine." In E. Hsü 2001: 99–120.

Harrell, Stevan, ed. 1995. *Cultural Encounters on China's Ethnic Frontiers*. Seattle: University of Washington Press.

Harvey, David. 1990. *The Condition of Postmodernity*. London: Blackwell.

Hayes, James. 1985. "Specialists and Written Materials in the Village World." In Johnson, Nathan, and Rawski, eds. 1985.

He, Liankui. 1979. *Zhongguo lisu yanjiu* (Research on Chinese rituals and customs). Taibei: Zhonghua.

Henderson, John B. 1984. *The Development and Decline of Chinese Cosmology*. New York: Columbia University Press.

——. 1991. *Scripture, Canon and Commentary: A Comparison of Confucian and Western Exegesis*. Princeton, NJ: Princeton University Press.

Herman, John H. 1997. "Empire in the Southwest: Early Qing Reforms to the Native Chieftain System." *Journal of Asian Studies* 56.1 (February): 47–74.

Hevia, James. 1995. *Cherishing Men from Afar: Qing Guest Ritual and the Macartney Embassy of 1793*. Durham, NC: Duke University Press.

——. 1998. "Polemical Historiography: A Response to Joseph W. Esherick." *Modern China* 24.2 (July): 319–27.

Higgins, Charlie. 1997. "The Hexagram and the Kabbalah." Available at: http://www.mension.com/del_3.htm.

Hightower, James R. 1970. *The Poetry of T'ao Ch'ien*. Oxford: Clarendon Press.

Hinsch, Bret. 2004. "Myth and the Construction of Foreign Ethnic Identity in Early and Medieval China." *Asian Ethnicity* 5.1: 81–103.

Ho, Peng Yoke. 1972. "The System of the *Book of Changes* and Chinese Science." *Japanese Studies in the History of Science* 11: 23–39.

——. 1991. "Chinese Science: The Traditional Chinese View." *Bulletin of the London School of Oriental and African Studies* 54: 506–19.

——. 2003. *Chinese Mathematical Astrology: Reaching Out to the Stars*. London: RoutledgeCurzon.

Ho, Ping-ti. 1967. "The Significance of the Ch'ing Period in Chinese History." *Journal of Asian Studies* 26.2 (February): 189–95.

——. 1998. "In Defense of Sinicization: A Rebuttal of Evelyn Rawski's 'Reenvisioning the Qing.'" *Journal of Asian Studies* 57.1 (February): 123–55.

Hofmeyer, Isabel. 2004. *The Portable Bunyan: A Transnational History of The Pilgrim's Progress*. Princeton, NJ: Princeton University Press.

Hon, Tze-ki. 2005a. "Constancy in Change: A Comparison of James Legge's and Richard Wilhelm's Interpretations of the *Yijing*." *Monumenta Serica* 53: 315–36.

——. 2005b. *The Yijing and Chinese Politics: Classical Commentary and Literati Activism in the Northern Song Period, 960–1127*. Albany, NY: State University of New York Press.

Hong, Pimo and Jiang Yuzhen. 1990. *Zhongguo gudai suanming shu* (The divinatory arts of ancient China). Shanghai: Shanghai renmin chubanshe.

Hostetler. Laura. 2001. *Qing Colonial Enterprise: Ethnography and Cartography in Early Modern China*. Chicago: The University of Chicago Press.

Howland, D.R. 1996. *Borders of Chinese Civilization: Geography and History at Empire's End*. Durham, NC: Duke University Press.

Hsiao, Kung-ch'üan. 1967. *Rural China: Imperial Control in the Nineteenth Century*. Seattle: University of Washington Press.

Hsiung, Shih-i. 1948. *The Life of Chiang Kai-shek*. London: Peter Davies.

Hsu, Elizabeth, ed. 2001. *Innovation in Chinese Medicine*. Cambridge: Cambridge University Press.

Hsü, Immanuel C.Y. 1960. *China's Entrance into the Family of Nations: The Diplomatic Phase, 1858–1880*. Cambridge, MA: Harvard University Press.

Hsu, Mei-ling. 1997. "An Inquiry into Early Chinese Atlases through the Ming Dynasty." In Wolter and Grim, eds. 1997: 31–50.

Hu, Pu'an, comp. 1968. *Zhongguo quanguo fengsu zhi* (A treatise on Chinese national customs). Taibei: Qixin shuju. Originally published in Shanghai, 1923.

Hu, Wenhui. 2000. *Zhongguo zaoqi fangshu yu wenxian congkao* (Collected investigations into the predictive and prescriptive arts and literature of early China). Guangzhou: Zhongshan daxue chubanshe.

Hu, Xinsheng. 1998. *Zhongguo gudai wushu* (The magical arts of ancient China). Jinan: Shandong renmin chuban she.

Huang, Chin-shing. 1995. *Philosophy, Philology, and Politics in Eighteenth Century China: Li Fu and the Lu-Wang School under the Ch'ing*. Cambridge: Cambridge University Press.

Huang, Chun-chieh. 2007. "The Idea of 'Zhongguo' and Its Transformation in Early Modern Japan and Contemporary Taiwan." *Journal of Kanbun Studies* 2: 398–408.

Huang, Chunyao [aka Huang Yunsheng]. n.d. [late Ming dynasty]. *Yijing yijian nengjie* (Understanding the *Classic of Changes* at a glance). VL, Borg Cin 384 (1). This work was revised and reprinted repeatedly from the time of Huang's death in 1645 to 1901. See R. Smith 2011.

Huang, Pei. 2011. *Reorienting the Manchus: A Study of Sinicization, 1583–1795*. Ithaca, NY: East Asia Program, Cornell University.

Huang, Yinong [Huang Yi-Long]. 1994. "Shao Fuzhong Tianxue shuo xiaokao" (A brief analysis of Shao Fuzhong's *Tianxue shuo*). *Guoli zhongyang tushuguan guankan* 27.2: 163–66.

——. [Huang Yi-Long]. 1996. "Tongshu–Zhongguo chuantong tianwen yu shehui di jiaorong" (Almanacs: the blending of traditional Chinese astronomy and society). *Hanxue yanjiu* 14.2.

——. [Huang Yi-Long]. 1998. *Zhongguo ren di meng* (The dreams of the Chinese people). Shanghai: Shanghai guji chuban she.

Huang, Yizhou. 1893. *Lishu tonggu* (A general study of ritual classics). Dinghai [?]: Huangshi shiguan.

Huang Yunsheng. See under Huang, Chunyao.

Hucker, Charles. 1961. *The Traditional Chinese State in Ming Times*. Tucson, AZ: University of Arizona Press.

Hummel, Arthur, ed. 1943–44. *Eminent Chinese of the Qing Period 1644–1912*. Washington, DC: Library of Congress. Two vols.

Iwami, Kiyohiro. 2009. "Silk Trade and Gongxian Tribute System under the Tang Dynasty." *The Memoirs of the Toyo Bunko* 67: 31–48.

Jacobs, Bruce. 1982. "The Concept of *Guanxi* and Local Politics in a Rural Chinese Cultural Setting." In Greenblatt, *et al.*, eds. 1982: 209–36.

Javary, Geneviève. 1991. "Le Père Bouvet, a-t-il retrouvé Pythagore en Chine?" *Hexagrammes* 6: 113–22.

Jiang, Xiaoyuan. 1992. *Xingzhan xue yu chuantong wenhua* (Astrology and traditional culture). Shanghai: Shanghai guji chuban she.

Jiang, Yong. 1810. *Lishu gangmu* (Outline of works on ritual). n.p.

——. 1989. *He Luo jingyun* (The quintessence of the *Hetu* and the *Luoshu*). Beijing: Xueyuan chubanshe. Edited by Sun Guozhong.

Jiao, Xun. 1970. *Yixue sanshu* (Three works on the *Changes*). Taibei: Guangwen shuju.

Jin Liangnian, ed. 1994. *Zhongguo shenmi wenhua baike zhishi* (An encyclopedia of China's mystical culture). Shanghai: Shanghai wenhua chuban she.

Joachim, Christian. 1992. "Confucius and Capitalism: Views of Confucianism and Economic Development." *Journal of Chinese Religions* 2: 135–71.

Johnson, David. 2009. *Spectacle and Sacrifice: The Ritual Foundations of Village Life in North China*. Cambridge, MA: Harvard University Asia Center.

——, Andrew J. Nathan and Evelyn Rawski, eds. 1985. *Popular Culture in Late Imperial China*. Berkeley, CA: University of California Press.

Johnston, R.F. 1910. *Lion and Dragon in Northern China*. New York: E.P. Dutton.

Jones, William C., trans. 1994. *The Great Qing Code*. Oxford: Clarendon Press.

Jordan, David K. 1972. *Gods, Ghosts and Ancestors: The Folk Religion of a Chinese Village*. Berkeley, CA: University of California Press.

—— and Daniel L. Overmyer. 1986. *The Flying Phoenix: Aspects of Chinese Sectarianism in Taiwan*. Princeton, NJ: Princeton University Press.

Jullien, François. 1995. *The Propensity of Things: Toward a History of Efficacy in China*. Trans. Janet Lloyd. New York: Zone Books.

Kalinowski, Marc. 1990. "La litterature divinatoire dans le Daozang." *Cahiers d'Extrême-Asie* 5: 85–114.

——. 1991. *Cosmologie et divination dans la Chine ancienne: Le Compendium des Cinq Agents* (*Wuxing dayi, VIe siècle*). Paris: École Française d'Extrême-Orient.

——. 1998–1999. "The Xingde Texts from Mawangdui." *Early China* 23–24: 125–202. Trans. Phyllis Brooks.

——. 2002. "The Interaction between Popular Religion and Shushu Culture in the Dunhuang Manuscripts." Paper for the Third International Conference on Sinology, History Section, Institute of History and Philogy, Academia Sinica, Taipei, Taiwan.

——, ed. 2003. *Divination et société dans la Chine médiévale: Étude des manuscrits de Dunhuang de la Bibliothèque Nationale de France et de la British Library*. Paris: Bibliothèque Nationale de France.

——. 2004. "Technical Traditions in Ancient China and Shushu Culture in Chinese Religion." In Lagerwey, ed. 2004: 223–48.

——. 2011. *Wang Chong: Balance de discours. Destin, providence et divination*. Paris: Belles Lettres (Bibliothèque chinoise).

Kang, C.H. and Ethel R. Nelson. 1979. *The Discovery of Genesis: How the Truths of Genesis Were Found Hidden in the Chinese Language*. St. Louis, MO: Concordia.

Kang, David C. 2010. *East Asia Before the West: Five Centuries of Trade and Tribute*. New York: Columbia University Press.

Kelly, Liam. 2003. "Vietnam as a 'Domain of Manifest Civility'." *Journal of Southeast Asian Studies* 34.1: 63–76.

Kern, Martin, ed. 2005. *Text and Ritual in Early China*. Seattle: University of Washington Press.

Kim, Hyun Jin. 2009. *Ethnicity and Foreigners in Ancient Greece and China*. London: Gerald Duckworth.

King, Geoff. 1996. *Mapping Reality: An Exploration of Cultural Cartographies*. New York: St. Martin's Press.

Kiong, Simon. 1906. *Quelques mots sur la politesse chinoise*. Shanghai: Imprimerie de la mission catholique.

Kleeman, Terry F. 1994. *A God's Own Tale: The Book of Transformations of Wenchang, the Divine Lord of Zitong*. Albany, NY: State University of New York Press.

Kohn, Livia, ed. 2000. *Daoism Handbook*. Leiden: E.J. Brill.

Konings, Patricia. 1990. "Astronomical Reports Offered by Ferdinand Verbiest S.J. to the Chinese Emperor." Paper delivered at the Sixth International Conference on the History of Science in China. Cambridge University, Cambridge, England, August 2–7.

Kuhn, Philip. 1980. *Rebellion and Its Enemies in Late Imperial China*. Cambridge, MA: Harvard University Press.

——. 1990. *Soulstealers: The Chinese Sorcery Scare of 1768*. Cambridge, MA: Harvard University Press.

Kunst, Richard A. 1985. "The Original 'Yijing': A Text, Phonetic Transcription, Translation, and Indexes, with Sample Glosses." Ph.D. dissertation, University of California, Berkeley.

Kuo, Li-ying. 1994. "Divination, jeux de hasard et purification dans le bouddhisme chinois: autour d'un sûtra apocryphe chinois, le Zhanchajing." In Fukui and Fussman 1994: 145–67.

Lackner, Michael. 1998. "Richard Wilhelm: A 'Sinicized' German Translator." In Alleton and Lackner, eds. 1998.

——. 2009. " 'The Other is Manifold': A Jesuit's Comparative Approach to China, Japan, and India 1583." In *Den Jadestein Erlangen: Festschrift für Harro von Senger*. Ed. Monika Gänßbauer. Frankfurt am Main: Lembeck Verlag: 121–39.

——. 2012. "The Last Stand of 'Chinese Science': Yuan Shushan, Traditional Divination and Western Knowledge in Republican China." Paper for the international workshop "Divinatory Traditions in East Asia: Historical, Comparative and Transnational Perspectives," Rice University, February 17–18.

——, Iwo Amelung and Joachim Kurtz, eds. 2001. *New Terms for New Ideas: Western Knowledge and Lexical Change in Late Imperial China*. Leiden: Brill.

Lagerwey, John, ed. 2004. *Religion and Chinese Society*. 2 vols. Hong Kong: Chinese University Press/Paris: École française de l'Extrême-Orient.

Lai, Guisan. 1994. *Jiao Xun Diaogu lou Yixue yan jiu* (Research on Jiao Xun's Study of the *Changes* in the *Diaogu lou*). Taibei: Liren shuju.

Landry-Deron, Isabelle. 1995. *Les leçons des sciences occidentales de l'empereur de Chinese Kangxi (1662–1722): Texte des journaux des Pères Bouvet et Gerbillion*. Paris: E.H.E.S.S.

Lary, Diana, ed. 2007. *The Chinese State at the Borders*. Vancouver, BC: University of British Columbia Press.

Lee, Jeong Mi. 2008. "Cultural Expressions of Tokugawa Japan and Choson Korea: An Analysis of the Korean Embassies in the Eighteenth Century." Ph.D. Dissertation, East Asia Studies, University of Toronto.

Lee, Jung Young. 1982a. "The *Book of Change* and Korean Thought." In Phillips and Yu (1982).

——. 1982b. "The Origin and Significance of the *Chongyok* or Book of Correct Change." *Journal of Chinese Philosophy* 9: 211–41.

——. 1994. *Embracing Change: Postmodern Interpretations of the I Ching from a Christian Perspective*. Scranton, PA: University of Scranton Press.

Legge, James, trans. 1963. *The I Ching* [originally rendered *Yi King*]. New York: Dover Publications. Reprint of 1899 edition.

Leonard, Jane Kate. 1984. *Wei Yuan and China's Rediscovery of the Maritime World*. Cambridge, MA: Harvard East Asian Monographs.

Lewis, Mark Edward. 1999. *Writing and Authority in Early China*. Albany, NY: State University of New York Press.

——. 2006. *The Construction of Space in Early China*. Albany, NY: State University of New York Press.

Li, Cunyang. 2004. "Qin Huitian and His General Study on the Five Rites." *Sungkyun Journal of East Asian Studies* 4.1: 156–83.

Li, Fuyao. 1891. *Duli congchao* (Research notes on the ritual classics). n.p.

Li, Guangdi, ed. 1998. *Zhouyi zhezhong* (A balanced [edition of the] *Zhou Changes*). Chengdu: Ba Shu shushe, 1998. Edited by Liu Dajun; originally published in 1715.

Li, Jinyu. 1986. "The Politics of Propriety." MA thesis, Rice University.

Li, Ling. 2000a. *Zhongguo fangshu kao* (An examination of the Chinese predictive and prescriptive arts). Beijing: Dongfang chuban she. Revised edition.

——. 2000b. *Zhongguo fangshu xukao* (Continuation of an examination of Chinese predictive and prescriptive arts). Beijing: Dongfang chuban she. Revised edition.

—— and Liu Lexian, eds. 1993. *Zhongguo fangshu gaiguan* (A general survey of the predictive and prescriptive arts of China). Beijing: Renmin Zhongguo chuban she.

Li, Shenglong, ed. 1993. *Zhanxing shu* (The astrological arts). Haikou: Hainan chuban she.

Li, Wenchao and Hans Poser, eds. 2000. *Das Neueste über China: G.W. Leibnizens Novissima Sinica von 1697*. Stuttgart: Franz Steiner.

Li, Xiaocong. 1997. *Ouzhou shoucang bufen zhongwen gu ditu xulu* (A descriptive catalogue of pre-1900 maps seen in Europe). Beijing: Guoji wenhua.

Li, Yan. 1984. *Zhongguo suanxue shi* (A history of Chinese mathematics). Shanghai: Shanghai shudian.

Li, Yunquan. 2004. *Chaogong zhidu shilun: Zhongguo gudai duiwai guanxi tizhi yanjiu* (A history of the tributary system: Research on China's ancient institutions for [the conduct of] foreign relations). Beijing: Xinhua chubanshe.

Li, Zhongjun 1994. *Xiangshu Yixue fazhan shi* (A history of the development of [the school of] images and numbers in *Changes* learning). 2 vols. Jinan: Qi Li shushe.

Liang Zhaotao. 1989. *Zhongguo gudai wushu: Zongjiao di qiyuan he fazhan* (China's ancient magical arts: the origins and development of religion). Guangzhou: Zhongshan daxue chuban she.

Liao, Hsien-Huei. 2005. "Exploring Weal and Woe: The Song Elite's Mantic Beliefs and Practices." *T'oung Pao* 91.4–5: 347–95.

Liao, Mingchun. 1991. *Zhouyi yanjiu shi* (History of research on the *Zhou Changes*). Changsha: Hunan chubanshe.

Lin, Fu-shih. 1995. "Religious Taoism and Dreams: An Analysis of the Dream-Data Collected in the *Yün-chi ch'i-ch'ien*." *Cahiers d'Extrême-Asie* 8: 95–112.

Lin, Meirong. 1997. *Taiwan minjian xinyang yanjiu shumu* (Bibliography on research on Taiwan folk beliefs). Taibei: Zhongyang yanjiu yuan minzu xue yanjiu suo.

Ling, Shanqing. 1969. *Taiping tianguo yeshi* (An unofficial history of the Taiping heavenly kingdom). Taibei: Huawen.

Lippiello, Tiziana. 2000. "Interpreting Written Riddles: A Typical Chinese Way of Divination." In De Meyer and Engelfriet 2000: 41–52.

Liu, Dajun, ed. 2003. *Xiangshu Yixue yanjiu* (Research on [the school of] images and numbers in *Changes* learning). Chengdu: Ba Shu shushe

Liu, Daochao and Zhou Rongyi. 1994. *Shenmi di zeji* (The mystery of selecting auspicious [times]). Taibei: Shuquan chuban she.

Liu, Gang. 2007. "The Chinese Inventor of the Bi-Hemispherical World Map." *E-Perimetron* 2.3 (Summer): 185–93.

Liu, Hui-cheng. 1959. "An Analysis of Chinese Clan Rules: Confucian Theories in Action." In A. F. Wright, ed. 1959: 16–49.

Liu, Kwang-Ching. 1978. "The Ch'ing Restoration." In Fairbank, ed. 1978: 409–90.

——, ed. 1990a. *Orthodoxy and Heterodoxy in Late Imperial China*. Berkeley, CA: University of California Press.

——. 1990b. "Introduction." In Kwang-Ching Liu, ed. 1990a: 1–24.

——. 1990c "Socioethics as Orthodoxy: A Perspective." In Kwang-Ching Liu, ed. 1990a: 53–100.

—— and Richard Shek, eds. 2004. *Heterodoxy in Late Imperial China*. Honolulu: University of Hawaii Press.

Liu, Lydia H. 2004. *The Clash of Empires: The Invention of China in Modern World Making*. Cambridge, MA: Harvard University Press.

Liu, Shih-Shun, trans. 1975. *Vignettes from the Late Ch'ing*. Hong Kong: Chinese University Press.

Liu, Wenying. 2000. *Meng di mixin yu meng di tansuo* (Dream superstitions and the investigation of dreams). Beijing: Zhongguo shehui kexue chubanshe.

——. 2003. *Meng yu Zhongguo wenhua* (Dreams and Chinese culture). Beijing: Renmin chuban she.

Liu, Xicheng and Wang Wenbao, eds. 1991. *Zhongguo xiangzheng cidian* (Dictionary of Chinese symbols). Tianjin: Tianjin jiaoyu chuban she.

Liu, Yongming, ed. 1995. *Siku weishou shushu lei guji daquan* (Ancient books on the arts of calculation not included in the *Complete Collection of the Four Treasuries*). Hefei: Huangshan shushe.

Liu, Yujian. 1996. *Liang Han xiangshu Yixue yanjiu* (Research on the [school of] images and numbers in *Changes* studies in the Han dynasty). Nanning: Guangxi jiaoyu chubanshe.

Liu Zheng, 1993. "The Dilemma Facing Contemporary Research in the *I-ching*." *Chinese Studies in Philosophy* 24.4 (Summer): 47–64.

Liu, Zhiwan. 1974. *Zhongguo minjian xinyang lunji* (Collected essays on popular beliefs in China). Taibei: Zhongyang yanjiu yuan minzu xue yanjiu suo.

Lo, Yuet Keung. 2008. "Change Beyond Syncretism: Ouyi Zhixu's Buddhist Hermeneutics of the *Yijing*." *Journal of Chinese Philosophy* 35.2 (June): 273–95.

Loewe, Michael. ed. 1993. *Early Chinese Texts: A Bibliographical Guide*. Berkeley, CA: Society for the Study of Early China.

——. 1994. *Divination, Mythology and Monarchy in Han China*. Cambridge: Cambridge University Press.

—— and Edward Shaughnessy, eds. 1999. *The Cambridge History of Ancient China*. Cambridge: Cambridge University Press: 813–84.

—— and Michael Nylan, eds. 2009. *China's Early Empires: A Re-Appraisal*. Cambridge: Cambridge University Press.

Lu, Liangzhi. 1984. *Zhongguo ditu xue shi* (History of Chinese cartography). Beijing: Cehui chuban she.

Lü, Shaogang, ed. 1992. *Zhouyi cidian* (Dictionary of the *Zhou Changes*). Changchun: Jilin daxue chuban she.

Lundbaek, Knud. 1991. *Joseph de Prémare (1666–1736), S.J.: Chinese Philology and Figurism*. Aarhus, Denmark: Aarhus University Press.

——. 1993. "Shao Yong et les jésuites de l'ancienne mission de Chine." In *Actes du Ve Colloque internationale de Sinologie*. Chantilly, France: 175–190.

——. 1994. "Joseph Prémare and the Name of God in China." In Mungello, ed. 1994: 129–48.

Luo, Hongxian. 1579. *Guang yutu* (Enlarged terrestrial map). Reproduced in SKQS.

Luo, Lida. 1997. "Bai Jin yanjiu *Yijing* shishi jikao" (An examination of Joachim Bouvet's research on the *Classic of Changes*). *Hanxue yanjiu* 15.1: 173–85.

Luo, Yiqun. 1993. *Zhongguo Miaozu wushu toushi* (A perspective on the magical arts of the Miao people). Beijing: Zhongyang minzu xueyuan chuban she.

Lupke, Christopher, ed. 2005. *The Magnitude of Ming: Command, Allotment, and Fate in Chinese Culture*. Honolulu: University of Hawaii Press.

Lynn, Richard John. 1994. *The Classic of Changes: A New Translation of the I Ching as Interpreted by Wang Bi*. New York: Columbia University Press.

Ma Junliang. *c.* 1790. *Jingban tianwen quantu* (Capital edition of a complete map [of the world based on) astronomy). Woodson Research Center, Rice University. Available at: http://www.ruf.rice.edu/~asia/jingban.html.

Ma Xisha and Han Bingfang. 1992. *Zhongguo minjian zongjiao shi* (A history of Chinese popular religion). Shanghai: Shanghai renmin chuban she.

Ma, Xulun. 1941. "Shuoming" (Discussing fate). *Xuelin* 9: 15–34.

McCaffree, Joseph E. 1875–1876. *Bible and I Ching Relationships*. Hong Kong and Seattle: South Sky Book Co. Reprinted in 1982.

Mair, Victor. 1985 "Language and Ideology in the Written Popularizations of the *Sacred Edict*." in Johnson, *et al.*, eds. 1985: 325–59.

Malatesta, Edward. 1997. *The Society of Jesus and China: A Historical-Theological Essay*. St. Louis: Institute of Jesuit Sources.

Malek, R., ed. 2000. *Jews in China: From Kaifeng ... to Shanghai*. Sankt Augustin: Steyler.

Martin, W. A. P. 1897. *A Cycle of Cathay; or, China, North and South*. New York: F. H. Revell.

Masini, Frederico, ed. 1996. *Western Humanistic Culture Presented to China by Jesuit Missionaries (XVII–XVIII Centuries)*. Rome: Institutum Historicum S.I.

Mears, I. and L. E. Mears. 1931. *Creative Energy: Being an Introduction to the Study of the Yih King, or Book of Changes, with Translations from the Original Text*. London: John Murray.

Meserve, Ruth I. 1982. "The Inhospitable Land of the Barbarian." *Journal of Asian History* 16: 51–89.

Metzger, Thomas. 1977. *Escape from Predicament: Neo-Confucianism and China's Evolving Culture*. New York: Columbia University Press.

——. 1990. "Continuities between Modern and Premodern China: Some Neglected Methodological and Substantive Issues." In Cohen and Goldman, eds. 1990: 263–92.

Meyer, Jeffrey. 1991. *The Dragons of Tiananmen: Beijing as a Sacred City*. Columbia, SC: University of South Carolina Press.

Michael, Franz. 1971. *The Taiping Rebellion*. Seattle: University of Washington Press.

Millward, James A. 1998. *Beyond the Pass: Economy, Ethnicity, and Empire in Qing Central Asia, 1759–1864*. Stanford, CA: Stanford University Press.

——. 1999. "'Coming onto the Map': 'Western Regions' Geography and Cartographic Nomenclature in the Making of Chinese Empire in Xinjiang." *Late Imperial China* 20.2 (December): 61–98.

——, *et al.*, eds. 2004. *New Qing Imperial History: The Making of Inner Asian Empire at Qing Chengde*. London: RoutledgeCurzon.

Milne, William, trans. 1870. *The Sacred Edict*. Shanghai: American Presbyterian Mission Press.

Minamiki, George. 1985. *The Chinese Rites Controversy from Its Beginnings to Modern Times*. Chicago: Loyola University Press.

Ming, Kao. 1986. "Chu Hsi's Discipline of Propriety." In Wing-tsit Chan, ed., 1986.

Miu, Tianhua. 1977. *Chengyu dian* (Dictionary of proverbs). Taibei: Fuxing.

Mizuno, Norihito. 2003. "China in Tokugawa Foreign Relations: The Tokugawa Bakufu's Perception of and Attitudes toward Ming-Qing China." *Sino-Japanese Studies* 15: 108–44.

Morgan, Carole. 1980. *Le Tableau du bœuf du printemps: Étude d'une page de l'almanac chinois*. Paris: Collège de la France.

Mungello, David E. 1977. *Leibniz and Confucianism: The Search for Accord*. Honolulu: University of Hawaii Press.

——. 1989. *Curious Land: Jesuit Accommodation and the Origins of Sinology*. Honolulu: University of Hawaii Press.

——, ed. 1994. *The Chinese Rites Controversy: Its History and Meaning*. San Francisco, CA: Ricci Institute for Chinese-Western Cultural History.

——. 2009. *The Great Encounter of China and the West, 1500–1800*. Third edition. New York and Oxford: Rowman & Littlefield.

Naquin, Susan. 1976. *Millenarian Rebellion in China*. New Haven, CT: Yale University Press.

Nedostup, Rebecca. 2010. *Superstitious Regimes: Religion and the Politics of Chinese Modernity*. Cambridge, MA: Harvard East Asian Monographs.

Nevius, John. 1869. *China and the Chinese*. New York: Harper and Brothers.

Ng, Benjamin. See Ng, Wai-ming.

Ng, On-cho, ed. 2008. *The Imperative of Understanding: Chinese Philosophy, Comparative Philosophy, and Onto-Hermeneutics*. New York: Global Scholarly Publications.

Ng, Wai-ming. 2000. *The I Ching in Tokugawa Thought and Culture*. Honolulu: University of Hawaii Press.

Ngo, Van Xuyet. 1976. *Divination, magie et politique dans la Chine ancienne*. Paris: Presses Universitaires de France.

Nickerson, Peter. 1994. "Shamans, Demons, Diviners, and Taoists: Conflict and Assimilation in Medieval Chinese Ritual Practice (*c.* AD 100–1000)." *Taoist Resources* 5.1 (August): 41–66.

Nielsen, Bent. 2003. *A Companion to Yi jing Numerology and Cosmology: Chinese Studies of Images and Numbers from Han (202 BCE–220 CE) to Song (960–1279 CE)*. London: RoutledgeCurzon.

Oldstone-Moore, Jennifer. 2000. "The New Life Movement of Nationalist China: Confucianism, State Authority and Moral Formation." Ph.D. dissertation, University of Chicago.

O'Malley, John W., S.J., *et al.*, eds. 1999. *The Jesuits: Cultures, Sciences, and the Arts 1540–1773*. Toronto: University of Toronto Press.

Ong Eng Ann, Alexander. 2004. "Of 'Wei' and 'De': Revisiting the Yongle Emperor's Military and Cultural Offensives in Vietnam, 1406–1427." MA thesis, National University of Singapore.

Ong, Hean-Tatt. 1991. *The Chinese Pakua*, Selangor Darul Ehsan, Malaysia: Pelanduk Publications.

Ouyang, Chun. 1814. *Fengsui yi-shu* (Geomancy, book 1). n.p.

Overmyer, Daniel. 1976. *Folk Buddhist Religion*. Cambridge, MA: Harvard University Press.

Pankenier, David W. 1995. "The Cosmo-Political Background of Heaven's Mandate." *Early China* 20: 121–76.

Peck, Graham. 1967. *Two Kinds of Time*. Boston: Houghton Mifflin.

Perdue, Peter C. 2005. *China Marches West: The Qing Conquest of Central Eurasia*. Cambridge, MA: Harvard University Press.

Peterson, Willard. 1982. "Making Connections: 'Commentary on the 'Attached Verbalizations' of the *Book of Changes*." *Harvard Journal of Asiatic Studies* 42: 67–116.

Phillips, Earl and Eui-Young Yu, eds. 1982. *Religions in Korea: Beliefs and Cultural Values*. Los Angeles: Center for Korean-American and Korean Studies, California State University, Los Angeles.

Pines, Yuri. 2002. *Foundations of Confucian Thought: Intellectual Life in the Chunqiu Period, 722–453 BCE*. Honolulu: University of Hawaii Press.

——. 2005. "Beasts or Humans: Pre-Imperial Origins of the 'Sino-Barbarian' Dichotomy." In Amitai and Biran, eds. 2005: 59–102.

——. 2009. *Envisioning Empire: Chinese Political Thought of the Warring States Period (453–221 BCE)*. Honolulu: University of Hawaii Press.

Plaks, Andrew. 1976. *Archetype and Allegory in "Dream of the Red Chamber"*. Princeton, NJ: Princeton University Press.

Plopper, Clifford. 1926: *Chinese Religion Seen through the Proverb*. Shanghai: China Press.

Polachek, James M. 1992. *The Inner Opium War*. Cambridge, MA: Council on East Asian Studies, Harvard University.

Poo, Mu-chou. 1998. *In Search of Personal Welfare: A View of Ancient Chinese Religion*. Albany, NY: State University of New York Press.

——. 2005. *Enemies of Civilization: Attitudes toward Foreigners in Ancient Mesopotamia, Egypt, and China*. Albany, NY: State University of New York Press.

Poon, Shuk-wah. 2011. *Negotiating Religion in Modern China*. Shatin, Hong Kong: Hong Kong University Press.

Pritchard, Earl. 1943. "The Kowtow in the Macartney Embassy to China in 1793." *Far Eastern Quarterly* 2.2 (February): 163–203.

Ptak, Roderich. 1999. *China and the Asian Seas: Trade, Travel, and Visions of the Other (1400–1750)*. Farnham, Hampshire: Ashgate.

——. n.d. (*c*. 2005). "The Sino-European Map ("Shanhai yudi quantu") in the Encyclopedia *Sancai tuhui*." n.p. Available at: www.humanismolatino.online.pt/v1/pdf/C003-022.pdf: 1–17.

Puett, Michael. 2001. *The Ambivalence of Creation: Debates Concerning Innovation and Artifice in Early China*. Stanford, CA: Stanford University Press.

Qi, Yunshi. 1884. *Huang Qing fanbu yaolüe* (Essentials of the vassal [tribes] of the imperial Qing). 2 vols. Taibei: Wenhai chuban she (1965 reprint).

Qian, Mu. 1977. *Zhongguo lishi jingshen* (The spirit of Chinese history). Kowloon: Rensheng chuban she.

Qian, Nanxiu, Grace Fong and Richard J. Smith, eds. 2008. *Different Worlds of Discourse: Transformations of Gender and Genre in Late Qing and Early Republican China*. Leiden: Brill.

Qian, Shiming. 1990. *Yilin tongshuo* (A complete discussion of the Forest of Changes). Beijing: Huaxia chubanshe.

Qingshi bianzuan weiyuan hui, ed. 1961. *Qingshi* (History of the Qing). Taibei: Guofang yanjiu yuan.

Que, Weimin. 1996. "Zhongguo gudai zhishu ditu huizhi zhunze chutan" (A preliminary investigation into the standards for drawing in ancient Chinese gazetteers and maps). *Ziran kexue shi yanjiu* 15.4 (1996): 334–42.

Raphals, Lisa. 2001. "Divination and Medicine in China and Greece: A Comparative Perspective on the Baoshan Medical Divinations." Paper for the Conference on Divination in honor of Professor Ho Peng Yoke, Needham Research Institute, Cambridge University, December 7–8.

——. 2005. "Divination and Medicine in China and Greece: A Comparative Perspective on the Baoshan Illness Divinations." *East Asian Science, Technology and Medicine* 23: 26–53.

Rawski, Evelyn. 1996. "Reenvisioning the Qing: The Significance of the Qing Period in Chinese History." *Journal of Asian Studies* 55.4 (November): 829–50.

——. 1998. *The Last Emperors: A Social History of Qing Imperial Institutions*. Berkeley, CA: University of California Press.

Rees, Ronald. 1980. "Historical Links between Cartography and Art." *Geographical Review* 70: 60–78.

Reilly, Thomas H. "The Taiping Heavenly Kingdom." Available at: http://www.amazon.com/gp/reader/0295984309/ref=sr_1_5?p=random&ie=UTF8&qid=1327640990#reader_0295984309.

Ren, Jincheng. 1983. "Xibanya cang Ming ke *Gujin xingshi zhi tu*" (A Ming edition of the Map of advantageous terrain, past and present, archived in Spain). *Wenxian* 17: 213–21.

Rhind, D.W. and D. R. F. Taylor, eds. 1989. *Cartography: Past, Present and Future*. London: International Cartographic Association.

Rickett, Allyn trans. 1985. *Kuan-tzu: A Respository of Early Chinese Thought*. Hong Kong: Hong Kong University Press.

Robinet, Isabelle. 1995. "Les marches cosmiques et les carrés magique dans le Taoism." *Journal of Chinese Religions* 23: 81–94.

Robinson, Arthur. "Cartography as Art." In Rhind and Taylor 1989.

Ronan, Charles and Bonnie Oh, eds. 1988. *East Meets West: The Jesuits in China, 1582–1773*. Chicago: Loyola University Press.

Rong Zhaozu. 1969. *Mixin yu chuanshuo* (Superstitions and legends). Taibei: Fulu tushu gongsi.

Ropp, Paul S. 1981. *Dissent in Early Modern China*. Ann Arbor, MI: University of Michigan Press.

——, ed. 1990. *Heritage of China: Contemporary Perspectives on Chinese Civilization*. Berkeley, CA: University of California Press.

Rowe, William T. 1998. "Ancestral Rites and Political Authority in Late Imperial China: Chen Hongmou in Jiangxi." *Modern China* 24.4: 378–407.

——. 2001. *Saving the World: Chen Hongmou and Elite Consciousness in Eighteenth-Century China*. Stanford, CA: Stanford University Press.

——. 2009. *China's Last Empire: The Qing*. Cambridge, MA: Belknap Press.

Rule, Paul. 1987. *K'ung-tzu or Confucius?: The Jesuit Interpretation of Confucianism*. London: Allen & Unwin.

——. 1995. "Moses or China?" *Actes* 6: 303–31.

Rundstrom, Robert. 1991 "Mapping, Post-Modernism, Indigenous People, and the Changing Direction of North American Cartography." *Cartographia* 28.2 (Summer): 1–12.

Rutt, Richard, trans. 1996. *Zhouyi: The Book of Changes*. Richmond, Surrey: Curzon Press.

Ryan, James A. 1996. "Leibniz's Binary System and Shao Yong's *Yijing*." *Philosophy East and West* 46.1: 59–90.

Sahlins, Marshall. 1993. "Goodbye to Tristes Tropes: Ethnography in the Context of Modern World History." *Journal of Modern History* 65: 1–25.
Sakade, Yoshinobu. 2000. "Divination as Daoist Practice." In Kohn 2000: 541–66.
Sangren, P. Steven. 1987a. "Orthodoxy, Heterodoxy and the Structure of Value in Chinese Rituals." *Modern China* 13.3 (July): 63–89.
——. 1987b. *History and Magical Power in a Chinese Community*. Stanford, CA: Stanford University Press.
Sato, Masayuki. 2003. *The Confucian Quest for Order: The Origin and Formation of the Political Thought of Xun Zi*. Leiden: Brill.
Sawada, Janine Tasca. 2004. *Practical Pursuits: Religion, Politics, and Personal Cultivation in Nineteenth-Century Japan*. Honolulu: University of Hawaii Press.
Scarborough, William and C. Wilfred Allan, eds. 1926. *A Collection of Chinese Proverbs*. Shanghai.
Schulz, Larry. 1982. "Lai Chih-te (1525–1604) and the Phenomenology of the *Classic of Change* (*I-Ching*)." Ph.D. dissertation, Princeton University.
Schwartz, Benjamin, ed. 1972. *Reflections on the May Fourth Movement*. Cambridge, MA: Harvard East Asian Council.
Secret, François. 1979. "Quand la Kabbale expliquait le 'Yi king,' ou un aspect oublié du figuratisme du P. Joachim Bouvet." *Revue de l'histoire des religions* 195: 35–53.
Shanghai guji chuban she, ed. 1989. *Songben lidai dili zhizhang tu* (A Song dynasty edition of *Convenient Geographical Maps, Chronologically Organized*). Shanghai: Shanghai guji chuban she.
Shaughnessy, Edward L. 1983. "The Composition of the Zhouyi." Ph.D. dissertation, Stanford University.
——. 1997. "Commentary, Philosophy, and Translation: Reading Wang Bi's Commentary to the *Yi jing* in a New Way." *Early China* 22: 221–45.
——, trans. 1998. *I Ching: The Classic of Changes*. New York: Ballantine Books.
——. 2001. "The Fuyang *Zhou Yi* and the Making of a Divination Manual." *Asia Major*, third series, 14.1: 7–18.
——. 2002. "The Wangjiatai *Gui Cang*: An Alternative to *Yi Jing* Divination." In Cadonna and Bianchi 2002: 95–126.
——. Forthcoming. *Unearthing the Changes: Recently Discovered Manuscripts of the Yi Jing and Related Texts*. New York: Columbia University Press.
Shchutskii, Iulian. 1979. *Researches on the I Ching*. Trans. William MacDonald and Tsuyoshi Hasegawa. Princeton, NJ: Princeton University Press.
Shen, Peter 1998. *The Traditional Art of Chinese Fortune Reading*. Selangor Darul Ehsan, Malaysia: Pelanduk Publications.
Shen, Zhian. 1995. *Zhonghua xiangshu* (The physiognomic arts of China). Taibei: Wenjin chuban she.
Shi, Wei and Qiu Xiaobo, eds. 1994. *Zhouyi tushi da dian* (A dictionary explaining illustrations [related to the] *Zhou Changes*), Beijing: Zhongguo gongren chuban she.
Shih, Vincent. 1976. *The Taiping Ideology*. Seattle: University of Washington Press.
Shilling, Dennis R. 2009. *Yijing. Das Buch der Wandlungen*. Frankfurt am Main: Verlag der Weltreligionen.
Sima Fu. See Smith, Richard J. forthcoming.
Sivin, Nathan. 1990. "Science and Medicine in Chinese History." In Ropp, ed., 1990: 164–96.
——. 1995a. *Science in Ancient China: Researches and Reflections*. Aldershot: Variorum.
——. 1995b. *Medicine, Philosophy, and Religion in Ancient China: Researches and Reflections*. Brookfield, VT: Variorum.

——. 2005. "Why the Scientific Revolution Did Not Take Place in China—Or Did It?" PDF version available at ccat.sas.upenn.edu/~nsivin/scirev.pdf. Originally published in *Chinese Science* 5 (1982): 45–66.

—— and Michael Nylan. 1995. "The First Neo-Confucianism: An Introduction to Yang Hsiung's 'Canon of Supreme Mystery' (*T'ai Hsuan Ching, c.* 4 BC)." In Sivin 1995b: Part 3, 1–42.

Skinner, G.W., ed. 1977. *The City in Late Imperial China*. Stanford, CA: Stanford University Press.

Smith, Arthur. 1894. *Chinese Characteristics*. New York: F. H. Revell.

——. 1899. *Village Life in China*. New York: F. H. Revell.

Smith, Kidder. 1989. "*Zhouyi* Divination from Accounts in the *Zuozhuan*." *Harvard Journal of Asiatic Studies* 49.2: 424–63.

——. 1993. "The Difficulty of the *Yijing*." *Chinese Literature: Essays, Articles, Reviews* 15: 1–15.

——, *et al.* 1990. *Sung Dynasty Uses of the I Ching*. Princeton, NJ: Princeton University Press.

Smith, Richard J. 1975. "The Employment of Foreign Military Talent: Chinese Tradition and Late Ch'ing Practice." *Journal of the Hong Kong Branch of the Royal Asiatic Society* 1: 113–38.

——. 1988. "A Note on Qing Dynasty Calendars." *Late Imperial China* 9.1 (June): 123–45.

——. 1989. "The Future of Chinese Culture." *Futures* (October): 431–46.

——. 1990. "Ritual in Ch'ing Culture." In Kwang-Ching Liu, ed. 1990a: 281–310.

——. 1991. *Fortune-tellers and Philosophers: Divination in Traditional Chinese Society*. Boulder, CO: Westview Press. Revised paperback edition, 1993.

——. 1992. *Chinese Almanacs*. New York and Hong Kong: Oxford University Press.

——. 1993. "Divination in Ch'ing China." In Smith and Kwok 1993: 113–40.

——. 1994. *China's Cultural Heritage: The Qing Dynasty, 1644–1912*. Boulder, CO: Westview Press.

——. 1996. *Chinese Maps: Images of "All Under Heaven."* New York and Hong Kong: Oxford University Press.

——. 1997. "Flies and Fresh Air: Culture and Consumerism in Contemporary China." *Problems of Post-Communism* (March–April): 3–13.

——. 1998a. "Mapping China's World: Cultural Cartography in Late Imperial Times." In Yeh Wen-hsin, ed. 1998: 52–109.

——. 1998b. "The Teachings of Ritual and the Rectification of Customs: Echos of Tradition in the Political Culture of Modern China." In Yen-p'ing Hao and Hsiu-mei Wei, eds. 1998: 1173–216.

——. 1998c. "The Place of the *Yijing* (Classic of Changes) in World Culture: Some Historical and Contemporary Perspectives." *Journal of Chinese Philosophy* (Winter): 391–422.

——. 1998d. "The Languages of the *Yijing* and the Representation of Reality." *Oracle: The Journal of the I Ching Society* (Summer): 35–50.

——. 2001a. "Jesuit Interpretations of the *Yijing* (Classic of Changes) and Their Legacy." Paper for the conference on "Matteo Ricci and After: Four Centuries of Cultural Interactions between China and the West," Hong Kong, October 13–16.

——. 2001b. "Maps, Myths and Multiple Realities: Images of the 'Other' in Late Imperial China." Paper for a conference on "De l'image à l'action: la dynamique des représentations visuelles dans la culture intellectuelle et religieuse de la Chine." Paris, September 3–5.

——. 2002. "The Jesuits and Evidential Research in Late Imperial China: Some Reflections," *Ex/Change* (February): 7–12.

——. 2003. "The *Yijing* (Classic of Changes) in Comparative Perspective: The Value of Cross-Cultural Investigations." *International Journal of the Humanities* 1: 776–801.

——. 2006a. "Maps, Myths and Multiple Realities: Images of the 'Other' in Late Imperial China." Paper for a workshop on "Image of Maps/Maps of Imagination," Oxford University, Ashmolean Museum, May 12–13, 2006.

——. 2006b. "Knowing the Self and Knowing the 'Other': The Epistemological and Heuristic Value of the *Yijing* (Classic of Changes)." *Journal of Chinese Philosophy* 33.4: 465–77.

——. 2008a. *Fathoming the Cosmos and Ordering the World: The Yijing (I-Ching or Classic of Changes) and Its Evolution in China.* Charlottesville, VA: University of Virginia Press.

——. 2008b. "Divination in Late Imperial China: New Light on Some Old Problems." In On-cho Ng, ed. 2008: 273–315.

——. 2009. "Select Bibliography of Works on the *Yijing* Since 1985." *Journal of Chinese Philosophy*, Supplement to Volume 36: 152–63.

——. 2009–10. "The *Book of Changes* as a Mirror of the Mind." *Zhouyi Studies* (English version). 6.1: 41–64.

——. 2011. "Key Concepts of Fate and Prediction in the *Yijing* (Classic of Changes)." Paper for the conference on "Ming and Fatum—Key Concepts of Fate and Prediction in a Comparative Perspective," Erlangen, Germany, June 29–July 3. Available online at: http://chaocenter.rice.edu/Content.aspx?id=601.

——. 2012. *The I Ching: A Biography*. Princeton, NJ: Princeton University Press.

——. [Sima Fu]. Forthcoming. "Zhongguo Qingdai di zhanbu" [Divination in the Qing dynasty]. In Ping Yao and Cong Zhang, eds., *Zhishi lishi* [Intellectual History], Shanghai guji chubanshe.

—— and D.Y.Y. Kwok, eds. 1993. *Cosmology, Ontology and Human Efficacy: Essays in Chinese Thought.* Honolulu: University of Hawaii Press.

Spence, Jonathan. 1975. *Emperor of China: Self-Portrait of K'ang-hsi*. New York: Vintage.

Standaert, Nicholas. 1997. "New Trends in the Historiography of Christianity in China." *The Catholic Historical Review* 83.4 (October): 569–613.

——, ed. 2001. *Handbook of Christianity in China:* Vol. one, *635–1800*. Leiden: Brill.

Steinhardt, Nancy S. 1997. "Chinese Cartography and Calligraphy." *Oriental Art* 43.1: 10–20.

Strassberg, Richard E. 1994. *Inscribed Landscapes: Travel Writing from Imperial China*. Berkeley, CA: University of California Press.

——. 2008. *Wandering Spirits: Chen Shiyuan's Encyclopedia of Dreams*. Berkeley, CA: University of California Press.

Strickmann, Michel. 2002. *Chinese Magical Medicine*. Bernard Faure, ed. Stanford, CA: Stanford University Press.

Struve, Lynn A., ed. 2004. *The Qing Formation in World Historical Time*. Cambridge, MA: Harvard University Press.

Su, Ai. 1996. *Zhongguo gudai xiangshu yanjiu yu pipan* (Research and critique of ancient Chinese physiognomic arts). Changsha: Yuelu shushe.

Sun, Qifeng. 1884. *Chongkan jiali zhuo* (Notes on the republication of *Family Rituals*). n.p.

Sutton, Donald. 1981. "Pilot Surveys of Chinese Shamans, 1875–1945: A Spatial Approach to Social History." *Journal of Social History* 15.1: 39–50.

——. 2007a. "Introduction." In Sutton 2007c, ed.: 3–21.

——. 2007b. "Death Rites and Chinese Culture: Standardization and Variation in Ming and Qing Times." In Sutton 2007c, ed.: 125–53.

——, ed. 2007c. "Ritual, Cultural Standardization, and Orthopraxy in China: Reconsidering James J. Watson's Ideas." Special edition of *Modern China* 33.1 (January): 3–158.

Swisher, Earl. 1953. *China's Management of the American Barbarians: A Study of Sino-American Relations, 1841–1861, with Documents*. New Haven, CT: Yale University Press.

Swope, Kenneth M. 2002. "Deceit, Disguise, and Dependence: China, Japan, and the Future of the Tributary System, 1592–1596." *The International History Review* 24.4 (December): 757–82.

——. 2009. *A Dragon's Head and a Serpent's Tail: Ming China and the First Great East Asian War, 1592–1598*. Norman, OK: University of Oklahoma Press.

Szonyi, Michael. 2002. *Practicing Kinship: Lineage and Descent in Late Imperial China*. Stanford, CA: Stanford University Press.

——. 2007. "Making Claims about Standardization and Orthopraxy in Late Imperial China: Rituals and Cults in the Fuzhou Region in Light of Watson's Theories." In Sutton, ed. 2007c: 47–71.

Takata, Tokio, ed. 1995. *Inventaire sommaire des manuscrits et imprimés chinois de la Bibliothèque Vaticane*. Kyoto: Italian School of East Asian Studies: 1995. Expanded edition of Paul Pelliot's work of the same title, dated June 13–July 6, 1922.

Tambiah, Stanley. 1990. *Magic, Science, Religion, and the Scope of Rationality*. Cambridge: Cambridge University Press.

Tan, Qixiang. 1993a. *Qingren wenji: Dili lei huibian* (Collected writings by Qing authors: a compilation of geographical works). Vol. 3 (Ming–Qing). Jinan: Shandong jiaoyu chubanshe. Originally compiled by Wang Zhongmin.

——. 1993b. *Zhongguo lidai dili xuejia pingzhuan* (Critical biographies of Chinese geographers by period). Vol. 3 (Ming–Qing). Jinan: Shandong jiaoyu chubanshe.

Tang, Junyi. 1979. *Zhongguo wenhua zhi jingshen jiazhi* (The spiritual value of Chinese culture). Taibei: Zhengzhong . Reprint of 1953 edition.

Tang, Mingbang, *et al.*, eds. 1986. *Zhouyi zongheng lu* (Comprehensive account of the *Zhou Changes*). Wuhan: Hubei renmin chuban she.

Tat, Wei. 1977. *An Exposition of the I-Ching or Book of Changes*. Hong Kong: Dai Nippon Printing Co.

Taylor, Romeyn. 1990. "Official and Popular Religion and the Political Organization of Chinese Society in the Ming." In Kwang-Ching Liu, ed. 1990.

Teng, Emma Jinhua. 2004. *Taiwan's Imagined Geography: Chinese Colonial Travel Writing and Pictures, 1683–1895*. Cambridge, MA: Harvard University Press.

Teng, Ssu-yü and John K. Fairbank, eds. 1979. *China's Response to the West: A Documentary Survey, 1839–1923*. Cambridge, MA: Harvard University Press.

Thomas, Keith. 1971. *Religion and the Decline of Magic*. New York: Charles Scribner's Sons.

Tikhvinsky, S.L., and L.S. Perelomov, eds. 1981. *China and Her Neighbours from Ancient Times to the Middle Ages: A Collection of Essays*. Trans. Liv Tudge. Moscow: Progress.

Tong, Hollington. 1938. *Chiang Kai-shek: Soldier and Stateman*. London: Hurst and Blackett.

Topley, Marjorie, ed. 1967. *Some Traditional Chinese Ideas and Conceptions in Hong Kong Social Life Today*. Hong Kong: Royal Asiatic Society.

Tozer, Warren. 1970. "Taiwan's 'Cultural Renaissance.'" *China Quarterly* 43 (July–September).

Tsu, Timothy. 1997. "Geomancy and the Environment in Premodern Taiwan." *Asian Folklore Studies* 56: 65–77.

Tu, Wei-ming. 1972. "Li as a Process of Humanization." *Philosophy East and West* 22.2: 187–201.

——. 1979. "Ritual and Authority in Confucian Thought." Paper prepared for the annual meeting of the Association for Asian Studies, March 1979.

Turnbull, David. 1993–1994. "Local Knowledge and Comparative Scientific Traditions." *Knowledge and Policy* 6.3–4 (Fall/Winter): 29–54.

——. 1996. "Cartography and Science in Early Modern Europe: Mapping and the Construction of Knowledge Spaces." *Imago Mundi* 48: 5–24.

Unno, Kazutaka. 1977. "Concerning a MS Map of China in the Bibliothèque Nationale, Paris, Introduced to the World by Monsieur M. Destombes." *Memoirs of the Research Department of the Toyo Bunko* 35: 205–17.

Unschuld, Paul. 1985. *Medicine in China: A History of Ideas*. Berkeley, CA: University of California Press.

Van Lieu, Joshua. 2009. "The Politics of Condolence: Contested Representations of Tribute in late Nineteenth-Century Choson–Qing Relations." *Journal of Korean Studies* 14.1 (Fall): 83–115.

Vanderstappen, Harrie. 1988. "Chinese Art and the Jesuits in Peking." In Ronan and Oh, eds. 1988.

von Collani, Claudia. 1981. *Die Figuristen in der Chinamission*. Bern: Peter Lang.

——. 1982. "Chinese Figurism in the Eyes of European Contemporaries." *China Mission Studies (1550–1800) Bulletin* 4: 12–23.

——. 1985. *P. Joachim Bouvet S.J.: Sein Leben und sein Werk*, Nettetal: Steyler Verlag.

——. 1989. *Eine wissenschaftliche Akademie für China: Briefe des Chinamissionars Joachim Bouvet S.J. an Gottfried Wilhelm Leibniz und Jean Paul Bignon*. Wiesbaden: Steiner.

——. 1998. "Did Jesus Christ Really Come to China?" *Sino-Western Cultural Relations Journal* 20: 34–48.

——. 2000a. "Gottfried Wilhelm Leibniz and the China Mission of the Jesuits." In Wenchao Li and Hans Poser, eds. 2000.

——. 2000b. "Cabbala in China." In Malek, ed. 2000: 527–58.

——. 2007. "The First Encounter of the West with the *Yijing*: Introduction to and Edition of Letters and Latin Translations by French Jesuits from the 18th Century." *Monumenta Serica* 55: 227–387.

Wakeman, Frederic Jr. 1986. *The Great Enterprise*. Berkeley, CA: University of California Press.

——. 1994. *Policing Shanghai 1927–1937*. Berkeley, CA: University of California Press.

Waley-Cohen, Joanna. 2004. "The New Qing History." *Radical History Review* 88 (Winter): 193–206.

Wang, Aihe. 2000. *Cosmology and Political Culture in Early China*. Cambridge: Cambridge University Press.

Wang, Bugui. 1993. *Shenmi wenhua: Chenwei wenhua xintan* (Mystery culture: A new investigation into prophetic texts). Beijing: Zhongguo shehui kexue chubanshe.

Wang, Erh-min. 1993. "The Turn of Fortune (*Yun-hui*): Inherited Concepts and China's Response to the West." In Smith and Kwok 1993: 205–16.

Wang, Guiyuan. 1991. *Nüwu yu wushu* (Witches and magic). Shijiazhuang: Hebei renmin chubanshe renmin chuban she.

Wang, Hui. 1995. "The Fate of 'Mr. Science' in China: The Concept of Science and Its Application in Modern Chinese Thought." *positions* 3.1 (Spring): 1–67.

Wang, Nan. 1993. "Time and Change: A Comparative Study of Chinese and Western Almanacs." MA thesis, Rice University, Houston.

Wang, Qi. 1609. *Sancai tuhui* (Illustrated compilation of the Three Powers). Huayincao tang.

Wang, Qianjin. 1995. "*Huangyu quanlan tu cehui*" (The surveying and drawing of the *Huangyu quanlan tu*). In Dong, ed. 1995.

Wang, Shou-nan. 1987–88. "Chiang Kai-shek and the Promotion of the Chinese Cultural Renaissance Movement." *Chinese Studies in History* 20.2 (Winter): 82–90.

Wang, Yao and Chen Jian. 1987. *Tufan shiqi di zhanbu yanjiu* (Research on divination in early Tibet). *Xianggang Zhongwen daxue Zhongguo wenhua yanjiu suo zhuankan* 9: 1–161.

Wang, Yude 1994a. *Shenmi di fengshui* (Mysterious fengshui). Taibei: Shuquan chuban she.

——. 1994b. *Fengshui shu zhuping* (Notes and commentaries on the art of fengshui). Taibei: Yunlong chuban she.
——. 1994c. *Shenmi di bagua* (The mysterious eight trigrams). Taibei: Shuquan chuban she.
——. *et al.*, eds. 1993. *Zhonghua shenmi wenhua* (The mystical culture of China). Changsha: Hunan chuban she.
Wang, Yunwu, ed. 1985–1991. *Congshu jicheng chubian* (The complete collectanea; first series). Beijing: Zhonghua shuju. Reprint of 1935 edition.
Wasserstrom, Jeffrey and Perry, Elizabeth, eds. 1992. *Popular Protest and Political Culture in Modern China*. Boulder, CO: Westview Press.
Watson, Burton, trans. 1963. *Hsün Tzu: Basic Writings*. New York: Columbia University Press.
Watson, James. 1985. "Standardizing the Gods: The Promotion of T'ien Hou ('Empress of Heaven') along the South China Coast, 960–1960." In Johnson, Nathan, and Rawski, eds. 1985.
——. 2007. "Orthopraxy Revisited." In Sutton, ed. 2007c: 154–58.
—— and Evelyn Rawski, eds. 1988. *Death Ritual in Late Imperial and Modern China*. Berkeley, CA: University of California Press.
Wei, Shaosheng. 1992. *Zhongguo gudai zhanbu shu* (The divinatory arts of ancient China). Zhengzhou: Zhongzhou guji chuban she.
Wei, Tat. 1977. *An Exposition of the I-Ching*. Hong Kong: Dai Nippon Printing Co.
Wei, William. 1985. *Counterrevolution in China: The Nationalists in Jiangxi during the Soviet Period*. Ann Arbor, MI: University of Michigan Press.
Wei, Zhengtong. 1974. *Zhongguo di zhihui* (The wisdom of China). Taibei: Shuiniu.
——. 1981. *Zhongguo wenhua gailun* (Introduction to Chinese culture). Taibei: Shuiniu. Reprint of 1968 edition.
Weller, Robert. 1987. *Unities and Diversities in Chinese Religion*. Seattle: University of Washington Press.
Weng, Naijian. 1993. *Zhongguo gudai xiangshu yanjiu zong pipan* (A general critique of research on the ancient physiognomic arts of China). Wuhan: Wuhan gongye daxue chuban she.
White, Douglass Alan. 1976. "Interpretations of the Central Concept of the *I-Ching* during the Han, Sung and Ming Dynasties." Ph.D. dissertation, Harvard University.
Whyte, Martin King. 1974. *Small Groups and Political Rituals in China*. Berkeley, CA: University of California Press.
Wieger, Leon. 1913. *Moral Tenets and Customs in China*. Trans. L. Davrout. Ho-kien Fu: Catholic Mission Press.
Wilhelm, Richard trans. 1967. *The I Ching or Book of Changes*. Princeton, NJ: Princeton University Press.
Williams, E.T. 1913. "The State Religion of China during the Manchu Dynasty." *Journal of the North China Branch of the Royal Asiatic Society* 44: 11–45.
Wills, John E., Jr. 1988. "Tribute, Defensiveness, and Dependency: Uses and Limits of Some Basic Ideas about Mid-Qing Dynasty Foreign Relations." *American Neptune* 48: 225–29.
——. 1995. "How We Got Obsessed with the 'Tribute System' and Why It's Time to Get Over It." Paper for the Association for Asian Studies Annual Meeting, Washington, DC.
——. *et al.*, eds. 2011. *China and Maritime Europe, 1500–1800: Trade, Settlement, Diplomacy, and Missions*. Cambridge: Cambridge University Press.
Wing-tsit. Chan, trans. 1967. *Reflections on Things at Hand: The Neo-Confucian Anthology*. New York: Columbia University Press.
Witek, John S.J. 1982. *Controversial Ideas in China and Europe: A Biography of Jean-François Foucquet S.J. (1665–1741)*. Rome: Institutum Historicum S.I.
Wolf, Arthur. 1974a. "Gods, Ghosts, and Ancestors." In Wolf, ed. 1974b.

——, ed. 1974b. *Religion and Ritual in Chinese Society*. Stanford, CA: Stanford University Press.

Wolter, J. A. and Grim, R. E., eds. 1997. *Images of the World: The Atlas through History*. Washington, DC: Library of Congress.

Wong, Lawrence Wang-chi. 2010. "'Chouban yiwu (Handling the Affairs of the Barbarians)': Westerners as Barbarians in the Translation History of 19th Century China." Paper for the Fourth Asian Translation Traditions Conference: Appropriation and Transformation in Asian Translation Traditions, December 15–17.

Woo, David. 1981. "Map as Expression: A Study of Traditional Chinese Cartographic Style." M.A. thesis, University of Washington, Seattle.

Wood, Denis, with John Fels. 1992. *The Power of Maps*. New York: Guilford Press.

Woodward, David. 1985. "Reality, Symbolism, Time, and Space in Medieval World Maps." *Annals of the Association of American Geographers* 75.4 (December): 510–21.

——, *et al*. 1994. "Concluding Remarks." In Harley and Woodward, eds. 1994.

Wright, Arthur F. ed. 1959. *Confucianism and Chinese Civilization*. Stanford, CA: Stanford University Press.

——. 1960. "The Study of Chinese Civilization." *Journal of the History of Ideas* 21.2 (April–June): 234–35.

——. 1963. "On the Uses of Generalization in the Study of Chinese History." In Gottschalk, ed. 1963: 36–58.

Wright, Mary C. 1962. *The Last Stand of Chinese Conservatism: The T'ung-chih Restoration, 1862–74*. Stanford, CA: Stanford University Press.

Wu, Hua, ed. 1993. *Zhouyi da cidian* (Great dictionary of the *Zhou Changes*).

Wu, Lin. ed. 1995. *Yixiang tushuo* (Illustrated discussions on the images of the *Changes*). Taibei: Songlin chubanshe.

——, 1996. *Yigua: Xiangshu yili* (The hexagrams of the *Changes*: images and numbers and meanings and principles). Taibei: Sunglin chubanshe.

Wu, Silas. 1970. "Emperors at Work." *Tsing Hua Journal of Chinese Studies*, n.s. 8, nos. 1–2 (August): 210–27.

——. 1979. *Passage to Power: K'ang-hsi and His Heir Apparent, 1661–1722*. Cambridge, MA: Harvard University Press.

Wyatt, Don J. 1996. *The Recluse of Loyang: Shao Yung and the Moral Evolution of Early Sung Thought*. Honolulu: University of Hawaii Press.

Xiao, Ai. 1996. *Zhongguo gudai xiangshu yanjiu yu pipan* (Research and criticism on the ancient physiognomic arts of China). Changsha: Yuelu shushe.

Xiao, Yuan, *et al*., eds. 1992. *Zhouyi cidian* (Dictionary of the *Zhou Changes*). Beijing: Zhongguo gongren chuban she.

Xie, Gui'an. 1998. *Zhongguo chenyao wenhua yanjiu* (Research on the culture of prophecy in China). Haikou: Hainan chuban she.

Xinwenfeng chuban gongsi, bianji bu. 1986. *Congshu jicheng xinbian* (A new edition of the complete collectanea). Taibei: Xinwenfeng chuban gongsi.

Xiong, Victor Cunrui. 2002. "Astrological Divination at the Tang Court." Paper for the Princeton Tang Conference, Princeton, New Jersey, April 18–20. Available at: http://unix.cc.wmich.edu/~xiong/Tang%20Divination.pdf: 1–33.

Xiong Yuezhi. 1994. *Xixue dongjian yu wan Qing shehui* (The Eastern dissemination of Western learning and Late Qing society). Shanghai: Shanghai renmin chuban she.

Xu, Dishan. 1940. *Fuji mixin di yanjiu* (Research on the superstition of spirit-writing). Changsha: Shangwu yinshu guan.

Xu, Qianxue. 1696. *Duli tongkao* (Research on the study of ritual). n.p.

Yan, Ping, *et al*., eds. 1995. *Zhonghua gu ditu jizhen* (A collection of treasures from Chinese ancient maps). Xi'an: Xi'an ditu chubanshe.

Yan, Yunxiang. 1996. *The Flow of Gifts: Reciprocity and Social Networks in a Chinese Village*. Stanford, CA: Stanford University Press.

Yang, C.K. 1961. *Religion in Chinese Society*. Berkeley, CA: University of California Press.

Yang, Chang and Song Chuanyin. 1992. *Gudai cezi shu zhuping* (Critical annotations of works on the ancient art of fathoming characters). Beijing: Beijing shifan daxue chubanshe.

Yang, Hongsheng. 1995. *Bentu yu yuwai:Yixue di xiandaihua yu shijiehua* (The native land and beyond: The modernization and globalization of *Changes* studies). Shanghai: Shanghai shehui kexue yuan.

Yang, L. S. 1957. "The Concept of *Pao* as a Basis for Social Relations in China." In Fairbank, ed. 1957: 291–309.

Yang, Li. 1998. *Book of Changes and Traditional Chinese Medicine*. Beijing: Beijing Science and Technology Press.

Yang, Mayfair Mei-hui. 1994. *Gifts, Favors and Banquets: The Art of Social Relationships in China*. Ithaca, NY: Cornell University Press.

Yang Shao-Yun. 2007. "Becoming Zhongguo, Becoming Han: Tracing and Re-Conceptualizing Ethnicity in Ancient North China, 770 BC–AD 581." MA thesis, National University of Singapore.

Yang, Shufang. 1978. Qingdai zhongyang zhengzhi zhidu (Central governent institutions of the Qing dynasty). Taibei: Shangwu yishu guan.

Yang, Zong, *et al.*, eds. 1991. *Zhongguo shiyong jinji daquan* (A great compendium of Chinese taboos for practical use). Shanghai: Shanghai wenhua chuban she.

Yao Weijun. 1994. *Shenmi di zhanmeng* (Mysterious dream divination). Shuquan chuban she.

Yates, Robin. 2001. "The History of Military Divination in China." Paper for the Conference on Divination in honor of Professor Ho Peng Yoke, Needham Research Institute, Cambridge University, December 7–8.

——. 2005. "The History of Military Divination in China." *East Asian Science, Technology, and Medicine* 24: 15–43.

Yee, Cordell. 1992. "A Cartography of Introspection: Chinese Maps as Other Than European." *Asian Art* 5.4 (Fall): 29–45.

——, *et al.* 1996. *Space and Place: Mapmaking East and West*. Annapolis, MD: St. John's College Press.

Yeh, Jonathan. 2011. "Bestseller Almanac Continues to Influence Chinese after Thousands of Years." *Culture.tw*. April 1. Available at: http://culture.tw/index.php?option=com_content&task=view&id=2033&Itemid=157.

Yeh, Wen-hsin, ed. 1998. *Landscape, Culture and Power in Chinese Society*. Berkeley, CA: University of California, Center for East Asian Studies.

Yen, Hsiao-pei. 2005. "Body Politics, Modernity and National Salvation: The Modern Girl and the New Life Movement." *Asian Studies Review* 29.2 (June): 165–86.

Yi, Ding, *et al.* 1996. *Zhongguo gudai fengshui yu jianzhu xuanzhi* (Geomancy and architectural site selection in ancient China). Shijiazhuang: Hebei kexue jishu chuban she.

Yin, Xiaolin, ed. 1993. *Hanyu shumu cidian* (Dictionary of Chinese numerical expressions). Beijing: Zhonghua shuju.

Ying, Dingcheng. 1990. *Zhongguo wenhua zhi benyuan* (The origins of Chinese culture). Xianning: Hubei renmin chuban she Hubei renmin chuban she.

Yoon, Hong Key. 1992. "The Expression of Landforms in Chinese Geomantic Maps." *Cartographic Journal* 29 (June): 12–15.

You Yi shufang zhuren. 2003. *Cong tuxiang kan Yijing* (Looking at the *Classic of Changes* from the standpoint of images). Taibei: Chuntian chubanshe.

Young, John D. 1975. "An Early Confucian Attack on Christianity: Yang Kuang-hsien and his *Pu-te-i*," *Journal of the Chinese University of Hong Kong* 3: 155–86.

Yu Dong. 1996. *Catalogo delle opere cinese missionarie della Biblioteca Apostolica Vaticana*, vols XVI–XVIII. Vatican City: Biblioteca Apostolica Vaticana.

Yu, Xiaoqun. 1994. *Shushu tanmi* (An investigation into the secrets of the computational arts). Beijing: Sanlian shudian.

Yuan, Shushan. 1919. *Mingli tanyuan* (An investigation into the origins of destiny). Shanghai: Runde shuju. Four vols.

——. 1928. *Shu bushi xingxiang xue* (An account of divination studies). Shanghai: Runde shuju. Four vols.

——. 1947. *Mingpu* (Records of destiny). Shanghai: Runde shuju.

——. 1948a. *Zhongguo lidai buren zhuan* (Biographies of diviners in China by period). Shanghai: Runde shuju.

——. 1948b. *Zhong Xi xiangren tanyuan* (An investigation into the origins of Chinese and Western physiognomy). Hong Kong: Runde shuju.

——. 1961a. *Da Liuren tanyuan* (An investigation into the origins of the great Six Yang Waters). 2 vols. Hong Kong: Runde shuju.

——. 1961b. *Xuanji tanyuan* (An investigation into the origins of auspicious choices). 2 vols. Hong Kong: Runde shuju.

——. n.d. *Xin Mingli tanyuan* (A new investigation into the origins of destiny). Hong Kong: Runde shuju.

Yuan Zhongzhe. n.d. *Shenxiang quanbian* (Complete guide to spiritual observation). Reprinted in the TSJC.

Yun, Peter I. 1998. "Rethinking the Tribute System: Korean States and Northeast Asian Interstate Relations, 600–1600." Ph.D. dissertation, University of California, Los Angeles.

——. n.d. "Rhetoric and Reality of the Tribute System: Interstate Relations in the 10th–11th Century East Asia."

Zeitlin, Judith. 1998. "Spirit-Writing and Performance in the Work of You Tong (1618–1704)." *T'oung Pao* 84: 102–35.

Zhai, Tingpu. 2000. *Zhouyi yu Huaxia wenming* (The *Zhou Changes* and Chinese civilization). Shanghai: Shanghai renmin chuban she.

Zhang, Guiguang. *Zhouyi zhanbu cidian* (Dictionary of *Zhou Changes* divination). Taibei: Jieyou chuban she.

Zhang, Huang. 1613. *Tushu bian* (A compilation of illustrations and writings). Reproduced in SKQS.

Zhang, Longxi. 1998. "Shenme shi '*Huairou yuanren*?' Zhengming, kaozheng yu houxiandai shi shixue" (What is '*Huairou yuanren*?' The rectification of names, evidential scholarship and postmodern historiography). *Ershiyi shiji* 45 (February).

Zhang, Qicheng. ed. 1996. *Yijing yingyong da baike* (A practical encyclopedia of the *Classic of Changes*). 2 vols. Taibei: Dijing qiye gufen youxian gongsi.

——. 2003. *Xiangshu Yixue* (Images and numbers in *Changes* learning). Beijing: Zhongguo shudian.

Zhang, Ronghua. 1991. *Zhongguo gudai minjian fangshu* (The predictive and prescriptive arts of the common people in ancient China). Anhui renmin chuban she.

Zhang, Rongming. 2000. *Fangshu yu Zhongguo chuantong wenhua* (The predictive and prescriptive arts and the traditional culture of China). Shanghai: Xuelin chubanshe.

Zhang, Shanwen, ed. 1992. *Zhouyi cidian* (Dictionary of the *Zhou Changes*). Shanghai: Shanghai guji chuban she.

Zhang, Shiming. 2006. "A Historical and Jurisprudential Analysis of Suzerain-Vassal State Relationships in the Qing Dynasty." *Frontiers of History in China* 1: 124–57.

Zhang, Weiwen. 2008. "Religious Daoist Studies of the *Book of Changes* (*Yi jing*) and Their Historical and Contemporary Influence." *Contemporary Chinese Thought* 39.3 (Spring): 74–97.

Zhang, Yanyuan. 1983. *Lidai minghua ji* (Records of famous Chinese paintings by period). Beijing.

Zhang, Yaowen. n.d. [1975?] *Wushu zhanbu quanshu* (Complete book on the five arts of divination). Taibei: Dahua.

Zhang, Yongtang, ed. 1992a. *Guancang Zhongguo shushu chubian* (Index to archives on the arts of calculation in China, first series). Taiwan: Qinghua daxue.

——, ed. 1992b. *Zhongguo lidai shushu chubian—zhengshi pian* (Index to the arts of calculation in China by period, first series—orthodox histories). Taipei: Qinghua daxue.

Zhao, Rong and Yang Zhengtai. 1998. *Zhongguo dili xue shi* (History of Chinese geography). Beijing: Shangwu yinshu guan.

Zheng, Wan'geng. 1989. *Taixuan jiaoshi* (A collated and annotated commentary on the [*Classic of*] *Great Mystery*). Beijing: Beijing Shifan daxue chubanshe.

——. 1997. *Yixue yuanliu* (The origin and development of *Changes* studies). Shenyang: Shenyang chubanshe.

——. 2008. "Process Thinking in *The Book of Changes* (*Yi jing*)." *Contemporary Chinese Thought* 39.3 (Spring): 59–73.

Zheng, Xiaojiang. 1993. *Zhongguo shenmi shu daguan* (An overview of the mystical arts of China). Nanchang: Baihua zhou wenyi chuban she.

Zhongguo cehui kexue yanjiu yuan, comp. 1998. *Zhonghua gu ditu zhenpin xuanji* (A selection of ancient Chinese map treasures). Harbin: Ha'erbin ditu chubanshe.

Zhou, Lin'gen. 1971. *Zhongguo jindai lijiao shi* (A history of the teachings of ritual in modern China). Taibei: Haiyang hsueh-yuan.

Zhu, Bokun. 1995. *Yixue zhexue shi* (A history of the philosophical aspects of the *Changes*). 4 vols. Beijing: Huaxia chuban she.

——. ed. 1996. *Zhouyi zhishi tonglan* (An intellectual reference book on the *Zhou Changes*). Jinan: Qi Lu shushe.

Zhuang, Wanshou [Chuang Wan-shou]. 2011. "Zhongguo minzu zhuyi tanyuan: Zhongguo Hanzi guji sheng xiong yu yi di shu xie zhi jiegou" (On the origins of Chinese nationalism: deconstruction of [the categories of] "saint" and "evil" and "Chinese" and "barbarians" from ancient Chinese literature written in "Han characters"). *Changrong* [*Chang Jung*] *daxue xuebao* 15.1: 31–57.

Zhuo, Xinping. 1998. "Suoyin pai yu Zhong Xi wenhua rentong" (The Figurists and the acknowledgment of similarities in Chinese and Western culture). *Dao feng Hanyu shenxue xuekan* 8 (1998). Reprinted in *Jidu zongjiao lun* n.d.: 292–317.

Ziporyn, Brook. 1995–1996. "Space, Time, and Eighty-One 'Kinds' in Yang Xiong's *Taixuan Jing*." *Medieval China* 2: 1–24.

Zito, Angela. 1984. "Re-Presenting Sacrifice: Cosmology and the Editing of Texts". *Ch'ing-shih wen-t'i* 5.2 (December): 47–78.

——. 1987. "City Gods, Filiality, and Hegemony in Late Imperial China." *Modern China* 13.3 (July): 333–71.

——. 1993. "Ritualizing *Li*: Implications for Studying Power and Gender." *positions* 1.2 (Fall): 321–48.

——. 1997. *Of Body and Brush: Grand Sacrifice as Text/Performance in 18th Century China*. Chicago: University of Chicago Press.

Zogner, Lothar. 1983. *China Cartographica*. Berlin: Staatsbibliothek Preussischer Kulturbesitz.

Zuesse, Evan. 1987. "Divination." In Mircea Eliade, ed. *The Encyclopedia of Religion*. New York: Macmillan.

Zürcher, Erik. 1994. "Jesuit Accommodation and the Chinese Cultural Imperative." In Mungello, ed. 1994: 31–64.

INDEX

Note: In the interest of space, I have eliminated from this index the proper names of most people and places, as well as the titles of most books and essays.